Progress in

MEDICAL GENETICS

Edited by

ALEXANDER G. BEARN, M.D.

Senior Vice-President,
Medical and Scientific Affairs,
Merck Sharp & Dohme International,
Rahway, New Jersey, and
Professor of Medicine,
Cornell University Medical Center,
New York, New York

ARNO G. MOTULSKY, M.D.

Professor of Medicine and Genetics and
Director, Center for Inherited Diseases,
University of Washington,
Seattle, Washington

BARTON CHILDS, M.D.

Professor Emeritus of Pediatrics,
The Johns Hopkins University,
Baltimore, Maryland

Progress in

MEDICAL GENETICS

Volume VI

GENETICS OF NEUROLOGICAL DISORDERS

PRAEGER SPECIAL STUDIES • PRAEGER SCIENTIFIC

New York • Philadelphia • Eastbourne, UK
Toronto • Hong Kong • Tokyo • Sydney

Library of Congress Cataloging in Publication Data
Main entry under title:

Genetics of neurological disorders.

(Progress in medical genetics ; v. 6)
Includes index.
1. Nervous system--Diseases--Genetic aspects.
2. Neurogenetics. I. Bearn, Alexander G., 1923-
II. Motulsky, Arno G., 1923- III. Childs, Barton.
IV. Series: Progress in medical genetics ; new ser., v. 6.
[DNLM: 1. Nervous System Diseases--familial & genetic.
W1 PR6709 New ser., v. 6 / WL 100 G328]
RB155.P7 n.s., vol. 6 616'.042 s 85-508
[RC346] [616.8'0442]
ISBN 0-03-001769-6 (alk. paper)

Published in 1985 by Praeger Publishers
CBS Educational and Professional Publishing, a Division of CBS Inc.
521 Fifth Avenue, New York, NY 10175 USA

56789 052 987654321

Printed in the United States of America on acid-free paper

INTERNATIONAL OFFICES

Orders from outside the United States should be sent to the appropriate address listed below. Orders from areas not listed below should be placed through CBS International Publishing, 383 Madison Ave., New York, NY 10175 USA

Australia, New Zealand
Holt Saunders, Pty, Ltd., 9 Waltham St., Artarmon, N.S.W. 2064, Sydney, Australia

Canada
Holt, Rinehart & Winston of Canada, 55 Horner Ave., Toronto, Ontario, Canada M8Z 4X6

Europe, the Middle East, & Africa
Holt Saunders, Ltd., 1 St. Anne's Road, Eastbourne, East Sussex, England BN21 3UN

Japan
Holt Saunders, Ltd., Ichibancho Central Building, 22-1 Ichibancho, 3rd Floor, Chiyodaku, Tokyo, Japan

Hong Kong, Southeast Asia
Holt Saunders Asia, Ltd., 10 Fl, Intercontinental Plaza, 94 Granville Road, Tsim Sha Tsui East, Kowloon, Hong Kong

Manuscript submissions should be sent to the Editorial Director, Praeger Publishers, 521 Fifth Avenue, New York, NY 10175 USA

Foreword

Volume VI of *Progress in Medical Genetics* is devoted to neurological disorders. Genetic factors in the causation of neurological disease have been recognized at least since the beginning of the century, but it has been only recently that the importance of genetic influences has become fully appreciated. In this volume the editors have brought together geneticists and neurologists in an attempt to focus on certain diseases of particular interest to the practicing neurological clinician.

An introductory chapter by Thomas Bird provides an overall view of the broad influence of genes on neurological disease. In some instances the presence of a single mutant gene appears sufficient to account for the clinical syndrome; in others where heterogeneity is evident, more than one gene can be implicated. In many genetically influenced diseases, environmental influences, difficult to identify specifically, play a role.

Epilepsy has long been known to "run in families," but the evident clinical and genetic heterogeneity has led to much confusion. V. Elving Anderson and W. Allen Hauser provide an admirable summary of this field and discuss the extremely important problem of genetic counseling in this group of disorders.

Muscular dystrophy remains a collection of somewhat similar syndromes that—although the gene products remain elusive—are now being sorted out both clinically and genetically. Peter Harper reviews new knowledge in this field.

Huntington's disease has always been the classic example of a neurological disease caused by a single gene that expresses itself clinically in mid-life. Richard Myers, Miriam Schoenfeld, and Edward Bird review the recent development of a disease first described more than 100 years ago. Recent information using recombinant DNA technology indicates that the gene is a chromosome. This information will have powerful implications for genetic counseling.

The hereditary ataxias, like the muscular dystrophies, are extremely heterogeneous. Håvard Skre, Tor Haugstad, and Kåre Berg explore this confusing subject extensively. This chapter will serve as an important source for future students of the subject as well as a valuable contemporary summary.

The recognition that resistance and susceptibility to infectious agents are strongly conditioned by genetic factors was first alluded to by J. B. S. Haldane about 35 years ago. Only recently has his suggestion been capable of rigorous analysis—largely as a result of the development in our understanding of the structure and function of viruses. The development of molecular virology and the increased understanding of how viruses multiply within the cells they infect are illuminating *resistance* in a way not imaginable a decade ago. Raymond Roos examines these topics.

This volume closes with a vivid and stimulating chapter by Nancy Wexler, pointing to the human implications and complexities of detecting carriers of genetic disease and diagnosing disease before it becomes clinically manifest. For all practicing physicians, not only neurologists, this chapter raises many questions and has the modesty not to offer ultimate solutions.

Taken together, these chapters manifest the importance of genetic factors in the etiology of neurological disease. They also illustrate that the clarification of how genes cause disease, the precise chromosomal location of the gene, the nature of the mutational event, and the probing of the gene using recombinant technology will illuminate an area of neurological disease previously hidden from view by eponymous names or all-inclusive tortured and bewildering classifications. There is, of course, still much to do, but this volume surely underscores the importance of genetic factors in our understanding, at both the clinical and molecular level, of neurological disease.

Contents

1 Medical Genetics and Clinical Neurology: The Common Ground

Thomas D. Bird

Department of Medicine, Divisions of Neurology and Medical Genetics, University of Washington School of Medicine, and Veterans Administration Medical Center, Seattle, Washington

This introductory chapter emphasizes the common ground shared by medical geneticists and clinical neurologists. Neurologists often lack the basic information required for genetic counseling, and medical geneticists are not usually trained for making or assessing the adequacy of neurological diagnosis. Collaboration and cooperation between these two specialties provides a rewarding experience for both and is highly beneficial to their patients. Furthermore, research in genetics and the neurosciences is progressing at a rapid rate; so there is an obvious need for further collaboration at all basic science and clinical levels.

Medical geneticists are frequently called upon to evaluate families with neurological diseases. Eldridge (1980) has estimated that as many as 500 single-gene disorders display major neurological symptoms. Table 1.1 shows the frequency with which families with neurological disorders were seen during a recent three-year period (1979 through 1981) at the adult Medical Genetics Clinic at the University of Washington Medical Center. This facility has separate pediatric medical genetics and muscular dystrophy clinics, so that pediatric disorders and conditions covered by the Muscular Dystrophy Association (including muscular dystrophies, hereditary neuropathies, and Friedreich's ataxia) are underrepresented in the table. Nevertheless, 43.5 percent of the total clinical population represented families with neurological conditions. In fact, this heavy load of neurological cases had previously led to the establishment of a "neurogenetics day" held in the clinic on a monthly basis. The data in Table 1.1 are similar to those reported in the mid-1970s from the same clinic (Bird and Hall, 1977) except that most families with Down's syndrome are now counseled at the prenatal or pediatric Medical Genetics Clinics. The four most common groups of neurogenetic disorders represented at the adult genetics clinic were Huntington's disease, muscular dystrophies, hereditary spinocerebellar degenerations, and hereditary neuropathies. Appropriately, each of these disorders constitutes the subject of a separate chapter in this

TABLE 1.1. Neurological Disorders at the Adult Medical Genetics Clinic, University of Washington, 1979–81

Disorder	Number of Families	Percentage of Clinic Population
Huntington's disease	40	10.1
Muscular dystrophies	33	8.3
Hereditary spinocerebellar degenerations	22	5.5
Hereditary neuropathies	17	4.3
Mental retardation	13	3.3
Neurofibromatosis	12	3.0
Congenital CNS anomalies	9	2.3
Miscellaneous (tuberous sclerosis, porphyria, epilepsy, schizophrenia, manic-depression, Alzheimer's, MS, ALS, dystonia, Tourette's, others)	27	6.8
Total neurological	173	43.5
Total clinic	398	100.0

CNS = central nervous system
MS = multiple sclerosis
ALS = amyotrophic lateral sclerosis

volume. Although our medical center has a special interest in neurogenetics, and other medical genetics units will attract somewhat different patient populations, unquestionably neurological disorders form an important proportion of every medical genetics counseling service.

Neurologists should also recognize that many of their patients have diseases requiring genetic counseling. Bird and Hall (1977) found that 22 percent of hospitalized *pediatric* neurology patients had disorders with chromosomal, single-gene, or multifactorial-polygenic implications. In the same study approximately 4 percent of hospitalized *adult* neurology patients had genetic diseases, and at least an additional 5 percent of this adult population had a family history suggesting a neurogenetic disorder. Since 26 percent of the medical records in the survey had no adequately recorded family history, many potential neurogenetic conditions went unrecognized by the clinicians.

Further epidemiological data concerning clinical neurogenetics are presented in Table 1.2. This table lists six of the most common disorders evaluated by neurologists (stroke, epilepsy, mental retardation, dementia, Parkinson's syndrome, and multiple sclerosis). The table also shows the estimated number of persons in the United States affected with each condition. The first five categories are clearly heterogeneous. In some cases each disorder may be triggered primarily by environmental conditions; on other occasions each may be caused by single-genes, multiple-genes, or gene-environment interactions. (Multiple sclerosis is likely to be a single disease with a well-documented increased familial incidence.) Table 1.2 also cites references reviewing potential genetic factors in each disorder. If by conservative estimate 5 percent of the patient population indicated in Table 1.2 represents disorders with a strong genetic influence, this neurogenetic group would include 300,000 individuals.

Table 1.3 shows some advances in medical genetics of demonstrated or potential importance to clinical neurogenetics. Neurologists should become familiar with the basic principles of these advances because they are—or soon will be—pertinent to the practice of clinical neurology. For this reason Table 1.3 also lists articles providing a general review of each category.

Examples of recent advances in medical genetics of interest to neurologists include the assignment of the loci for four hereditary neurological disorders to specific human chromosomes: one form of hereditary ataxia on chromosome 6 (Jackson et al., 1977), one form of hereditary motor/sensory neuropathy on chromosome 1 (Bird, Ott and Giblett, 1982), myotonic muscular dystrophy on chromosome 19 (Eiberg et al., 1981) and Huntington's disease on chromosome 4 (Gusella et al., 1983). The latter work of Gusella and colleagues is especially important because it represents the linkage of a DNA polymorphism to a devastating, autosomal dominant, degenerative brain disorder of delayed onset and unknown cause. Advances in chromosome banding techniques have led to the identification of a fascinating form of X-linked mental retardation (the so-called fragile X syndrome), which may occur as often as 1 in every 1,000 male births (Herbst and Miller, 1980; Jacobs et al., 1980; Blomquist et al., 1982). Furthermore, practicing physicians should know that the prenatal diagnosis of sickle-cell anemia and β-thalassemia has been accomplished through a combination of the first four advances listed

TABLE 1.2. Common Multifactorial Neurological Syndromes with Proven or Suspected Genetic Influences

Neurological Syndrome	Estimated Affected Population (prevalence, U.S.)	Reference to Genetic Factors
Cerebrovascular disease	1.8×10^6	Marshall, 1971 Rice et al., 1980
Epilepsy	$1.0–1.5 \times 10^6$	Newmark and Penry, 1980 Jennings and Bird, 1981 Anderson et al., 1982
Mental retardation (severe)	1.0×10^6	Milunsky, 1975 Crawfurd, 1982
Dementia	$0.5–1.0 \times 10^6$	Goudsmit et al., 1981 Heston et al., 1981
Parkinson's syndrome	0.5×10^6	Martin, Young, and Anderson, 1973 Barbeau and Pourcher, 1982
Multiple sclerosis	$0.6–1.4 \times 10^5$	Tiwari et al., 1980 Williams et al., 1980 Roberts and Bates, 1982

TABLE 1.3. Advances in Medical Genetics of Importance to Clinical Neurogenetics

Advance	Review Article
Gene mapping/linkage studies (including somatic cell genetics)	Sparkes et al., 1980
Recombinant DNA and DNA cloning techniques	Davies, 1981 Miller, 1981
Defining the molecular mechanisms of the hemoglobinopathies	Winter, Hanash, and Rucknagel, 1979 Antonarakis, Phillips, and Kazazian, 1982
Prenatal diagnosis	Dumars, Dalrymple, and Murray, 1976 Milunsky, 1976 Golbus et al., 1979
Chromosome banding techniques	Yunis and Chandler, 1977 Therman, 1980
Elucidation of inborn errors of metabolism	Rosenberg, 1981 Kolodny and Cable, 1982

DNA = deoxyribonucleic acid

in Table 1.3 (Antonarakis, Phillips, and Kazazian, 1982; Orkin et al., 1982). These techniques will continue to be applied to an increasing number of genetic disorders, including those affecting the nervous system. A final example comes from the field of inborn errors of metabolism where it is now recognized that adults may be affected with rare autosomal recessive diseases (such as lipid storage disorders) previously assumed to be confined to the pediatric population (Johnson, 1981; Longstreth et al., 1982).

Table 1.4 cites advances in neurology of importance to clinical neurogenetics. Again, general review articles are listed for each category for the

TABLE 1.4. Advances in Neurology of Importance to Clinical Neurogenetics

Advance	Review Article
Discovery of latent transmissible CNS diseases	Masters et al., 1979 Johnson, 1982
Identification of central neurotransmitters and related disorders	Sourkes, 1981 Iverson, 1982 Moore, 1982
Advances in neuroimaging (CAT, NMR, PET)	Oldendorf, 1981 Pykett, 1982
Evoked potential techniques (visual, auditory, somatosensory)	Giblin, 1981 Sokol, 1980 Stockard, Stockard, and Sharbrough, 1980
Discovery of muscle receptor antibody in myasthenia gravis	Drachman, 1981 Lisak and Barchi, 1982

CAT = computerized axial tomography
NMR = nuclear magnetic resonance
PET = positron emission tomography

benefit of the interested medical geneticist. Jakob-Creutzfeldt disease, for example, has been demonstrated to be caused by a slow transmissible agent, and this rare disorder shows unexplained geographic and familial clustering (Masters et al., 1979). Slow transmissible agents may be important in the understanding of other neurological disorders with increased familial incidence including, but not limited to, the dementias. The discovery that Parkinson's disease is a treatable disorder of central neurotransmission has greatly increased the interest of neuroscientists in other similar degenerative brain processes such as Huntington's disease. Recent advances in neuroimaging and the recording of evoked potentials will expand our understanding of which areas of the nervous system are affected by mutant genes, increase the accuracy of neurological diagnosis, and potentially lead to the presymptomatic diagnosis of some neurogenetic disorders (Bird and Crill, 1981; Kuhl et al., 1982). Furthermore, the discovery of muscle receptor antibodies in myasthenia gravis has shown that humoral antibodies can produce a chronic neurological disorder, has helped to explain the transient neonatal occurrence of the muscle weakness, and is of special interest to medical geneticists because of recognized genetic factors in autoimmune diseases (Lisak and Barchi, 1982).

It is beyond the scope of this chapter to review basic concepts in genetics for neurologists, and neurology for geneticists. However, it is appropriate to cite several references that will serve as general introductions to these topics for the two groups of specialists. The text by Thompson and Thompson (1980) provides a good review of medical genetics for clinicians. More detailed volumes covering human genetics and hereditary metabolic disorders, respectively, are those of Vogel and Motulsky (1979) and Stanbury and coauthors (1982). Many neurologists are unaware of the catalog of genetic disorders authored by McKusick (1983), which serves as an annotated list of all hereditary diseases and an entrance to the recent literature for any given disorder.

Medical geneticists can find valuable discussions of neurological diseases and syndromes in the texts by Adams and Victor (1981), Gilroy and Meyer (1979), Baker and Baker (1975), and Menkes (1980). The *Handbook of Clinical Neurology* (Vinken and Bruyn, 1970–81) is an extensive, detailed, multiauthored, 43-volume compendium of neurology that may be found in many medical center libraries. The last two volumes of this series compose a neurogenetics directory. Both neurologists and medical geneticists will find helpful clinical information and useful wide-ranging bibliographies in Baraitser's *The Genetics of Neurological Disorders* (1982) and *Genetics in Neurology* (Ionasescu and Zellweger, 1983).

In summary, neurogenetic disorders are common, and affected families are likely to be evaluated by both medical geneticists and clinical neurologists. These two disciplines have much to gain from increased familiarity with recent advances in their respective fields. Further close collaboration between the two groups is inevitable. It will be beneficial to medical science and improve patient care.

REFERENCES

Adams, R. C. and Victor, M. 1981. *Principles of Neurology*. McGraw-Hill, New York.

Anderson, V. E., Hauser, W. A., Penry, J. K., Sing, C. F. (Eds.). 1982. *Genetic Basis of the Epilepsies*. Raven Press, New York.

Antonarakis, S. E., Phillips, H. A., Kazazian, H. H. 1982. Genetic diseases: diagnosis by restriction endonuclease analysis. *J. Ped.*, 100:845–856.

Baker, A. B. and Baker, L. H. 1975. *Clinical Neurology*. Harper & Row, Hagerstown (3 volumes, updated annually).

Baraitser, M. 1982. *The Genetics of Neurological Disorders*. Oxford Press, Oxford.

Barbeau, A. and Pourcher, E. 1982. New data on the genetics of Parkinson's disease. *Canad. J. Neurol.*, 9:53–60.

Bird, T. D. and Crill, W. E. 1981. Pattern reversal visual evoked potentials in hereditary ataxias and spinal degenerations. *Ann. Neurol.*, 9:243–250.

Bird, T. D. and Hall, J. G. 1977. Clinical neurogenetics; a survey of the relationship of medical genetics to clinical neurology. *Neurology*, 27:1057–1060.

Bird, T. D., Ott, J., Giblett, E. R. 1982. Evidence for linkage of Charcot-Marie-Tooth neuropathy to the Duffy locus on chromosome one. *Am. J. Hum. Gen.*, 34:388–394.

Blomquist, H. K., Gustavson, K. H., Holmgren, G., Nordenson, I., Sweins, A. 1982. Fragile site X chromosomes and X-linked mental retardation in severely retarded boys in a northern Swedish county. A prevalence study. *Clinical Genetics*, 21:209–214.

Crawfurd, J. 1982. Severe mental handicap: pathogenesis, treatment and prevention. *Brit. Med. J.*, 285:762–766.

Davies, K. E. 1981. The application of DNA recombinant technology to the analysis of the human genome and genetic disease. *Hum. Genet.*, 58:351–357.

Drachman, D. B. 1981. The biology of myasthenia gravis. *Ann. Rev. Neurosci.*, 4:195–225.

Dumars, K. W., Dalrymple, G. T., Murray, A. K. 1976. Prenatal diagnosis and genetic counseling. *West. J. Med.*, 124:377–387.

Eiberg, H., Mohr, J., Nielson, L. S., Simonsen, N. 1981. Linkage relationships between the locus for C3 and 50 polymorphic systems; assignment of C3 to the DM-SE-LU linkage group. 6th World Congress of Genetics, Jerusalem (abstract).

Eldridge, R. 1980. Clinical neurogenetics: needs versus resources. *Neurology*, 30:860–863.

Giblin, D. R. 1980. Scalp-recorded somatosensory evoked potentials. In Aminoff, M. J. (Ed.), *Electrodiagnosis in Clinical Neurology*. Churchill Livingstone, New York.

Gilroy, J. and Meyer, J. S. 1979. *Medical Neurology*. Macmillan, New York.

Golbus, M. S., Loughman, W. D., Epstein, C. J., Halbasch, G., Stephens, J. D., Hall, B. C. 1979. Prenatal genetic diagnosis in 3000 amniocenteses. *New Eng. J. Med.*, 300:157–163.

Goudsmit, J., White, B. J., Weitkamp, L. R., Keats, B. J. B., Morrow, C. H., Gajdusek, D. C. 1981. Familial Alzheimer's disease in two kindreds of the same geographic and ethnic origin. *J. Neurol. Sci.*, 49:79–89.

Gusella, J. F., Wexler, N. S., Conneally, P. M., Naylor, S. L., Anderson, M. A., Tanzi, R. E., Watkins, P. C., Ottina, K., Wallace, M. R., Sakaguchi, A. Y., Young, A. B., Soulson, I., Bonilla, E., Martin, J. B. 1983. A polymorphic DNA marker genetically linked to Huntington's disease. *Nature*, 306:234–238.

Herbst, D. S. and Miller, J. R. 1980. Nonspecific X-linked mental retardation. II: The frequency in British Columbia. *Am. J. Med. Gen.*, 7:461–469.

Heston, L. L., Mastri, A. R., Anderson, V. E., White, J. 1981. Dementia of the Alzheimer type. Clinical genetics, natural history and associated conditions. *Arch. Gen. Psychiat.*, 38:1085–1090.

Ionasescu, V. and Zellweger, H. 1983. *Genetics in Neurology*. Raven Press, New York.

Iverson, L. 1982. Neurotransmitters and CNS disease. *Lancet*, 2:914–918.

Jackson, J. F., Currier, R. D., Terasaki, P. I., Morton, N. E. 1977. Spinocerebellar ataxias and HLA linkage. *New Eng. J. Med.*, 296:1138–1141.

Jacobs, P. A., Glover, T. W., Mayer, M., Fox, P., Gerrard, J. W., Dunn, H. G., Herbst, D. S. 1980. X-linked mental retardation: a study of 7 families. *Am. J. Med. Gen.*, 7:471–489.

Jennings, M. T. and Bird, T. D. 1981. Genetic influences in the epilepsies: a review of the literature with practical implications. *Am. J. Dis. Child.*, 135:450–457.

Johnson, R. T. 1982. The contribution of virologic research to clinical neurology. *New Eng. J. Med.*, 307:660–662.

Johnson, W. G. 1981. The clinical spectrum of hexosaminidase deficiency. *Neurology*, 31:1453–1456.

Kolodny, E. H. and Cable, W. J. L. 1982. Inborn errors of metabolism. *Ann. Neurol.*, 11:221–232.

Kuhl, D. E., Phelps, M. E., Markham, C. H., Metter, E. J., Reige, W. H., Winter, J. 1982. Cerebral metabolism and atrophy in Huntington's disease determined by ^{18}FDG and computed tomographic scan. *Ann. Neurol.*, 12:425–434.

Lisak, R. P. and Barchi, R. L. 1982. *Myasthenia Gravis*. Saunders, Phila.

Longstreth, W. T., Daven, J. R., Farrell, D. F., Bolen, J. W., Bird, T. D. 1982. Adult dystonic lipidosis: clinical, histological and biochemical findings of a neurovisceral storage disease. *Neurology*, 32:1295–1299.

Marshall, J. 1971. Familial incidence of cerebrovascular disease. *J. Med. Gen.*, 8:84–89.

Martin, W. E., Young, W. I., Anderson, V. E. 1973. Parkinson's disease: a genetic study. *Brain*, 96:495–506.

Masters, C. L., Gajdusek, D. C., Gibbs, C. J., Bernoulli, C., Asher, D. M. 1979. Familial Creutzfeldt-Jakob disease and other familial dementias: an inquiry into possible modes of transmission of virus-induced familial disease. In Prusiner, S. B., Hadlow, W. J. (Eds.), *Slow Transmissible Diseases of the Nervous System*, Vol. I. Academic Press, New York, pp. 143–194.

McKusick, V. A. 1983. *Mendelian Inheritance in Man*. Johns Hopkins Press, Baltimore.

Menkes, J. H. 1980. *Textbook of Child Neurology*. Lea and Febiger, Phila.

Miller, W. 1981. Recombinant DNA and the pediatrician. *J. Ped.*, 99:1–15.

Milunsky, A. 1975. *The Prevention of Genetic Disease and Mental Retardation*. Saunders, Phila.

Milunsky, A. 1976. Prenatal diagnosis of genetic disorders. *New Eng. J. Med.*, 295:377–380.

Moore, R. Y. 1982. Catecholamine neuron systems in brain. *Ann. Neurol.*, 12:321–327.

Newmark, M. E. and Penry, J. K. 1980. *Genetics of Epilepsy: A Review*. Raven Press, New York.

Oldendorf, W. H. 1981. Nuclear medicine in clinical neurology: an update. *Ann. Neurol.*, 10:207–213.

Orkin, S. H., Little, R. F. R., Kazazian, H. H., Boehm, C. D. 1982. Improved detection of the sickle mutation by DNA analysis. Application to prenatal diagnosis. *New Eng. J. Med.*, 307:32–36.

Pykett, I. L. 1982. NMR imaging in medicine. *Scientific American*, 246:78–88.

Rice, G. P. A., Boughner, D. R., Stiller, C., Eber, G. C. 1980. Familial stroke syndrome associated with mitral valve prolapse. *Ann. Neurol.*, 7:130–134.

Roberts, E. F. and Bates, D. 1982. The genetic contribution to multiple sclerosis: evidence from Northeast England. *J. Neurol. Sci.*, 54:287–293.

Rosenberg, R. N. 1981. Biochemical genetics of neurologic disease. *New Eng. J. Med.*, 305:1181–1193.

Sokol, S. 1980. Visual evoked potentials. In Aminoff, M. J. (Ed.), *Electrodiagnosis in Clinical Neurology*. Churchill Livingstone, New York.

Sourkes, T. L. 1981. Parkinson's disease and other disorders of the basal ganglia. In Seigel, G., Albers, R. W., Agranoff, B. W., Katzman, R. (Eds.), *Basic Neurochemistry*. Little, Brown, Boston, pp. 719–736.

Sparkes, R. S., Spence, M. A., Mohandas, T., Shapiro, L. J., Funderburk, S. J. 1980. Human gene mapping, genetic linkage and clinical applications. *Ann. Int. Med.*, 93:469–479.

Stanbury, J. B., Wyngaarden, J. B., Fredrickson, D. S., Goldstein, J. L., Brown, M. S. 1982. *The Metabolic Basis of Inherited Disease*. McGraw-Hill, New York.

Stockard, J. J., Stockard, J. E., Sharbrough, R. W. 1980. Brainstem auditory evoked potentials in neurology. In Aminoff, M. J. (Ed.), *Electrodiagnosis in Clinical Neurology*. Churchill Livingstone, New York.

Therman, E. 1980. *Human Chromosomes: Structure, Behavior, Effects*. Springer-Verlag, New York.

Thompson, J. S. and Thompson, M. W. 1980. *Genetics in Medicine*. Saunders, Phila.

Tiwari, J. L., Hodge, S. E., Terasaki, P. I., Spence, M. A. 1980. HLA and the inheritance of multiple sclerosis: linkage analysis of 72 pedigrees. *Am. J. Hum. Gen.*, 32:103–111.

Vinken, P. J. and Bruyn, G. W. (Eds.). 1970–81. *Handbook of Clinical Neurology*, (43 volumes). North-Holland/American Elsevier, New York.

Vogel, R. and Motulsky, A. G. 1979. *Human Genetics: Problems and Approaches*. Springer-Verlag, New York.

Williams, A., Eldridge, R., McFarland, H., Houff, S., Krebs, H., McFarlin, D. 1980. Multiple sclerosis in twins. *Neurology*, 30:1139–1147.
Winter, W. P., Hanash, S. M., Rucknagel, D. L. 1979. Genetic mechanisms contributing to the expression of the human hemoglobin loci. In Harris, H., Hirschhorn, K. (Eds.), *Advances in Human Genetics*, Vol. 19. Plenum, New York, pp. 229–292.
Yunis, J. J. and Chandler, M. E. 1977. The chromosomes of man: clinical and biologic significance. *Am. J. Path.*, 88:466–495.

2 The Genetics of Epilepsy

V. Elving Anderson

Professor, Genetics and Cell Biology, University of Minnesota, Minneapolis, Minnesota

W. Allen Hauser

Associate Professor, Neurology and Public Health, G. H. Sergievsky Center, Faculty of Medicine, Columbia University, New York, New York

---

INTRODUCTION

Epilepsy is a common medical problem, and familial differences in susceptibility have long been recognized. Nevertheless, our understanding of the genetic basis has been slow in developing, and there are many inconsistencies in the conclusions drawn from various studies.

Part of the problem arises from inadequacies in the way data were presented and analyzed in many of the earlier studies, but the complexity of the nervous system, especially in its developmental aspects, also has presented formidable obstacles. There are recent developments, however, that permit considerable optimism that some significant advances may be possible in the near future.

Within the past few years there have been several comprehensive reviews of the relationship between family history and epilepsy (Doose, 1981; Hauser, Annegers, and Anderson, 1983). In an earlier volume Newmark and Penry (1980) summarized the clinical and family history studies through 1976. A neurogenetic directory edited by Myrianthopoulos (1981) provides a compilation of descriptions of a number of the clinical seizure syndromes. Finally, the major approaches required for thorough study were the subject of an interdisciplinary conference (Anderson et al., 1982).

Recent advances in the neurosciences at the molecular and cellular levels are beginning to demonstrate the power of genetic methodologies (Breakefield, Pintar, and Rosenberg, 1982). From the phenotypic point of view, methods for the clinical evaluation of brain structure and function will assist in defining heterogeneity.

This review is selective rather than exhaustive. Emphasis is placed on papers published since 1975. When a series of papers have emerged from an ongoing study, we have usually cited only the most recent or most comprehensive report.

GENERAL CONSIDERATIONS

Epidemiological Data

The evaluation of the contribution of genetic and environmental factors to the occurrence of seizures requires appropriate epidemiological data (Baumann, 1982; Ward, 1982; Hauser, Annegers, and Anderson, 1983). For each seizure type the comparison between the base rate in the population and the risk among siblings or offspring of probands provides the basic information needed to estimate the etiological importance of genetic factors. Similarly, in genetic counseling families need to know both the base rate and the empiric risk to relatives of probands.

Leviton and Cowan (1982) have reviewed the epidemiological features of seizure disorders in children, with appropriate comments on genetics.

Neugebauer and Susser (1979) have emphasized the areas that need more extensive and careful analysis.

These papers carefully outline the important distinction among estimates of incidence, prevalence, and cumulative risk. It is unfortunate that many interpretations of family histories concerning epilepsy have been flawed by the inappropriate choice of comparison data from the general population. For example, the cumulative incidence rate for epilepsy by age 40 from the Rochester, Minnesota, study (Figure 2.1) is 1.7 percent, but the prevalence at that age is only 0.6 percent. It is the former, and not the latter, that should be used for comparison with the empiric risk that relatives of probands will develop epilepsy by that age.

The possibility that racial and environmental factors play a role in the development of seizures must be taken into account. Higher rates for the incidence of febrile seizures in Japan and the Mariana Islands have been noted when compared with studies in Europe and North America. Japan and other countries (Mathai et al., 1968; Tsuboi, 1982). For most other types of seizures

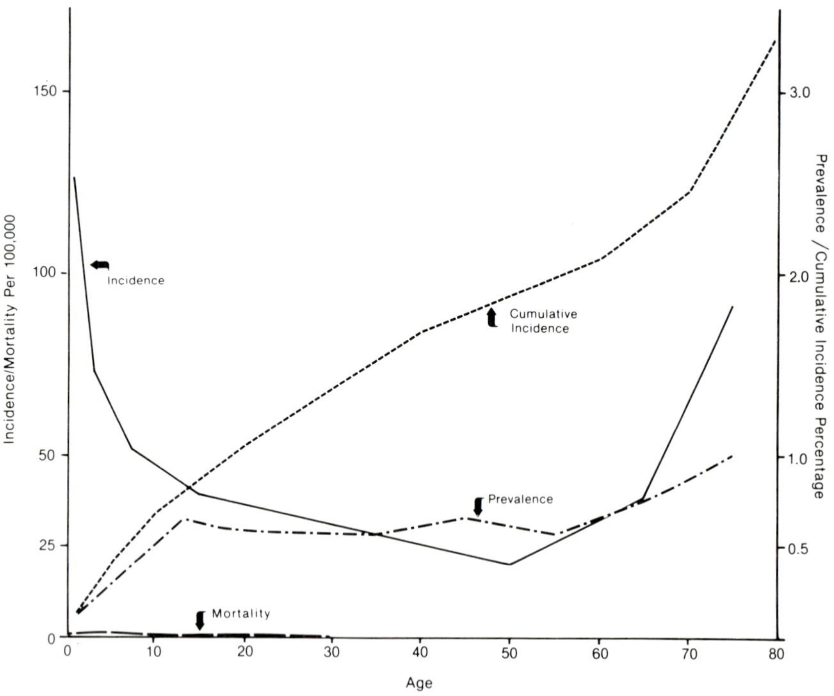

FIGURE 2.1 Incidence, prevalence, and cumulative incidence rates for epilepsy in Rochester, Minnesota, 1935–74. (Courtesy of J. F. Annegers, personal communication, 1983.)

it is difficult to know whether the differences among various reports reflect differences in populations or in methods of study (Hauser, Annegers, and Anderson, 1983).

One of the few studies providing data on racial variation in risks for epilcpsy was carried out in the New Haven area (Shamansky and Glaser, 1979). The cumulative incidence rates for epilepsy to age 15 were 19.63 per 1,000 for black males, 19.51 for black females, 9.53 for white males, and 9.10 for white females. During the later years of the study, there were higher rates of epilepsy in black and white children living in lower (as compared with higher) socioeconomic areas.

Classification of Seizures

As in any other genetic study it is important that the seizure phenotype be understood thoroughly (Hauser, 1982). Over the years considerable changes have occurred in the approach to the classification of seizures – in part because no one classification scheme has been suitable for all purposes. Recent revisions in the classification of epileptic seizures incorporate details of the clinical manifestations and electroencephalogram (EEG) patterns in patients with epileptic seizures. This classification is of use in the clinical management of patients but has not yet been demonstrated to be of particular use in the categorization of cases for genetic purposes. A long-term goal would be to develop a taxonomy that is sensitive to underlying differences in etiological mechanisms.

We have utilized a composite and somewhat simplified approach (Hauser et al., 1982). Seizures are categorized based upon the presence (symptomatic) or absence (idiopathic) of an historical event felt to reasonably predispose an individual to seizures. Those probands with symptomatic seizures are further categorized by the temporal relationship of the inducing or provoking insult to the seizure (acute versus remote). Another distinction involves the number of seizures (single versus recurrent).

The major considerations used in preparing the most recent International Classification of Seizures are reviewed in a committee report (Dreifuss, 1981), and a helpful interpretation from the genetic point of view is found in a paper by Delgado-Escueta, Treiman, and Enrile-Bacsal (1982).

Electroencephalographic Data

Some features of the EEG are of particular importance for genetic considerations. Vogel (1970) reported extensively on a number of characteristics of the EEG, finding a remarkable similarity of EEG frequency spectra among siblings. More recently he has explored the possibility that genetic factors influence the development and maturation of the brain (Vogel et al., 1982).

Studies by Doose and his coworkers in Kiel, Germany, have shown a strong family occurrence for certain specific EEG patterns. A theta EEG pattern (bilateral, synchronous, monomorphous waves of 4–7 Hz) seen maximally in the parietal area is found most commonly in children aged 2 to 6 years (Doose and Gundel, 1982). Among siblings of probands with this theta pattern, up to 30 percent of 3- to 4-years olds will show the same EEG pattern. Doose and Gundel argue that theta waves may serve as an indicator of a genetic predisposition for seizures. Other studies have been reported on occipital 2-4/s rhythms, spike and wave patterns, and the photoconvulsive reaction (Gerken and Doose, 1972, 1973; Doose and Gerken, 1973).

One of the problems for genetic studies is that generalized spike and wave (GSW) patterns, as well as other epileptiform patterns identified in probands with epilepsy, have rather specific age distributions (Doose, 1981; Kellaway, 1982). For example, even in individuals with epilepsy, interictal GSW patterns occur in the highest proportion of cases between the ages of 5 to 15 years are quite uncommon at younger and older ages. Thus it would be unlikely that parents or younger or older siblings of probands would demonstrate a spike and wave EEG pattern even if one had been expressed at some time in their lives. If, as Doose expects, theta patterns are an equivalent of GSW, it would be important to pay particular attention to these wave forms, that might otherwise be treated as normal variations.

A further problem in the use of the EEG as a marker of epilepsy is the time constraint in the recording itself. Some epilepsy probands, for example, show a GSW only at night or during sleep (Kellaway, 1982). While activating procedures such as hyperventilation, sleep, or photic stimulation are used to enhance the likelihood of identifying abnormal potentials, the development of epileptiform abnormalities with such provocative techniques may also provide unique information independent of any spontaneously occurring patterns. Thus the electroencephalographic evaluation of probands and relatives in a genetic study must be planned with great care and with attention to the best of current methods and instrumentation. It is also important that we have data about the frequency of electroencephalographic patterns of interest in the nonepileptic population, such as those provided by Eeg-Olofsson, Petersén, and Selldén (1971).

Basic Seizure Mechanisms

Seizures are the clinical manifestation of an abnormal and uncontrolled discharge among a set of neurons. Hauser, Annegers, and Anderson (1983) pointed out that the optimal state for neuronal excitability lies at an intermediate level between either extreme. In parallel with other physiological systems that function best in intermediate ranges, we would expect to find a complex control mechanism with a number of steps under independent genetic control. Thus epilepsy might arise from a defect in one or more of three major areas: neuronal inhibition, inactivation of excitatory neural transmitters, and feedback control.

From another point of view, a distinction can be made between seizure initiation and seizure propagation. At a basic level this involves neuronal excitability and synchrony and discharging neural circuits (Noebels, 1979). These considerations are highly relevant for the major subdivisions in the International Classification of Seizures.

Partial elementary seizures are the manifestation of a seizure discharge limited to a circumscribed area of the brain. A seizure state is initiated, but it does not spread widely.

Partial complex seizures involve a wider spread and deeper brain nuclei, but a limited proportion of the brain is involved in the discharge. This wider spread may be related to the anatomic site of origin of the discharge and/or to a different substrate in which propagation is facilitated. The involvement of deeper gray matter is thought to explain the memory deficit and confusion that accompany such seizures.

Generalized onset seizures often—perhaps always—involve a focal origin or localized site of initiation. In this case, however, there is a rapid generalization to involve the entire brain. Deeper brain structures mediate the generalization but are not necessarily the site of initiation.

It is possible, then, to envision the contribution of genetic factors either to the neuronal excitability that initiates seizures or to the mechanisms that facilitate the propagation of seizures to a wider area or throughout the brain.

GENETIC APPROACHES

Mendelian Traits

Evidence for the etiological heterogeneity of seizures is seen in the large number of single-gene disorders that can increase the risk of seizures (Table 2.1). An understanding of these traits is of obvious importance for the clinical evaluation and treatment of seizure patients (Jennings and Bird, 1981; Baraitser, 1982). Furthermore, it would appear that these "experiments of nature" have much to tell us about etiological pathways that can produce seizures.

For many of these individual conditions, however, it is difficult to find in the literature information regarding the following questions:

1. What proportion of individuals with this trait develop seizures?
2. What is the distribution by age at onset of seizures?
3. What types of seizures are observed?
4. What variations in EEG patterns have been found?
5. Which anticonvulsants are most effective and least effective in controlling the seizures associated with this trait?

6. Is there significant variation from family to family in the frequency and clinical manifestation of seizures?

TABLE 2.1. Mendelian Traits Associated with Seizures, with and without Mental Retardation

		With Seizures					
		Seizures and Retardation		Seizures Only		Total	
Mode of Inheritance	Total Number*	Number	Percent	Number	Percent	Number	Percent
Autosomal dominant	1,489	8	0.5	20	1.3	28	1.9
Autosomal recessive	1,117	61	5.5	27	2.4	88	7.9
X-linked recessive	205	10	4.9	7	3.4	17	8.3
Total	2,811	79	2.8	54	1.9	133	4.7

*From McKusick 1978.
Source: Anderson, 1982.

It must be emphasized that there is no one-to-one correspondence between seizures and any of these traits. In few of the traits are seizures found in all of the affected individuals. Furthermore, it is not yet clear whether a comprehensive review of the Mendelian traits will provide insight into the etiological pathways involved for the manifestation of seizures in general.

Chromosomal Syndromes

Essentially all of the detectable changes in the autosomes have some effect upon the nervous system, usually in the form of mild to severe mental retardation. The frequency of epilepsy in the various chromosomal syndromes has not been well defined, but the attack rate appears to be quite variable.

Among individuals with Down's syndrome, the frequency of epilepsy has been reported to vary from 3 to 9 percent (Pueschel and Rynders, 1982). In a survey of 1,654 hospitalized Down's syndrome patients around London and in southeast England, Veall (1974) found a prevalence of 5.8 percent. (*Epilepsy* was defined as having had a fit during the two years prior to the survey or having been on regular anticonvulsants during that period for fits that had appeared earlier.) Veall acknowledged that an institutionalized sample is biased from several points of view and that this definition of *epilepsy* excludes those who have stopped having seizures or those who have died shortly after the onset of seizures. Nevertheless, the data are useful as a first approximation to estimates of prevalence. In the age group from 0 to 19, the prevalence of epilepsy was 1.9 percent. In the age groups from 20 to 59 there were some variation, but the overall average prevalence was 6.5 percent. In the much smaller sample from ages 60 and above, the prevalence was 15.4 percent. Thus the general form of this prevalence curve is very similar to that observed in the Rochester, Minnesota, population (Figure 2.1), although about fivefold higher in older age groups.

In a study of 111 patients with Down's syndrome in a hospital serving the city of Glasgow, 11 (8.1 percent) had epilepsy (MacGillivray, 1967). In general, their seizures were milder than for the other retarded patients. In another series of 128 patients in a hospital in Lancaster, England, 13 (10.2 percent) had a history of epilepsy. Only 5 of these had onset of seizures prior to age 25 (Tangye, 1979).

The diagnosis of tuberous sclerosis was at first considered for an infant girl with scattered vitiliginous spots and multiple hyperpigmented spots. The presence of dysmorphic features of the head and face, however, led to a chromosome study that showed a ring chromosome 14 (Schmidt et al., 1981). A review of this and six previously reported cases showed that seizures occurred in six out of seven. These were generalized tonic-clonic seizures and myoclonic jerks, which were fully controlled with anticonvulsant treatment.

Another syndrome with a high frequency of seizures is found in those cases with an extra inverted duplicated chromosome 15 (Schreck et al., 1977; Wisniewski et al., 1979; and Maraschio et al., 1981). Of the 28 cases reported in these three papers, 21 (75 percent) had seizures. In contrast with many of the chromosomal syndromes, there were no major physical abnormalities.

In the clinical descriptions of patients with various chromosomal problems, most of the attention is given to dysmorphic features. The general level of mental and motor development usually is mentioned, but the presence or absence of seizures may or may not be reported. The data needed for an appropriate epidemiological analysis are almost never available. A comprehensive review of the presently available literature and careful attention in planning new surveys would be helpful.

Statistical Analysis

As indicated above, there are a few families with single-gene traits in which the pattern of inheritance can be stated without much question. For other subgroups of the epilepsies, claims have been made for a specific mode of inheritance, usually autosomal dominant or polygenic. Most of these claims are premature, and it is only recently that the data in a form suitable for analysis are becoming available.

Within the past decade considerable attention has been given to the study of genetics for the common chronic disorders (Sing, Hanis, and Moll, 1982). A number of analytical strategies have been developed so that different genetic models can be applied.

The choice of genetic models to be tested has strong implications for the sampling process (Moll, 1982). Moll illustrated this point by a reference to *kifafa*, an interesting disorder found in Africa (Jilek-Aall, Jilek, and Miller, 1979). In addition to seizures (which may be triggered by photic stimuli), the affected individuals show Parkinsonism, transient psychotic episodes, facial edema, and dry skin. Jilek-Aall and coworkers found the segregation ratio to be 0.20 when both parents were normal and 0.26 when one parent was affected. In a limited test of the data Moll was not able to draw any conclusions as to mode of inheritance, but her general approach to the problem was instructive.

It is further possible to apply the concept of multiple threshold models to data on the occurrence of seizures (Cloninger et al., 1982). These authors also discussed the use of genetic analysis to test heterogeneity within epilepsy and drew attention to the fact that many studies report the highest recurrence risk of epilepsy among daughters of epileptic mothers, an observation that does not fit the usual assumptions under multifactorial inheritance.

In view of the probability of a high level of heterogeneity for epilepsy, Anderson, Chern, and Schwanebeck (1981) have carried out computer simulation tests to evaluate the potential usefulness of multiplex sibships—those in which two or more siblings are affected. This approach was recommended for biochemical and other intensive studies of common disorders with a large degree of heterogeneity.

Twin Studies

The overall concordance rate for epilepsy in six major twin studies was 60.2 percent for monozygotic pairs and 13.2 percent for dizygotic pairs (Tsuboi, 1980). In addition, there have been a number of reports each including one or a few twin pairs concordant for specific types of epilepsy.

One of the larger population-based twin studies involved a follow-up of all twins born in Denmark from 1870 to 1910 (Harvald and Hauge, 1965). Among the monozygotic pairs, 10 out of 27 (37.0 percent) were concordant for epilepsy. Among the dizygotic pairs (pooling same-sex and opposite-sex) 10 out of 100 (10.0 percent) were concordant.

One of the larger series of twins identified through clinical practice is the one studied carefully by Lennox and Lennox (1960). When analysis is restricted to chronic seizures (epilepsy), the results are as follows. When there was evidence of an abnormality on clinical examination suggesting a brain lesion in the index case, concordance was seen in 4 out of 37 (10.8 percent) monozygotic pairs and in 5 out of 67 (7.5 percent) dizygotic pairs. When the clinical information favored an intact brain, concordance was 33 out of 47 (70.2 percent) monozygotic pairs as contrasted with 3 out of 54 (5.6 percent) dizygotic pairs. For many of the individuals there was adequate information about the EEG records, and these also were carefully analyzed.

In a smaller series Inouye (1960) identified 40 twin pairs from outpatients and inpatients. In reanalyzing the data, we excluded cases with Mendelian syndromes (a pair with ichthyosis and mental retardation, a pair with phenylketonuria, a pair with epiloia) and a pair first identified at 79 years of age. Furthermore, we did not count febrile convulsions alone as epilepsy. Among 24 monozygotic pairs, 12 were concordant for epilepsy. Among the 12 discordant pairs, the cotwin had an abnormal EEG in seven cases. Among 12 dizygotic pairs, none were concordant for epilepsy, although in one pair the cotwin had febrile seizures. There was EEG information on 8 of the dizygotic cotwins, and 2 showed an abnormal EEG.

Eaves (1982) has prepared an excellent review of the twin study methods that are available for use in planning new studies of epilepsy and seizures.

Association and Linkage

A number of medical conditions have been shown to have an association with a specific HLA haplotype, and a few conditions appear to show a genetic linkage with the HLA locus. Studies of these two phenomena with respect to epilepsy, however, have not yet produced consistent results. Smeraldi et al. (1975) reported an increased incidence of HLA-B7 in patients with astatic-myoclonic epilepsy. Cazzullo and Canger (1979) found a significant elevation of HLA-B7 (with a relative risk of 4.38) in 22 patients with Lennox-Gastaut syndrome. A significant excess of the A1-B8 haplotype was found in two studies of absence epilepsy (Fichsel and Kessler, 1982; Rivas, 1983). On the other hand, Eeg-Olofsson, Safwenberg, and Wigertz (1982) found a significant *deficiency* of the A1-B8 haplotype in probands with "benign epilepsy of childhood" and their parents. Monaghan, O'Sullivan, and O'Donohoe (1982) reported no significant differences in 19 patients with cryptogenic myoclonic epilepsy.

Comparisons among these studies are difficult since they involve small samples of patients and since few reports were based on the same types of seizures. Information about EEG patterns generally was not provided, although most of the syndromes investigated are associated with an increased frequency of GSW patterns. Further studies of this type are certainly desirable, although caution in interpretation is advised in view of the statistical difficulties that have been encountered in HLA studies of other conditions.

In the future the available number of marker loci will be reaching the point where linkage studies of selected epilepsy pedigrees will be justified. White and Skolnick (1982) have outlined the important contribution of DNA sequence polymorphisms to this task.

ANIMAL STUDIES

It may seem that *genetic heterogeneity* is a term that is introduced to cover inadequate data or reasoning. How can one rigorously test hypotheses of genetic heterogeneity? What is the evidence that specific seizure or EEG features can be conditioned by genes? Can one characteristic of seizures be under different genetic control than others?

The presence of precipitating factors might be assumed to be sufficient to cause seizures in and of themselves. Are there any animal models that suggest the possibility of genetic variation in response to such factors?

In the search for biochemical explanations for seizure susceptibility, is there good evidence that different mechanisms exist between and within species? Are any new biochemical hypotheses arising in animals that can be tested in the human?

In animals, as in humans, there are a number of Mendelian traits associated with seizures. In animals it will be easier to determine whether specific anatomical and biochemical correlates are involved in the pathways from genes to seizures. The clinical variability that is seen when a given mutant gene is placed in different genetic backgrounds may give some clue as to the variability that is seen in humans within and among families.

It is fortunate that there are a number of animal models for epilepsy and that the genetic and biochemical aspects have been reviewed recently (Consroe and Edmonds, 1979; Seyfried, 1979, 1982a, 1982b; Jobe and Laird, 1981; van Gelder, 1981). These epilepsies may not have precisely the same precipitating factors, seizure pattern, and EEG features as human epilepsies, but they can provide models and hypotheses for testing in human samples. Some of the recent studies will help to illustrate the usefulness of animal models.

Audiogenic Seizures

A number of different genetic hypotheses have been proposed to explain susceptibility to audiogenic seizures (ASs) in the mouse. Seyfried (1982b) approached this problem by studying 21 recombinant inbred (RI) strains produced by inbreeding the F_2 generations derived from strains with high and low susceptibility. If the susceptibility is controlled by a single genetic locus, the 21 strains would be sorted into two distinct groups, some clearly high and others low. Instead, the seizure scores ranged from high to low, thus clearly excluding a single factor mode of inheritance and demonstrating the presence of multiple factors with a threshold for expression.

One of the RI strains (BXD-15) was highly susceptible at both young and adult ages (Seyfried, 1983). When the BSD-15 strain was crossed with the original AS-susceptible strain, the hybrids were susceptible only at the young age. On the other hand, when the BXD-15 strain was crossed with the original AS-resistant line, the hybrids were susceptible only at the adult age. Furthermore, the early onset susceptibility was strongly associated with variation at the *Ah* locus that controls the induction of aryl hydrocarbon hydroxylase activity, whereas there was no association between the *Ah* locus and adult onset AS susceptibility. These findings strongly suggest that the genetic mechanism responsible for early onset AS susceptibility is different from that responsible for late onset AS susceptibility.

Others studies provide further evidence for genetic heterogeneity. It is known that susceptibility to ASs can be induced in some initially resistant strains of mice by exposure to an initial auditory stimulus (acoustic priming). Maxson, Cowen, and Sze (1977) found strain differences that suggest at least two neuromechanisms of acoustic priming, each with its own genetic basis but both requiring corticosteroids for expression. McCall and Frierson (1981) studied the response of AS-susceptible and -resistant strains to a different type of stimulus: increasing pressure in helium/oxygen atmospheres. Differences in seizure threshold pressure were indeed found, mainly the result of a major locus (possibly linked to the H2 locus) and a minor locus (probably unlinked).

Some evidence also is emerging about possible biochemical explanations for AS susceptibility. Rosenblatt, Lauter, and Trams (1976) found that Ca^{2+}-stimulated ATPase activity was significantly lower in membrane-enriched preparations from the brains of AS-prone (DBA) mice than in control (C57 and C3H) mice. A subsequent study of cell cultures showed that the reduction of ecto-ATPase activity was found in glia cells (Trams and Lauter, 1978). Palayoor and Seyfried (1983) studied congenic mice and found that the Ca^{2+}-ATPase activity is not controlled by the Ias^b gene, which inhibits the spread of ASs. Furthermore, among 13 BXD recombinant inbred strains varying in AS susceptibility, a highly significant correlation was found between the seizure susceptibility and the Ca^{2+}-ATPase activity (but not the total, Mg^{2+}-, or Na^+K^+-ATPase activities). This result provides strong evidence that the difference in Ca^{2+}-ATPase activity is tightly associated with the seizure susceptibility and is not merely the incidental result of homozygous fixation in the original susceptible inbred strain.

Mendelian Traits

Among the nearly 150 identified neurological mutants in the mouse, there are alleles at more than 13 different loci for which recurrent spontaneous seizures are part of the manifestation (Noebels, 1982). The variations in EEGs and in seizure patterns are of particular interest to clinicians. Noebels points out, however, that heterogeneity at this genetic level does not preclude the possibility that several of the related molecular defects might occur at separate biochemical steps of a single neurotransmitter receptor system or cellular pathway. Noebels (1979) also noted that seizure disorders of a similar general pattern can result from mutations at different loci.

The electrocorticogram in the mutant *tottering* (*tg/tg*) mouse shows abnormal 6/s bursts of bilaterally synchronous spike-waves resembling the 3/s spike-wave complex that is seen in humans (Kaplan, Seyfried, and Glasser, 1979; Noebels, 1982). These cortical discharges are accompanied by a sudden arrest of movement, twitching of the vibrissae or jaw, and single myoclonic jerks of the head or a limb, thus resembling seizures of the centrencephalic type, both myoclonic and "absence." The *tg/tg* mouse also shows a highly stereotyped pattern of focal motor seizures, which provides a strong reminder that genetically conditioned seizures do not have to be generalized in form.

The brain of the *tg/tg* mouse (Levitt and Noebels, 1981) has a significant increase in the number of noradrenergic axons in those brain terminal fields that are innervated by the nucleus locus ceruleus. A concomitant rise in epinephrine levels was found in those same areas including hippocampus, cerebellum, and dorsal lateral geniculate. On the other hand, the number and size of locus ceruleus cell somata were identical in both wild type and *tg/tg* mice. Seyfried et al. (1981) found that *tg/tg* mice have a reduced concentration of gangliosides in the cerebellum, probably secondary to a loss or shrinkage of neuronal membranes.

In a further study of homozygous *tg/tg* mice, Heller, Dichter, and Sidman (1983) explored the antiepileptic effects of several drugs. The absence seizures were significantly reduced by ethosuximide, diazepam, and phenobarbital but not by phenytoin. This specificity is similar to that observed for absence seizures in humans and for seizures induced in animals by pentylenetetrazol. The authors hypothesized that the drugs effective against *tg/tg* seizures may share a common mechanism through their affects on gamma-aminobutyric acid (GABA). Further evidence in support of this hypothesis is the fact that aminooxyacetic acid, a drug known to raise GABA levels, is also very effective against absence seizures in the *tg/tg* mouse.

Among those mice homozygous for epileptiform (*epf*), 5 percent die at three to six months of age in a posture suggesting status epilepticus as the cause of death (Hare and Hare, 1979). The seizures can be induced by mild stimuli (physical, visual, or auditory). After a seizure there is a refractory period that lasts in individual mice from a few minutes to several hours.

Just at or after puberty, 25 percent of the homozygotes for spontaneous seizures (*sps*) show behavioral arrest and spontaneous generalized seizures (Maxson et al., 1983). The behavioral arrest is associated with 1-2 Hz high-voltage spikes in the neocortex, and the generalized convulsions are associated with paroxysmal activity. The affected mice do not show any gross motor abnormality such as tremors, spasms, ataxias, or other motor incoordinations. They are hyperreactive to tactile stimuli, but this does not trigger the behavioral arrest or generalized convulsions. It is also of some interest that the homozygous mice develop crusty sores of the skin, tiny growths around the eyes, and cataracts (which may indicate some metabolic defect).

Alcohol Withdrawal Seizures

There have been no adequate genetic studies of alcohol withdrawal seizures in the human. It seems appropriate, therefore, to review the evidence for marked strain differences in response to alcohol withdrawal in the mouse (Griffiths and Littleton, 1977; Kakihana, 1979). In DBA/2 mice the withdrawal syndrome is much more severe and includes spontaneous convulsions and some deaths, whereas C57BL mice show little response. Griffiths and Littleton reported that DBA (but not C57) mice showed a significant reduction in brain GABA and proline concentrations and an increase in aspartate concentrations during alcohol withdrawal. Kakihana found that plasma corticosterone levels were significantly elevated in DBA mice on an alcohol diet and increased further during withdrawal, whereas C57BL mice showed no significant changes.

In mice from a heterogeneous stock (HS), the most obvious and consistent signs of alcohol withdrawal have been seizures and loss of hypothalamic regulation of body temperature (Hutchins et al., 1981). (At low room temperature the mice show marked hypothermia after ethanol withdrawal, while hyperthermia is observed at 34° Centigrade.) On several of the measures females were significantly less affected by withdrawal even though they consumed significantly more ethanol. Further studies in HS mice showed that alcohol preference is not related to severity of withdrawal symptoms (Allen, Fantom, and Wilson, 1982), suggesting that alcohol preference and alcohol dependence may be mediated through more or less separate biological systems.

Starting with a foundation population of 200 genetically heterogeneous mice, a selective breeding program has been initiated, using a multivariate index of the ethanol withdrawal syndrome as a criterion (McClearn et al., 1982). After five generations the results indicate that selection is proceeding successfully (Allen et al., 1983), thus providing additional evidence for genetic factors controlling signs of alcohol withdrawal.

Finally, in a study of handling-induced convulsions after ethanol withdrawal in 17 recombinant inbred lines, 8 resembled the C57BL parent, 8 resembled the DBA parent, and only 1 showed variable results (Crabbe et al., 1983), thus indicating the possible influence of a major gene on ethanol withdrawal severity.

Other Studies

Jobe and Laird (1981) reviewed the neurochemical mechanisms proposed to account for seizure susceptibility and intensity in the several animal genetic models of epilepsy. In the epileptic rat there is evidence for a defect in noradrenergic transmission but little evidence for dopaminergic influence (Jobe et al., 1982). In contrast, dopaminergic transmission appears to have a strong modulatory effect on seizure activity in the DBA/2J strain of mice. The authors concluded that there is unlikely to be a common neurotransmitter defect that underlies all genetic seizure disorders in man or other animals. A biochemical heterogeneity appears much more likely.

Van Gelder (1981) reviewed the association between seizures and alterations of amino acid metabolism and concluded that changes in the compartmentalized metabolism of glutamic acid (either alone or as a consequence of an excessive release of taurine from tissues) may predispose toward epilepsy. In a further test of the potential role of taurine, Bonhaus, Laird, and Huxtable (1982) found that the high-affinity taurine uptake into platelets and into P_2 synaptosomal brain fractions were significantly lower in seizure-susceptible rats than in resistant strains.

Feline generalized penicillin epilepsy exhibits many of the EEG, behavioral, and pharmacological features that are characteristic of human primary generalized epilepsy. Gloor (1982) interpreted the data as suggesting that the spike and wave discharges result from the enhancement of the excitability of cortical neurons rather than from a breakdown of inhibition. Such a distinction may have important implications for the formulation and testing of genetic models for spontaneous epilepsy.

FAMILY STUDIES

Two General Strategies

Two general strategies have been used for the selection of probands for family studies in epilepsy and other common conditions. In the first of these the selection is as broad as possible, covering all ages and all etiologies. Information about seizure type, age at onset, family history, and EEG patterns is used as a part of the analysis of the data. The second strategy is to define a specific subtype of seizures as a basis for selecting the probands. Both of these strategies are of value, although the questions being asked are somewhat different.

One current illustration of the first strategy is seen in studies based at the Mayo Clinic in Rochester, Minnesota. Hauser and Kurland (1975) reported an extensive epidemiological analysis of the occurrence of seizures and epilepsy in Olmsted County over a 30-year period. More recently Annegers, Hauser, and Anderson (1982) have initiated a genetic study selecting probands from within that population. This approach has the advantage of providing an essentially complete ascertainment of cases of epilepsy as well as accurate information about the population base rates. The analysis reported was based on 196 probands who were under 15 years of age at initial diagnosis of idiopathic epilepsy while residing in Rochester, Minnesota, from 1935 through 1974. Among 489 siblings, 12 cases of epilepsy were observed (2.5 percent) as compared with an expected number of 4.8 (1.0 percent) calculated from the age-specific incidence rates in the general population. The cumulative risks for epilepsy and for all seizures are shown in Table 2.2. (The higher values observed among the children are based on small numbers.)

The second approach is illustrated by the studies initiated by Metrakos and Metrakos (1960) and Metrakos and Metrakos (1961) and amplified in subsequent papers (Metrakos and Metrakos, 1970; Metrakos and Metrakos, 1974). They focused their attention on centrencephalic epilepsy. They defined

TABLE 2.2. Cumulative Risk of Epilepsy and Seizure Disorders (in percent)

To Age	Siblings of Probands	Children of Probands	Rochester Population
Epilepsy			
5	0.7	2.7	0.4
10	1.2	4.9	0.7
20	2.7	10.6	1.1
40	3.6	–	1.7
All seizure disorders			
5	8.1	6.7	3.1
10	9.0	8.8	3.5
20	10.8	14.3	4.1
40	11.6	–	5.1

Source: From Annegers, Hauser, and Anderson, 1982.

their probands using three criteria: seizure history (recurrent petit mal and/or grand mal seizures), no obvious neuropathology, and a generalized spike and wave EEG pattern. Their reports included an analysis of the frequency of a generalized spike and wave EEG pattern or seizures in relatives and will be reviewed in a later section.

Symptomatic Seizures

Neonatal Convulsions

Neonatal convulsions is a term applied to seizures occurring in the first month of life (Blume, 1982; Gomez and Klass, 1983). About 60 percent of the cases have no further seizures and do not demonstrate evidence of neurologic deficit at follow-up. Some decades ago the majority of cases were associated with metabolic insults such as hypocalcemia or hypoglycemia. With better management of newborns generally (and in particular those with low birth weight), neonatal convulsions attributable to these metabolic causes have been all but eliminated, and the remainder are primarily associated with anoxic insults and/or intracranial hemorrhage. Some of the cases of neonatal seizures (some genetic and some apparently not) are associated with mental retardation and other serious outcomes.

With this background it is of interest that there are some families with a number of cases of neonatal seizures that are benign in outcome. Bjerre and Corelius (1968) reported 14 family members in five generations who had frequent convulsions within the first few weeks of life. Vitamin B_6 dependency was excluded, and there were no disturbances of electrolyte or glucose balance and no problems with amino acid metabolism. There was no retardation, and the outcome was favorable.

Quattlebaum (1979) also reported a family with 15 affected members in four generations. The outcome was benign, and the risk of subsequent epilepsy was low. In both of these families the pattern of inheritance was compatible with an autosomal dominant trait.

Febrile Convulsions

Convulsions associated with febrile illness are one of the common acute neurological disturbances seen in childhood (Hauser, 1981). In the United States one third of the patients with one febrile convulsion will experience a recurrence of febrile seizures. Furthermore, patients with febrile seizures have a three- to sixfold increase in the risk of epilepsy when compared with the general population.

The detailed studies on febrile convulsions have provided clear evidence for an increased risk among siblings of probands (Lennox-Buchthal, 1973; Tsuboi, 1982). Furthermore, there are large differences in population base rates and in the risks for febrile convulsions among siblings of probands (Table 2.3).

The two studies listed from Japan (Fukuyama, Kagawa, and Tanaka, 1979; Tsuboi, 1982) show similar sibling risks of about 20 percent and a population rate of about 7 percent. The former report used a literature population rate of 2.9 percent for statistical analysis, but this is clearly inappropriate since the authors quoted higher rates for Japan (much closer to the population and control rates from Tsuboi). The sample sizes for these studies were large enough to permit partitioning the data by parental history of febrile seizures and by the number of febrile seizures experienced by the probands (Table 2.4).

The three other reports in Table 2.3 (including data from the United States and Denmark) involve epidemiological studies as a basis for estimating population frequencies and for selecting probands. The population rates and sibling risks are much lower than for the Japanese studies, but the results from all these studies are consistent with (although not proof of) multifactorial inheritance. For the Japanese data an estimate of the heritability of liability (based on the graph in Smith [1970]) would be about 0.65 or 0.70, depending on whether the control or population rates, respectively, are used for comparison. For the pooled data from the United States and Denmark the estimate is about 0.55, which is not appreciably different. Thus the two sets of data are internally consistent, but the differences in population base rates between the two sets remain unexplained.

Several twin studies also have been reported. Lennox-Buchthal (1971) analyzed the data for the twin series collected by W. G. Lennox and found 13 out of 19 pairs (68 percent) concordant for febrile convulsions. It was noted, however, that these twins had a higher proportion with a severely abnormal birth history and with subsequent epilepsy than did the population-based series studied by Frantzen et al. (1970).

TABLE 2.3. Risk of Febrile Convulsions (FC) among Siblings of Probands with FC

Study	Siblings of Probands			Population			Control		
	Total Number	With FC		Total Number	With FC		Total Number	With FC	
		Number	Percent		Number	Percent		Number	Percent
Fukuyama, Kagawa, and Tanaka, 1979	287	57	19.9						
Tsuboi, 1982	1,094	227	20.7	16,806	1,123	6.7	1,252	105	8.4
Total	1,381	284	20.6						
Frantzen et al., 1970	303	33	10.9	9,656	292	3.0			
Van den Berg, 1974	323	29	9.0	6,956	144	2.1	489	6	1.2
Hauser, Annegers, and Kurland, 1982	1,042	82	7.9			2.3[a]			
Total	1,668	144	8.6			2.4[b]			

FC = febrile convulsions

[a]Expected proportion calculated from age-specific population rates.

[b]Pooled estimate weighted by number of siblings.

TABLE 2.4. Risk of Febrile Convulsions (FC) in Siblings of Probands, by Parental History of FC and Number of Episodes of FC in the Proband

	Fukuyama, Kugawa, and Tanaka 1979			Tsuboi, 1982		
		With FC			With FC	
	Total Siblings	Number	Percent	Total Siblings	Number	Percent
Parental history						
Both with FC	4	2	50.0	18	7	38.9
One with FC	73	17	23.3	210	66	31.4
Neither with FC	210	38	18.1	866	154	17.8
Total	287	57	19.9	1,094	227	20.7
Number of episodes of FC in proband						
1-4	160	26	16.2			
5-19	82	18	22.0			
10 or more	45	13	28.8			
Total	287	57	19.9			

FC = febrile convulsions

Schiøttz-Christiensen (1972) searched the records for twins born in the Copenhagen area from 1950 to 1965. There were 66 index pairs with at least one having febrile convulsions, but two pairs refused to participate, leaving 64. The incidence of febrile convulsions among the twin-born individuals was estimated at 3.0 percent. Among the monozygotic pairs 8 out of 26 (31 percent) were concordant, whereas among the dizygotic pairs 5 out of 37 (14 percent) were concordant. Since the ascertainment of probands was carefully described, it was possible to estimate the risk of febrile convulsions to be 44 percent for monozygotic cotwins of probands and 20 percent for dizygotic cotwins.

The monozygotic-dizygotic differences in the Schiøttz-Christiensen study were not statistically significant, and the results have been cited as evidence against genetic factors. It seems more reasonable to conclude that the results were indefinite because of the relatively small sample size and that both genetic and nongenetic factors are involved. This interpretation is strengthened by Tsuboi's (1982) report of concordance rates of 46 percent (25/54) for monozygotic pairs and 13 percent (12/89) for dizygotic pairs in his study.

Posttraumatic Epilepsy

Anecdotal observations have led to the suggestion that individuals with seizures following head trauma have a genetic predisposition for seizures, but data to support these contentions are scanty and contradictory. The proportion of cases with posttraumatic seizures was greater in patients with a family history of seizures in the studies of Caveness (1963) and Jennett (1975), but the difference was not significant in either study. Phillips (1954) reported a family history of epilepsy in 12.6 percent of patients with seizures following closed head trauma. He concluded that this was no higher than expected among any group of epileptic patients but presented no comparable information from the general population or from a head trauma cohort. Evans (1962) reported a family history of seizures in 7 percent of trauma patients with seizures and in only 2 percent of trauma patients without seizures.

Seizures occurring within the first few days of a head trauma episode presumably result from different mechanisms than seizures that begin months or years following the episode. Thus recent studies of posttraumatic epilepsy consider early seizures (within the first one to two weeks) separately from late seizures (after the first two weeks). With this distinction in mind, it has been suggested that patients with a family history of epilepsy are at greater risk for late posttraumatic epilepsy than trauma victims without a family history. In Jennett's series patients with a family history of epilepsy did not have a higher frequency of early seizures than those without a family history. In Oslo, patients with a family history of seizures were thought to have an increased frequency of early seizures (Kollevold, 1978), although the proportion of patients with a family history of seizures was 6 percent, actually less than the 8 percent reported by Jennett.

Other Symptomatic Seizures

Infantile spasms is a term for a unique type of seizure that usually starts in the first year of life and that generally responds to adrenocorticotropic hormone (ACTH) therapy. Infantile spasms are a symptom complex associated with a number of diseases to the nervous system, including tuberous sclerosis and phenylketonuria (Dyken and Krawiecki, 1982). (The separate and distinct syndrome of infantile spasms, hypsarrhythmia, and mental retardation is sometimes called *West syndrome*.) Feinberg and Leahy (1977) reported a pedigree of infantile spasms with five affected males from three generations in a pattern compatible with an X-linked recessive mode of inheritance. The EEG at three months showed hypsarrhythmia (a pattern seen in the majority of cases with infantile spasms) and at four months showed multiple left midtemporal and right parietal focal epileptiform spikes without hypsarrhythmia. The proband at 13 months was seizure free on ACTH therapy. The four affected relatives died at ages from nine months to six years.

Rimoin and Metrakos (1964) studied two groups of children with hemiplegia: 98 with a history of at least one convulsion and 60 with no history of convulsions. Among the parents of the former group, 2.6 percent had a history of convulsions, and 15 percent had an epileptiform EEG. The comparable figures for the parents of the latter group were 1.5 percent and 3 percent. A similar difference was found for the siblings: in the first group, 5.6

percent of the siblings had a convulsion and 30 percent an epileptiform EEG, whereas in the latter group 3.6 percent had convulsions and 16 percent an epileptiform EEG. The results were interpreted as evidence for individual variation in seizure threshold that may be triggered by environmental stimuli, but heterogeneity in the etiological mechanisms leading to hemiplegia would be an alternative explanation.

As part of the Collaborative Perinatal Research Project, Hirtz, Nelson, and Ellenberg (1983) identified 39 children who developed seizures within a two-week period following a childhood immunization. All but one of these seizures were associated with a fever. The records collected before the birth of the children showed that 9 out of the 39 (23 percent) had a family history of a febrile or nonfebrile convulsion in a near relative. A similar family history had been obtained for 14 percent of the 1,706 children who developed febrile convulsions and in only 7 percent of the remainder of the 50,000 pregnancies.

Generalized Onset Epilepsy

The electroencephalographical hallmark of the generalized epilepsies is the interictal GSW. Thus studies of the generalized epilepsies are by definition studies of GSW, but GSW may also occur in individuals who have other forms of epilepsy—or for that matter in asymptomatic individuals. Thus patients having generalized epilepsies with GSW must be considered highly selected groups, and the results from studies of such patients cannot be generalized to all individuals with GSW.

Eisner, Paulin, and Livingston (1959) reviewed the data for a number of seizure patients and then focused their attention on 321 probands having a diagnosis of idiopathic major motor (generalized tonic-clonic) epilepsy. The data for siblings were presented in the form of cumulative risks, which can be presented graphically (Figure 2.2). When the onset of major motor seizures in the proband occurred before age 4, the cumulative risk for major motor epilepsy in the siblings was 7.5 percent by age 20. When the proband had onset at ages 4 to 15, the cumulative risk for siblings was 4.3 percent by age 20 and 8.0 percent by age 40. The comparable risk for the control group was 2.3 percent by age 40.

The study carried out by Metrakos and Metrakos (1970) has served as a benchmark for comparison with later studies and thus deserves a more extended discussion. The probands were identified at Montreal Children's Hospital and thus had onset in childhood. The classification of probands follows the proposal by Penfield and Jasper (1954), which places seizures into three main groups (localized, unlocalized, and centrencephalic), depending upon the presumed origin of the epileptic discharge. The term *centrencephalic* persists in the literature, although Gastaut (1964) suggested that it be replaced by "epilepsy of subcortical origin," and Gloor (1968) preferred the term "cortico-reticular epilepsy."

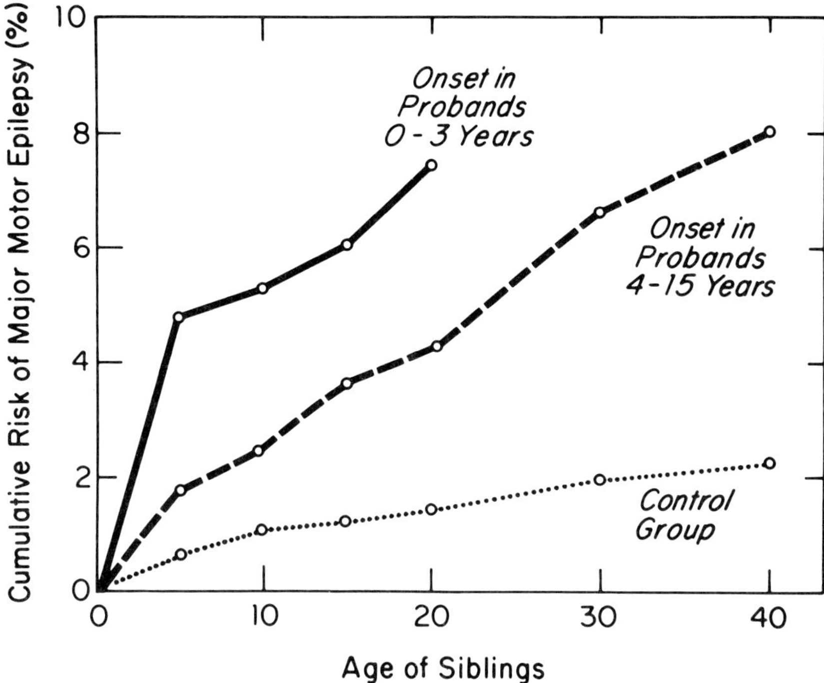

FIGURE 2.2 Cumulative risk for major motor epilepsy in siblings of probands with epilepsy. (From Anderson, 1982, using data from Eisner, Pauli, and Livingston, 1959.)

Based on the EEG tracings, the probands were divided into those with typical and atypical patterns (Metrakos and Metrakos, 1961). The prevalence rates of convulsions among near relatives of these two categories of probands were not different; thus the data were pooled.

The probands had either recurrent petit mal (absence) or grand mal seizures (not tabulated separately) with no obvious neuropathology that would account for their seizures. The EEGs for the probands and relatives were considered positive for a GSW pattern whether such a pattern occurred during the resting period, during hyperventilation, or during photic stimulation. The category of "spike-wave abnormalities" referred to all types of generalized paroxysmal abnormalities, both epileptiform and nonepileptiform (Andermann, 1982).

Control probands were collected by examining every twentieth admission to the same hospital and selecting those who had never had a convulsion, whose illness was not considered to be neuropathologic, and whose EEG was within normal or borderline normal limits.

The history of seizures among siblings of the centrencephalic and control probands is shown in Table 2.5. The information about EEG abnormalities among siblings also was evaluated carefully. Of particular interest was the fact that the percentage of siblings showing a centrencephalic EEG was highest in the age range of 4 to 16 years. Follow-up EEG studies were done on 94 of the siblings (Metrakos and Metrakos, 1974). At the time of their first EEG, the mean age was 7.6 years, and 29 percent had a centrencephalic EEG. The mean age at the time of the second EEG was 15.3 years, and at that time 28 percent had the EEG trait. However, 45 percent exhibited the trait in their first and/or second EEG. These data were used to support the hypothesis of an autosomal dominant inheritance for the EEG trait, that is thought to underlie the seizures.

Metrakos and Metrakos (1970) point out that there is no clear line of demarcation between focal and generalized forms of epilepsy. They suggest that centrencephalic epilepsy, focal epilepsy, and febrile convulsions all share the common genetically controlled predisposing factor—namely, the centrencephalic EEG trait. They furthermore point out that among those siblings of probands with centrencephalic epilepsy, who themselves have the EEG trait, about 25 percent have clinical seizures.

TABLE 2.5. Risk of Seizures and Epilepsy among Siblings of Probands with Epilepsy

Study probands	Total Number of Siblings	With Epilepsy		All Seizures	
		Number	Percent	Number	Percent
Metrakos and Metrakos, 1961					
Centrencephalic epilepsy (GSW)	519			66	12.7
Controls	322			15	4.7
Doose et al., 1973					
Absence (GSW)	448			30	6.7
Doose et al., 1970					
Myoclonic-astatic petit mal (GSW)	95			12	12.6
Matthes, 1969					
Absence (3/sec GSW)	240	9	3.8	24	10.0
Akinetic (slow GSW)	131	5	3.8	9	6.9
Tsuboi and Christian, 1973					
Impulsive petit mal	705	31	4.4		
Gerken et al., 1976					
Focal epilepsy only (FSW)	157			3	1.9
Generalized epilepsy only (FSW)	66			2	3.0
Multiple and other (FSW)	54			3	5.6
Total	277			8	2.9
Andermann, 1982					
Focal (mostly temporal)	229			11	4.8
Controls	458			18	3.9
Jensen, 1975					
Temporal lobe	171	5	2.9	13	7.6

GSW = generalized spike and wave
FSW = focal sharp wave

Gloor et al. (1982) elaborated on their concept of the genetic background for GSW patterns. This pattern represents an oscillation between a short period of increased excitation (the spike) and a following longer period of increased inhibition (the wave). They assume that the three-per-second spike and wave phenomenon is the result of the activity of a major gene. Furthermore, they propose that this genetic predisposition may be common to most epileptogenic conditions.

Another EEG pattern of interest is the appearance on the EEG of GSW discharges during stimulation with intermittent light (photoconvulsive response [PCR]) (Doose et al., 1969). The EEG pattern is essentially similar to the GSW pattern that occurs spontaneously. PCR has attracted a good deal of attention since it is the most common stimulus-sensitive brain phenomenon thus far identified in the human. In most other species an auditory stimulus is the most effective in producing such changes. PCR shows developmental fluctuations with age, with the highest manifestation generally being in the 5 to 15 years of age range. This, of course, will create difficulty in doing family studies since parents will be in an age range at which a lower percentage of manifestation will be expected.

In a series of 743 normal Swedish children (ages 1 to 15) Eeg-Olofsson, Petersén, andSelldén (1970) found a PCR in 12.3 percent of girls and 5.2 percent of boys. This bias toward a higher expression in females is a usual finding in other studies. In one of the earlier papers on this topic, Watson and Marcus (1962) observed a PCR in 13.2 percent of normal female subjects aged 18 to 30 years but in only 1.6 percent of males of the same age. The relatively high population rate should be taken into account when assessing reports of the association of PCR with febrile convulsions, psychomotor seizures, and some forms of focal epilepsy (Newmark and Penry, 1979).

A number of studies have shown a strong familial component in the manifestation of a PCR (Davidson and Watson, 1956; Hedenström, 1969; Jeavons and Harding, 1975; Takahashi, 1976). In some individuals, but not in all, a PCR occurs only if both eyes are stimulated (Procopis and Jameson, 1974).

When PCR declines with increasing age it may be replaced by some other EEG phenomenon. The photomyoclonic response (PMR) is a muscular response to intermittent light stimuli that seems physiologically different from the PCR. In a combined study of PCR and PMR, Rabending and Klepel (1970) studied relatives of 88 probands with both epilepsy and PCR. Among siblings of the probands, 22.0 percent showed PCR and 1.6 percent showed PMR. Among parents the proportions were reversed – 10.3 percent showing PCR and 19.2 percent showing PMR. Thus the two phenomena appeared to behave as age-dependent variations of the same underlying tendency (see also Klingler and Wessely, 1977).

At this point we can ask about the joint occurrence of GSW and PCR in epilepsy probands. Gerken and Doose (1973) studied 272 "mainly epileptic" probands with GSW in the resting and/or hyperventilation EEG. A second group was composed of 90 also mainly epileptic probands who had not shown a GSW in both the resting and hyperventilation EEG in at least three examinations at the age of 5 to 16 years. Finally, there was a group of 685 control children selected from the neighboring schools and kindergartens. The EEG findings for the siblings of probands are reported in Table 2.6. In a parallel paper Doose and Gerken (1973) looked at the PCR in the same series of probands. The findings in the tested siblings aged 5 to 16 years are shown in the same table. (The combinations of GSW and PCR in the siblings were not reported nor were these data related to the presence of seizures in the siblings.) The results suggest strongly that a GSW in the resting and hyperventilation EEG may be genetically different from GSW occurring only during photic stimulation.

Doose et al. (1973) studied 250 epileptic children who at some time had shown absences with generalized three-per-second spike-waves. Among 448 siblings there were 30 (6.7 percent) who had one or more seizures (Table 2.5). This rate was higher for sisters than brothers (9.1 percent versus 3.9 percent). Among 242 siblings for whom EEG tracings were available there were 54 (22.3 percent) with GSW and/or PCR. Of these, 15 showed GSW only, 33 showed PCR only, and 6 showed both.

Doose et al. (1970) also studied a smaller series of 51 probands with myoclonic-astatic petit mal of the centrencephalic type (Table 2.5). The onset is usually in the third and fourth year of life, and the course of the disease is generally unfavorable. Among 95 siblings there were 12 (12.6 percent) with a history of seizures.

Matthes (1969) carried out parallel studies (Table 2.5). Among 240 siblings of probands with absence epilepsy and three-per-second GSW, there were 9 (3.8 percent) with epilepsy and 15 (6.3 percent) with childhood convulsions, for a total of 10.0 percent. Among 131 siblings of probands with akinetic sei-

TABLE 2.6. Electroencephalographic Findings in Siblings of Probands and in Control Children

Probands			Gerken and Doose, 1973			Doose and Gerken, 1973		
			Total	With GSW		Total	With PCR	
Epilepsy	GSW	PCR	Siblings	Number	Percent	Siblings	Number	Percent
Yes	Yes	Yes	233	20	8.6	163	27	16.6
Yes	Yes	No	171	8	4.7	124	9	7.2
Yes	No	Yes	69	3	4.3	50	11	22.0
Yes	No	No	68	3	4.4	49	4	8.2
No	(control children)		685	12	1.8	662	33	5.0

GSW = generalized spike and wave (resting and/or hyperventilation EEG)
PCR = photoconvulsive response (GSW during photic stimulation with rhythmic flashes)

zures (but with a slow generalized spike and wave complex), there were 5 (3.8 percent) with epilepsy and an additional 4 (3.0 percent) with childhood convulsions, for a total of 6.9 percent.

Our interpretation of these results is that the GSW is an important electroencephalographic indicator of a genetically conditioned trait that affects susceptibility to seizures. However, the GSW appearing in response to photic stimulation should be recorded separately and may be genetically distinct. The story appears to be more complex than that reported by Metrakos and Metrakos, but the general importance of the electroencephalographic data remains unchallenged. The data are not yet sufficiently confirmed to determine the mode of inheritance.

One final category of generalized onset epilepsy is seen in the study of impulsive petit mal (or juvenile myoclonic epilepsy) carried out by Tsuboi and Christian (1973). The majority of the probands were aged 10 to 20 years at onset. In general the incidence of epilepsy was higher among relatives of female probands than among relatives of male probands. In addition, the risk of epilepsy was higher among female relatives than male relatives. This combination of findings does not fit the usual pattern for multifactorial inheritance and thus far has not received a satisfactory explanation. Among 705 siblings there were 31 (4.4 percent) who had developed epilepsy (Table 2.5).

The possible role of amino acids in human epilepsy was summarized by Janjua, Metrakos, and van Gelder (1982). They studied two groups of patients with epilepsy, one with a 3/sec. spike-wave EEG and a second group with other EEG abnormalities. The mean plasma levels of taurine (tau), glutamic acid (glu), and aspartic acid (asp) and the ratio tau/glu were found to be significantly altered in both of these groups of probands as well as in their near relatives, as compared to control probands. Goodman and Connolly (1982) reported evidence suggesting genetic variability in taurine excretion in the human population. Their calculations indicated that 21 out of 41 (51.2 percent) of epileptic patients appeared to be low excretors as compared with about 11 percent of the general population.

Quesney, Andermann, and Gloor (1981) showed that apomorphine can block the photosensitive response in patients with primary corticoreticular epilepsy. On the other hand, spontaneous spike and wave discharges were not blocked by apomorphine in any of the patients. The authors suggested that dopaminergic changes in primary generalized photosensitive epilepsy might be genetically determined.

Varma et al. (1982) reported that serum beta-glucuronidase activity was distinctly elevated in all of 30 patients with grand mal epilepsy, as compared with 30 normal subjects. There was no overlap in the ranges found for the two groups. This enzyme can degrade the glycosaminoglycans (acid mucopolysaccharides), which are components of the blood-brain barrier. The finding of such large differences in all of a group of seizure patients is surprising, and further reports will be awaited with interest.

Partial Epilepsies

The term *partial* (or *focal*) can be applied either to seizure manifestations or to EEG findings. Some of the cases of focal epilepsy may result from a structural epileptogenic focus such as a scar or mass. Others, however, may be related to the idiopathic centrencephalic epilepsies and may share a genetic predisposition. Bray and Wiser (1964, 1965) studied 40 probands with focal epilepsy whose EEGs showed paroxysmal sharp waves or spikes in the mid-temporal area, either unilaterally or bilaterally. Pedigree analysis was reported to show transmission as an autosomal dominant trait, although the details are not presented (except for a note that in 12 of the families at least one close relative had seizures associated with a focal temporal central spike abnormality on the EEG).

From the clinical EEG descriptions, many of the cases studied by Bray and Wiser appear similar to those studied by Heijbel, Blom, and Rasmuson (1975). The probands in the latter study had seizures and also Rolandic discharges on the EEG recording. Among the siblings at each age level one third showed Rolandic discharges. Of those with Rolandic discharges one half had seizures. This category of patients often is described as "benign epilepsy of childhood with centrotemporal EEG foci" and may represent a transition between generalized and partial epilepsies.

A larger series of 203 children with various forms of epilepsy were studied by Gerken et al. (1976). They were selected for having shown at least once focal sharp waves in the EEG (some in the Rolandic area). The sibling risk for seizures was considerably lower than in families of probands with spike and wave absences (Table 2.5).

One of the most carefully studied series is based on 60 patients with focal cerebral seizures (mostly temporal) who underwent surgical treatment for their epilepsy (Andermann, 1982). There was no significant increase in history of seizures among siblings as compared with control relatives (Table 2.5). There was, however, a significant increase in the prevalence of total EEG abnormalities, epileptiform EEG abnormalities, and focal and multiple diffuse (diffuse slow wave) abnormalities. Generalized spike-wave abnormalities, on the other hand, were not significantly increased.

Temporal lobe epilepsy, psychomotor epilepsy, and *complex partial seizures* are terms that have been used interchangeably over the years (Gastaut, 1982). However, the first term deals with the anatomical localization of the seizure onset, the second refers to a symptom complex, and the third implies propagation of the discharge to deeper brain structures. Thus, in the strict sense, these terms are not equivalent.

Barslund and Danielsen (1963) reported studies of three female monozygotic twin pairs. In two of the pairs both twins had similar temporal attacks. In the third pair one twin had dreamy states, and a daughter of the other twin had similar seizure episodes. In each of the three families epileptic seizures or temporal-spike foci were found among near relatives. In addition, other families were presented in which temporal epilepsy was present in several siblings.

A larger series of 74 patients who underwent unilateral temporal lobectomy for control of seizures was studied by Jensen (1975). Among 171 sibs there were 5 with epilepsy and 8 with other convulsions, for a total of 13 (7.6 percent) with seizure history (Table 2.5). The claim is made that the epilepsy rate among siblings is five to six times higher than expected in the general population, but this most likely results from the improper use of population prevalence values.

Lindsay, Ounsted, and Richards (1980) reported on a follow-up study of 100 children who were under 15 years of age at ascertainment of temporal lobe epilepsy. These probands who had siblings with convulsions appeared to have a more favorable remission rate.

A different approach was taken by Kristensen and Sindrup (1978). Among 96 probands who had partial complex epilepsy and also paranoid and hallucinatory signs, 16 (16.7 percent) had a family history of epilepsy. In contrast, 96 control probands selected for partial complex epilepsy without paranoid hallucinatory signs included 29 (30.2 percent) with a family history of epilepsy. The number and relationship of affected family members were not reported, however, nor was there any information about psychiatric disorders among relatives.

Studies of Offspring

Family Data

It is difficult to obtain a satisfactory sample of children of probands for analysis. Ideally, the proband should have been identified a generation ago to permit following the offspring to an age reasonably through the risk period.

Tsuboi and Endo (1977) analyzed their data about epilepsy and other seizures in children of probands from a variety of points of view (Table 2.7). The results are comparable with those from other studies of children and follow the trends demonstrated for siblings.

1. The risk was higher when the probands had idiopathic (as compared with symptomatic) epilepsy.
2. The risk was higher when the proband had an affected parent or sibling.
3. The risk was higher when the proband developed epilepsy at a younger age.
4. The risk was higher for children of affected mothers than for children of affected fathers.
5. Finally, the risk of seizures was considerably higher for those children who themselves showed a specific EEG abnormality.

TABLE 2.7. Risk of Seizures among Offspring of Epileptic Probands

Probands	Total Number	Offspring With Epilepsy Number	Offspring With Epilepsy Percent	Offspring With FC Number	Offspring With FC Percent	Offspring Total Affected Number	Offspring Total Affected Percent
Male probands	233	4	1.7	12	5.2	16	6.9
Female probands	273	8	2.9	22	8.1	30	11.0
Proband, idiopathic	382					42	11.0
Proband, symptomatic	124					4	3.2
Proband with affected parent or sibling	55					12	21.8
Other probands	451					34	7.5
Proband onset 0–9 years	99	3	3.0	11	11.1	14	14.1
Proband onset 10–29 years	250	7	2.8	16	6.4	23	9.2
Proband onset 30 or older	157	2	1.3	7	4.4	9	5.7
All probands combined	506	12	2.4	34	6.7	46	9.1
EEG in offspring GSW	38					11	28.9
No specific anomaly	64					3	4.7
Total offspring tested	102					14	13.7

FC = febrile convulsion
EEG = electroencephalograph
GSW = generalized spike and wave
Source: From Tsuboi and Endo, 1977.

An important ongoing study in Berlin is beginning to yield prospectively collected information about offspring of epileptic probands (Beck-Mannagetta and Janz, 1982; Janz and Beck-Mannagetta, 1982). The Rochester, Minnesota, data also have been studied from this perspective (Annegers, Hauser, and Anderson, 1982). The results are shown in Table 2.8 and indicate very similar findings.

Effects of Anticonvulsants

In the process of providing genetic counseling for women with epilepsy, it is important to consider the possibility that a fetus may be affected by genetic factors associated with the mother's predisposition for seizures or by the anticonvulsants that she may be taking. An evaluation of data from the Rochester, Minnesota, population-based study showed the rate of major malformations to be 2.4 percent for children born to mothers with epilepsy not taking anticonvulsant drugs during pregnancy, 10.7 percent for children of mothers who took anticonvulsants during pregnancy, and 3.8 percent among the children of men with epilepsy (Annegers et al., 1978).

In a prospective study at the Montreal Neurological Hospital the proportion with major malformations was 8.2 percent in live-born offspring of female patients and 1.1 percent in live-born offspring of male patients, as compared with 2.7 percent expected for the general population. In a retrospective study from the same hospital, the proportion of viable offspring having major congenital malformations was 15.9 percent for mothers taking anticonvulsant medication as compared with 6.5 percent for mothers not taking medication (Dansky, Andermann, and Andermann, 1982). The types of malformations seen were those generally associated with multifactorial etiology. This observation is compatible with the assumption that there is an interaction between a genetic predisposition and the effect of anticonvulsants.

The possibility that genetic variation in the fetus may affect teratogenic expression in the fetus is further suggested by information from multiple pregnancies. In one pair of dizygotic female twins (Loughnan, Gold, and Vance, 1973) one twin had a pulmonary atresia with patent ductus arteriosus, whereas the other was apparently unaffected. In a set of dizygotic triplets (one male and two female) all showed some malformation, but there was considerable variability in the nature of the changes (Bustamante and Stumpff, 1978). More recently we have the report of four-year-old dizygotic twin girls who may have been fathered by separate men of different races. One twin showed a variety of malformations, whereas the other showed only poor growth and a broad depressed nasal bridge (Phelan, Pellock, and Nance, 1982).

While there seems to be a high correlation between exposure to anticonvulsant drugs in utero and malformations, patients taking anticonvulsant drugs during pregnancy may be different from those not on such medication.

TABLE 2.8. Epilepsy and Febrile Convulsions in Offspring of Patients with Epilepsy

Study	Number of Probands	Total Offspring	With Epilepsy		With FC	
			Number	Percent	Number	Percent
Tsuboi and Endo, 1977	203	506	12	2.4	34	6.7
Janz and Beck-Mannagetta, 1982	414	768	26	3.4	36	4.7
Annegers, Hauser, and Anderson, 1982	196	93	4	4.3	3	3.2
Total		1,367	42	3.1	73	5.3

FC = febrile convulsions

Thus one could not exclude the possibility that the effect could be due in part to the epilepsy in the mother, per se, rather than to the anticonvulsants. (See also Janz, 1982; Annegers, Kurland, and Hauser, 1983.)

Finnell (1981) based an experimental model on mice of several inbred strains and on the three genotypes involving an autosomal recessive gene quaking (*qk*). This latter gene produces a spontaneous clonic-seizure disorder in homozygous mice that is treatable by the use of phenytoin. The results clearly showed that the occurrence of defects in the fetuses was correlated with maternal serum concentrations of phenytoin but not with the maternal or fetal genotype or the presence of a seizure disorder. Furthermore, there were no differences in rates of defects among the five lines that were tested. Finnell and Chernoff (1982) extended this model to test the effect of maternal seizure disorders on prenatal development. The results indicated that the maternal serum phenytoin level, and not the maternal seizure disorder, is the agent responsible for the malformations observed.

The American Academy of Pediatrics asked their Committee on Drugs to make a recommendation on this general point (Segal, 1979). The essence of their recommendation is as follows:

1. Whenever possible, a woman who has been seizure free for some years should be withdrawn from medication prior to pregnancy.
2. A woman who has epilepsy and requires medication may be advised that the risk of congenital malformations and/or mental retardation is two to three times greater than average.
3. For women who seek advice later than the first trimester of pregnancy, drug therapy generally should be continued because any major anatomical malformations presumably would have taken place already.
4. There is not yet sufficient information to recommend switching from one kind of anticonvulsant to another in the hope of ensuring a lower malformation rate.
5. The risk of seizures developing in the child, of course, would be established independently and would be over and above the possibility of malformation as described above.

GENETIC COUNSELING

The goal of genetic counseling is to provide individuals and families with the information and understanding they need in order to make informed choices about future reproduction. In the case of epilepsy close teamwork is required between the neurologist and geneticist, with the former evaluating the seizure history and EEG patterns and the latter helping to detect any genetic syndrome and estimating risks for siblings or children (Jennings and Bird, 1981; Reed, 1980; Baraitser, 1983a, 1983b; Ionasescu and Zellweger, 1983).

Major Factors in Evaluation

The extensive heterogeneity in epilepsy requires attention to a number of details. Our review of the literature shows that the following factors should be considered in the workup of counseling cases (Hauser, 1982).

1. *Age at onset*. In general, an earlier age at first idiopathic seizure in an index case is associated with a higher seizure risk for siblings. This observation parallels the findings for many other common disorders.
2. *Predisposing and precipitating factors*. The presence or absence of a history of anoxia, head trauma, cerebrovascular disorder, infection, tumor, or drug or alcohol withdrawal should be explored. There are age-specific changes in the relative contribution of these types of antecedent factors. For example, seizures associated with central nervous system (CNS) infections are seldom seen after five years of age. Nevertheless, there is little age-related change in the proportion of epilepsies that appear idiopathic (for which no antecedent factors are apparent).

   A distinction can be made between predisposing factors (that tend to increase the general hypersensitivity of neurons without necessarily triggering a seizure) and precipitating factors (that can more directly trigger a seizure episode). In some situations it may be possible to estimate whether transient or permanent brain damage has resulted. Furthermore, the time intervening between the presumed factor and the onset of seizures should be noted.
3. *Seizure type*. Some of the unique seizure syndromes discussed in the literature are not different entities but instead are age-specific variations in the manner in which seizures are manifested. A given individual may show distinct seizure syndromes at different ages. Furthermore, the manifestation of a specific seizure syndrome at a given age does not mean that the seizures necessarily will persist at older ages. Therefore, age-related changes in the seizure type appear to reflect the complex interaction of the nature and likelihood of specific antecedent factors, the developmental maturation of the nervous system, and genetic variation influencing the structure and biochemistry of the developing brain.

   Nevertheless, it is clear that some of the clinical seizure types may have different familial risks. Absence seizures, for example, have higher sibling risks than partial seizures. Excellent descriptions of the various seizure types and their relevance for genetic counseling are provided by Blume (1982) and Delgado-Escueta, Treiman, and Enrile-Bacsal (1982).
4. *EEG pattern*. The family studies of epilepsy probands (reviewed earlier) have established the importance of thorough EEG evaluation. Hyperventilation and photic stimulation should be used routinely within those age groups that are potentially responsive (about 4 to 19 years). Sleep also should be used when possible in order to increase the probability of eliciting a spike-wave pattern. The EEG response under each of these condi-

tions should be recorded separately, since the possibility of heterogeneity has not yet been explored fully.

5. *Associated findings*. The presence of malformations may suggest the desirability of a karyotype. Other physical and biochemical findings may contribute to the diagnosis of a specific Mendelian trait.
6. *Family history*. Specific questions should be directed toward the presence or absence of seizures in parents and siblings, with a more general inquiry about seizures in other relatives. If a relative is reported to have seizures, record should be made of the similarities and differences in comparison with the features observed in the index case. If there is a presumed precipitating cause (such as trauma) for the seizures in the index case but a close relative (such as one parent) also has a seizure history, the possibility of an underlying genetic predisposition should be considered.
7. *Possibility of a genetic syndrome*. Seizures may be observed as one manifestation of a large number of Mendelian traits (Table 2.1) and chromosomal syndromes. The total contribution of these specific conditions to the seizure caseload is quite small, but for the individual case a specific diagnosis significantly alters the counseling situation and may have other implications for the care of the patient. An abridged list of Mendelian traits is shown in Table 2.9, and a longer list is given in Anderson (1982), but the number of traits associated with seizures continues to expand. The most comprehensive discussion of these syndromes can be found in the volume edited by Myrianthopoulos (1981).
8. *Comparable population rate*. When a family is given an estimate of the risk for seizures in another child, they should have for comparison an estimate of the risk for the general population. Some discussions about genetic counseling for epilepsy have used incidence or prevalence rates. These are inappropriate values and should be replaced by the cumulative incidence rate. Since acute symptomatic seizures and seizures with onset at older ages are of less concern for counseling purposes, the cumulative risk for epilepsy (recurrent afebrile seizures) to age 40 seems reasonable. For the Rochester, Minnesota, population that figure is 1.7 percent, and the results are not significantly different for other white populations studied. The rates may be higher for black populations (Shamansky and Glaser, 1979). Furthermore, the population rates for febrile seizures are clearly higher in Japan than in Europe or the United States (Table 2.3).

Estimating Sibling Risks

In genetic counseling it has been customary to use the term *recurrence risk* to identify the estimated risk in siblings of probands. In discussions about seizures, however, the term *recurrence* generally means the repetition of a seizure event for a given person. Therefore, we find it useful to substitute the phrase *sibling risk*.

TABLE 2.9. Some Mendelian Traits Associated with Seizures

Autosomal Dominant	
14310*	Huntington disease
16220	Neurofibromatosis
17600	Porphyria, acute intermittent
19110	Tuberous sclerosis
Autosomal Recessive	
20420	Amaurotic family idiocy, juvenile type (Batten disease, Vogt-Spielmeyer disease, neuronal ceroid-lipofuscinosis)
20430	Amaurotic idiocy, adult type (Kufs' disease)
23050	Gangliosidosis, generalized GM_1, type I
23830	Hyperglycinemia, isolated nonketotic
24080	Hypoglycemia, leucine-induced
24350	Isovalericacidemia
24520	Krabbe's disease (globoid cell leukodystrophy)
24860	Maple syrup urine disease (branched-chain ketoaciduria)
25010	Metachromatic leukodystrophy, late infantile
25720	Niemann-Pick disease (sphingomyelin lipidosis)
26160	Phenylketonuria
26610	Pyridoxine dependency with seizures
27280	Tay-Sachs disease (GM_2-gangliosidosis, type I)
X-linked	
30800	Lesch-Nyhan syndrome
30940	Menkes syndrome
30950	Mental deficiency (Martin-Bell, or Renpenning type)
31160	Pelizaeus-Merzbacher disease

*The index numbers are from McKusick, 1978.
Source: Adapted from Jennings and Bird, 1981; Gomez and Klass, 1983.

Whenever a Mendelian or chromosomal syndrome is diagnosed, the problem becomes one of estimating the sibling risk for the syndrome (whether or not seizures develop in the sibling).

In the case of probands with febrile convulsions in Japan the sibling risk for febrile convulsions will be 20 percent as compared with a population rate of 7 to 8 percent (Table 2.3). Among Caucasians the sibling risk for febrile convulsions is 8 to 10 percent as compared with a population rate of 2 to 3 percent.

Estimates of sibling risk in other situations are based on the family studies that were reviewed earlier. The samples are not always defined in a comparable manner, and some types of seizures have not been well studied. A further problem is the failure to report the data separately for epilepsy and for other seizures (Table 2.5). Nevertheless, there is enough consistency to permit approximations for counseling purposes.

As a general figure we might start with the Rochester, Minnesota, study of probands with onset of idiopathic epilepsy at ages up to 15 years (Table 2.2). The cumulative risks for siblings were 3.6 percent for epilepsy or 4.7 percent when single idiopathic seizures (not febrile convulsions or seizures due to a presumed acute CNS insult) were included. With some rounding, this becomes 4 to 5 percent, as compared with 2 percent expected for the general population. The few studies in Table 2.5 that report epilepsy in siblings are very similar.

The risks for epilepsy in the offspring of epileptic probands from three studies (Table 2.8) are 2.4 percent, 3.4 percent, and 4.3 percent. In the first study the data were not reported separately for idiopathic and symptomatic probands. The figures for offspring of idiopathic epilepsy probands in the second and third studies are 3.8 percent and 4.3 percent (without age adjustment). These risk estimates are similar to those for siblings, but the cumulative risks (which involve age adjustment) may be somewhat higher for the offspring than for siblings.

The sibling risks for epilepsy will be somewhat lower (perhaps 3 percent) under the following conditions:

1. Proband with symptomatic epilepsy,
2. Proband with onset after 25 years, or
3. Proband with focal epilepsy.

For other questions the estimates from the current literature are inconsistent. Better data already exist in the records of large ongoing studies, and we need a collaborative comparison and pooling of the information. Meanwhile, the following estimates of sibling risk for epilepsy represent our effort to integrate the data presently available:

- 6 percent when the proband has absence or generalized clonic-tonic seizures and a generalized spike and wave EEG;
- 8 percent when, in addition to the preceding combination, the EEG also shows a photoconvulsive response and/or multifocal spikes;
- 8 percent when both the proband and one parent have idiopathic epilepsy (10 percent when either or both have an EEG with generalized spike and wave); and
- 12 to 15 percent for those tested siblings with generalized spike and wave when the proband also has a similar pattern.

It may be noted that the estaimates for sibling risks generally will be lower than the families may have feared. With the exception of the rarer genetic syndromes, sibling risks seldom will exceed 10 percent, as compared with a population base rate of 2 percent.

Genetic information may also be of help in estimating the prognosis for the index patient. In a study of 244 patients who presented for a first unprovoked seizure at any age, the cumulative risk of recurrence was 27 percent by 36 months after the initial seizure (Hauser et al., 1982). Among those idiopathic cases who had a sibling with seizures, the risk of recurrence was higher (35 percent) and earlier (within four months).

STRATEGIES FOR FUTURE RESEARCH

The potential role of genetic factors is disregarded in many ongoing studies of epilepsy, and the addition of genetic considerations could produce some valuable insights. In clinical studies the data could be sorted by the presence or absence of the history of epilepsy in a parent or a sibling. In experimental studies, tissues selected from genetically defined lines would strengthen the ability to interpret the data.

Wherever possible, methods to detect heterogeneity in etiology must be employed. On the other hand, there is a strong possibility of a convergent unity such that several distinct primary mechanisms may lead to epilepsy through a single final common pathway. The heterogeneity hypothesis can be studied in the large group of undefined idiopathic seizures, whereas the convergent unity hypothesis might be evaluated in a careful comparison of the mechanisms whereby the various Mendelian traits eventually increase susceptibility to seizures.

Insofar as possible, biochemical studies should be performed at the cellular level. Although we have no assurance that biochemical defects affecting the brain are also detectable in other body cells, the use of fibroblast cultures may prove useful for at least some seizure conditions.

Neuropeptides are known to be important for seizure susceptibility (Henriksen, 1982), and the possibility of genetic variation deserves careful attention (Bloom, 1982; Hahn, van Ness, and Chaudhari, 1982; Schmitt, 1982; Thoenen, 1982).

As indicated earlier, the use of restriction enzyme polymorphisms is rapidly improving the efficiency of genetic linkage studies (see also Housman and Gusella, 1982). Pedigrees with a number of affected individuals (five or more) in several generations are most suited for such a test. Clinicians who detect families of this type should bring them to the attention of a geneticist working on epilepsy.

Further advances in genetic analysis will require careful comparison of large samples from several different populations. This can be accomplished fairly easily if the major ongoing studies agree upon a common protocol for the tabulation of data in a form suitable for computer genetic analysis.

The advances suggested here will require a high degree of collaboration among persons with different backgrounds including clinical neurology, population genetics, molecular genetics, and neuroscience. Rapid developments in these several fields should make possible the resolution of some important questions in the near future.

Acknowledgments: This review was supported in part by Grant P50-NS-16308 from the National Institute of Neurological and Communicative Disorders and Stroke to the Minnesota Comprehensive Epilepsy Program and by Grant T32-NS-07153 from NINCDS to Columbia University.

REFERENCES

Allen, D. L., H. J. Fantom, and J. R. Wilson. 1982. Lack of association between preference for and dependence on ethanol. Drug Alcohol Depend. 9:119–125.

Allen, D. L., D. R. Petersen, J. R. Wilson, G. E. McClearn, and T. K. Nishimoto. 1983. Selective breeding for a multivariate index of ethanol dependence in mice: Results from the first five generations. In Alcoholism: Clin. Exp. Res. 7:443–447.

Andermann, E. 1982. Multifactorial inheritance of generalized and focal epilepsy. In Genetic Basis of the Epilepsies, Anderson, V. E., W. A. Hauser, J. K. Penry, and C. F. Sing (Eds.), pp. 355–374. New York, Raven Press.

Anderson, V. E. 1977. Genetic counseling for epilepsy. In Plan for Nation-wide Action on Epilepsy. Vol. II, Part 1, pp. 141-162. DHEW Publication No. (NIH) 78–311.

Anderson, V. E. 1982. Family studies of epilepsy. In Genetic Basis of the Epilepsies, Anderson, V. E., W. A. Hauser, J. K. Penry, and C. F. Sing (Eds.), pp. 103–112. New York, Raven Press.

Anderson, V. E., M. M. Chern, and E. Schwanebeck. 1981. Multiplex families and the problem of heterogeneity. In Genetic Research Strategies for Psychobiology and Psychiatry, Gershon, E. S., S. Matthysse, X. O. Breakefield, and R. D. Ciaranello (Eds.), pp. 341-351. Pacific Grove, California, Boxwood Press.

Anderson, V. E., W. A. Hauser, J. K. Penry, and C. F. Sing (Eds.). 1982. Genetic Basis of the Epilepsies. New York, Raven Press.

Annegers, J. F., W. A. Hauser, and V. E. Anderson. 1982. Risk of seizures among relatives of patients with epilepsy: Families in a defined population. In Genetic Basis of the Epilepsies, Anderson, V. E., W. A. Hauser, J. K. Penry, and C. F. Sing (Eds.), pp. 151–159. New York, Raven Press.

Annegers, J. F., W. A. Hauser, L. R. Elveback, V. E. Anderson, and L. T. Kurland. 1978. Congenital malformations and seizure disorders in the offspring of parents with epilepsy. Int. J. Epidemol. 7:241–247.

Annegers, J. F., L. T. Kurland, and W. A. Hauser. 1983. Teratogenicity of anticonvulsant drugs. In Epilepsy, Ward, A. A., Jr., J. K. Penry, and D. Purpura (Eds.), pp. 239–248. New York, Raven Press.

Baraitser, M. 1982. The Genetics of Neurological Disorders. New York, Oxford University Press.

Baraitser, M. 1983a. Recurrence risks in epilepsy. In Research Progress in Epilepsy, Rose, F. C. (Ed.), pp. 71–77, London, Pitman.

Baraitser, M. 1983b. Relevance of a family history of seizures. Arch. Dis. Child. 58:404–405.

Barslund, I. and J. Danielsen. 1963. Temporal epilepsy in monozygotic twins. Epilepsia 4:138–150.

Baumann, R. J. 1982. Classification and population studies of epilepsy. In Genetic Basis of the Epilepsies, Anderson, V. E., W. A. Hauser, J. K. Penry, and C. F. Sing (Eds.), pp. 11–20. New York, Raven Press.

Beck-Mannagetta, G. and D. Janz. 1982. Febrile convulsions in offspring of epileptic probands. In Genetic Basis of the Epilepsies, Anderson, V. E., W. A. Hauser, J. K. Penry, and C. F. Sing (Eds.), pp. 145–150. New York, Raven Press.

Bjerre, I. and E. Corelius. 1968. Benign familial neonatal convulsions. Acta Paediatr. Scand. 57:557–561.

Bloom, F. E. 1982. Prospects from retrospect. In Molecular Genetic Neuroscience, Schmitt, F. O., S. J. Bird, and F. E. Bloom (Eds.), pp. 467–475. New York, Raven Press.

Blume, W. T. 1982. Some seizure disorders affecting neonates and children. In Genetic Basis of the Epilepsies, Anderson, V. E., W. A. Hauser, J. K. Penry, and C. F. Sing (Eds.), pp. 35–48. New York, Raven Press.

Bonhaus, D., H. E. Laird, and R. J. Huxtable. 1982. Decreased taurine transport in epileptic rats. Proc. West. Pharmacol. Soc. 25:25–28.

Bray, P. F. and W. C. Wiser. 1964. Evidence for a genetic etiology of temporal-central abnormalities in focal epilepsy. New Engl. J. Med. 271:926–933.

Bray, P. F. and W. C. Wiser. 1965. Hereditary characteristics of familial temporal-central focal epilepsy. Pediatrics. 36:207–211.

Breakefield, X. O., J. E. Pintar, and M. B. Rosenberg. 1982. Current genetic approaches to the mammalian nervous system. In Molecular Approaches to Neurobiology, Brown, I. R. (Ed.), pp. 1–39. New York, Academic Press.

Bustamante, S. A. and L. C. Stumpff. 1978. Fetal hydantoin syndrome in triplets. Amer. J. Dis. Child 132:978–979.

Caveness, W. F. 1963. Onset and cessation of fits following craniocerebral trauma. J. Neurosurg. 20:570–582.

Cazzullo, C. L. and R. Canger. 1979. HLA and epilepsy. Epilepsia 20:179.

Cloninger, C. R., J. Rice, T. Reich, and P. McGuffin. 1982. Genetic analysis of seizure disorders as multidimensional threshold characters. In Genetic Basis of the Epilepsies, Anderson, V. E., W. A. Hauser, J. K. Penry, and C. F. Sing (Eds.), pp. 291–309. New York, Raven Press.

Consroe, P. and H. L. Edmonds, Jr. 1979. Genetic animal models of epilepsy: Introduction. Fed. Proc. 38:2397–2398.

Crabbe, J. C., A. Kosobud, E. R. Young, and J. S. Janowsky. 1983. Polygenic and single-gene determination of responses to ethanol in BXD/Ty recombinant inbred mouse strains. Neurobehav Toxicol Teratol 5:181–187.

Daly, R. F. and F. M. Forster. 1975. Inheritance of reading epilepsy. Neurology (Minneapolis) 25:1051–1054.

Dansky, L., E. Andermann, and F. Andermann. 1982. Major congenital malformations in the offspring of epileptic patients: Genetic and environmental risk factors. In Epilepsy, Pregnancy, and the Child, Janz, D., L. Bossi, M. Dam, H. Helge, A. Richens, and D. Schmidt (Eds.), pp. 223–233. New York, Raven Press.

Davidson, S. and C. W. Watson. 1956. Hereditary light sensitive epilepsy. Neurology (Minneapolis) 6:235–261.

Delgado-Escueta, A. V., D. M. Treiman, and F. Enrile-Bacsal. 1982. Phenotypic variations of seizures in adolescents and adults. In Genetic Basis of the Epilepsies, Anderson, V. E., W. A. Hauser, J. K. Penry, and C. F. Sing (Eds.), pp. 49–81. New York, Raven Press.

Doose, H. 1981. Genetic aspects of the epilepsies. Folia Psychiatr. Neurol. Jpn. 35:231–242.

Doose, H. 1982. Photosensitivity: Genetics and significance in the pathogenesis of epilepsy. In Genetic Basis of the Epilepsies, Anderson, V. E., W. A. Hauser, J. K. Penry, and C. F. Sing (Eds.), pp. 113–121. New York, Raven Press.

Doose, H. and H. Gerken. 1973. On the genetics of EEG-anomalies in childhood. IV. Photoconvulsive reaction. Neuropädiatrie 4:162–171.

Doose, H., H. Gerken, K. F. Hein-Völpel, and E. Völzke. 1969. Genetics of photosensitive epilepsy. Neuropädiatrie 1:56–73.

Doose, H., H. Gerken, T. Horstmann, and E. Völzke. 1973. Genetic factors in spike-wave absences. Epilepsia 14:57–75.

Doose, H., H. Gerken, R. Leonhardt, E. Völzke, and C. Völz. 1970. Centrencephalic myoclonic-astatic petit mal. Clinical and genetic investigations. Neuropädiatrie 2:59–78.

Doose, H. and A. Gundel. 1982. Rhythms of 4 to 7 CPS in the childhood EEG. In Genetic Basis of the Epilepsies, Anderson, V. E., W. A. Hauser, J. K. Penry, and C. F. Sing (Eds.), pp. 83–91. New York, Raven Press.

Dreifuss, F. E. 1981. Proposal for revised clinical and electroencephalographic classification of epileptic seizures. Epilepsia 22:489–501.

Dyken, P. and N. Krawiecki. 1983. Neurodegenerative diseases of infancy and childhood. Ann. Neurol. 13:351–364.

Eaves, L. J. 1982. The utility of twins. In Genetic Basis of the Epilepsies, Anderson, V. E., W. A. Hauser, J. K. Penry, and C. F. Sing (Eds.), pp. 249–276. New York, Raven Press.

Eeg-Olofsson, O., I. Petersén, and U. Selldén. 1970. The development of the electroencephalogram in normal children from the age of 1 through 15 years. Neuropädiatrie 2:375–404.

Eeg-Olofsson, O., J. Safwenberg, and A. Wigertz. 1982. HLA and epilepsy: An investigation of different types of epilepsy in children and their families. Epilepsia 23:27–34.

Eisner, V., L. L. Pauli, and S. Livingston. 1959. Hereditary aspects of epilepsy. Johns Hopkins Hosp. Bull. 105:245–271.

Evans, J. H. 1962. Post-traumatic epilepsy. Neurology (Minneapolis) 12:665–674.

Feinberg, A. P. and W. R. Leahy. 1977. Infantile spasms: Case report of sex-linked inheritance. Dev. Med. Child Neurol. 19:524–526.

Fichsel, H. and M. Kessler. 1982. HLA in primary generalized and partial epilepsies. In Advances in Epileptology: XIIIth Epilepsy International Symposium, Akimoto, H., H. Kazamatsuri, M. Seino, and A. Ward (Eds.), pp. 91–92. New York, Raven Press.

Finnell, R. H. 1981. Phenytoin-induced teratogenesis: A mouse model. Science 211:483–484.

Finnell, R. H. and G. F Chernoff. 1982. Mouse fetal hydantoin syndrome: Effects of maternal seizures. Epilepsia 23:423–429.
Frantzen, E., M. Lennox-Buchthal, A. Nygaard, and J. Stene. 1970. A genetic study of febrile convulsions. Neurology (Minneapolis) 20:909–917.
Fukuyama, Y., K. Kagawa, and K. Tanaka. 1979. A genetic study of febrile convulsions. Eur. Neurol. 18:166–182.
Gastaut, H. 1964. A proposed international classification of epileptic seizures. Epilepsia 5:297–306.
Gastaut, H. 1982. Temporal lobe epilepsy: Comments on its nosology and the resulting terminological problems. In Henri Gastaut and the Marseilles School's Contribution to the Neurosciences (EEG Suppl. No. 35), Broughton, R. J. (Ed.), pp. 127–134. Amsterdam, Elsevier Biomedical Press.
Gerken, H. and H. Doose. 1972. On the genetics of EEG-anomalies. II. Occipital 2-4/s rhythms. Neuropädiatrie 3:437–454.
Gerken, H. and H. Doose. 1973. On the genetics of EEG-anomalies in childhood. III. Spikes and waves. Neuropädiatrie 4:88–97.
Gerken, H., R. Kiefer, H. Doose, and E. Völzke. 1976. Genetic factors in childhood epilepsy with focal sharp waves. I. Clinical data and familial morbidity for seizures. Neuropädiatrie 8:3–9.
Gloor, P. 1968. Generalized cortico-reticular epilepsies. Epilepsia 9:249.
Gloor, P. 1982. Toward a unifying concept of epileptogenesis. In Advances in Epileptology: XIIIth Epilepsy International Symposium, Akimoto, H., H. Kazamatsuri, M. Seino, and A. Ward (Eds.), pp. 83–86. New York, Raven Press.
Gloor, P., J. Metrakos, K. Metrakos, E. Andermann, and N. van Gelder. 1982. Neurophysiological, genetic and biochemical nature of the epileptic diathesis. In Henri Gastaut and the Marseilles School's Contribution to the Neurosciences (EEG Suppl. No. 35), Broughton, R. J. (Ed.), pp. 45–56. Amsterdam, Elsevier Biomedical Press.
Gomez, M. R. and D. W. Klass. 1983. Epilepsies of infancy and childhood. Ann. Neurol. 13: 113–124.
Goodman, H. O. and B. M. Connolly. 1982. Taurine transport alleles: Dissecting a polygenic complex. In Genetic Basis of the Epilepsies, Anderson, V. E., W. A. Hauser, J. K. Penry, and C. F. Sing (Eds.), pp. 171–180. New York, Raven Press.
Griffiths, P. J. and J. M. Littleton. 1977. Concentrations of free amino acids in brains of mice of different strains during the physical syndrome of withdrawal from alcohol. Br. J. Exp. Pathol. 58:391–399.
Hahn, W. E., J. van Ness, and N. Chaudhari. 1982. Overview of the molecular genetics of mouse brain. In Molecular Genetic Neuroscience, Schmitt, F. O., S. J. Bird, and F. E. Bloom (Eds.), pp. 323–324. New York, Raven Press.
Hare, J. E. and A. S. Hare. 1979. Epileptiform mice, a new neurological mutant. J. Hered. 70:417–420.
Harvald, B. and M. Hauge. 1965. Hereditary factors elucidated by twin studies. In Genetics and the Epidemiology of Chronic Diseases, Neel, J. V., M. W. Shaw, and W. J. Schull (Eds.), pp. 61–76. Public Health Service Publication No. 1163.
Hauser, W. A. 1981. The natural history of febrile seizures. In Febrile Seizures, Nelson, K. B. and J. H. Ellenberg (Eds.), pp. 5–17. New York, Raven Press.
Hauser, W. A. 1982. Genetics and the clinical characteristics of seizures. In Genetic Basis of the Epilepsies, Anderson, V. E., W. A. Hauser, J. K. Penry, and C. F. Sing (Eds.), pp. 3–9. New York, Raven Press.
Hauser, W. A., V. E. Anderson, R. B. Loewenson, and S. M. McRoberts. 1982. Seizure recurrence after a first unprovoked seizure. N. Engl. J. Med. 307:522–528.
Hauser, W. A., J. F. Annegers, and V. E. Anderson. 1983. Epidemiology and the genetics of epilepsy. In Epilepsy, Ward A. A., Jr., J. K. Penry, and D. Purpura (Eds.), pp. 267–292. New York, Raven Press.
Hauser, W. A., J. F. Annegers, and L. T. Kurland. 1982. Risk of seizures among relatives of patients with febrile convulsions: A genetic epidemiologic study. Neurology (NY) 32:A185.
Hauser, W. A. and L. T. Kurland. 1975. The epidemiology of epilepsy in Rochester, Minnesota, 1935 through 1967. Epilepsia 16:1–66.
Hedenström, I. V. 1969. Sensitivity to photic stimulation in the relatives of epileptics. J. Amer. Med. Women's Assoc. 24:227–229.
Heijbel, J., S. Blom, and M. Rasmuson. 1975. Benign epilepsy of childhood with centrotemporal EEG foci: A genetic study. Epilepsia 16:285–293.

Heller, A. H., M. A. Dichter, and R. L. Sidman. 1983. Anticonvulsant sensitivity of absence seizures in the *tottering* mutant mouse. Epilepsia 25:25–34.
Henriksen, S. J. 1982. Neuroscience directions and the epilepsies: A prospective view. In Genetic Basis of the Epilepsies, Anderson, V. E., W. A. Hauser, J. K. Penry, and C. F. Sing (Eds.), pp. 225–236. New York, Raven Press.
Hirtz, D. G., K. B. Nelson, and J. H. Ellenberg. 1983. Seizures following childhood immunization. J. Pediatr. 102:14–18.
Housman, D. and J. Gusella. 1982. Molecular genetic approaches to neural degenerative disorders. In Molecular Genetic Neuroscience, Schmitt, F. O., S. J. Bird, and F. E. Bloom (Eds.), pp. 415–422. New York, Raven Press.
Hutchins, J. B., D. L. Allen, L. S. Cole-Harding, and J. R. Wilson. 1981. Behavioral and physiological measures for studying ethanol dependence in mice. Pharmacol. Biochem. Behav. 15:55–59.
Inouye, E. 1960. Observations on forty twin cases with chronic epilepsy and their co-twins. J. Nerv. Ment. Dis. 13:401–416.
Ionasescu, V. and H. Zellweger. 1983. Genetics in Neurology. New York, Raven Press.
Janjua, N. A., J. D. Metrakos, and N. M. van Gelder. 1982. Plasma amino acids in epilepsy. In Genetic Basis of the Epilepsies, Anderson, V. E., W. A. Hauser, J. K. Penry, and C. F. Sing (Eds.), pp. 181–197. New York, Raven Press.
Janz, D. 1982. Antiepileptic drugs and pregnancy: Altered utilization patterns and teratogenesis. Epilepsia 23 (Suppl. 1):S53–S63.
Janz, D. and G. Beck-Mannagetta. 1982. Epilepsy and neonatal seizures in the offspring of parents with epilepsy. In Genetic Basis of the Epilepsies, Anderson, V. E., W. A. Hauser, J. K. Penry, and C. F. Sing (Eds.), pp. 135–143. New York, Raven Press.
Jeavons, P. M. and G. A. Harding. 1975. Photosensitive Epilepsy. In Clinics in Developmental Medicine No. 56. Spastics International Medical Publications. Philadelphia, J. B. Lippincott.
Jennett, B. 1975. Epilepsy after Non-missile Head Injuries. Chicago, Wm. Heinemann.
Jennings, M. T. and J. D. Bird. 1981. Genetic influences in the epilepsies. Review of the literature with practical implications. Amer. J. Dis. Child. 135:450–457.
Jensen, I. 1975. Genetic factors in temporal lobe epilepsy. Acta Neurol. Scand. 52:381–394.
Jilek-Aall, L., W. Jilek, and J. R. Miller. 1979. Clinical and genetic aspects of seizure disorders prevalent in an isolated African population. Epilepsia 20:613–622.
Jobe, P. C. and H. E. Laird. 1981. Neurotransmitter abnormalities as determinants of seizure susceptibility and intensity in the genetic models of epilepsy. Biochem. Pharmacol. 30: 3137–3144.
Jobe, P. C., H. E. Laird II, K. Hoko, T. Ray, and J. W. Dailey. 1982. Abnormalities in monoamine levels in the central nervous system of the genetically epilepsy-prone rat. Epilepsia 23:359–366.
Kakihana, R. 1979. Alcohol intoxication and withdrawal in inbred strains of mice: Behavioral and endocrine studies. Behav. Neural Biol. 26:97–105.
Kaplan, B. J., T. N. Seyfried, and G. H. Glasser. 1979. Spontaneous polyspike discharges in an epileptic mutant mouse (*tottering*). Exper. Neurol. 66:577–586.
Kellaway, P. 1982. Maturational and biorhythmic changes in the elctroencephalogram. In Genetic Basis of the Epilepsies, Anderson, V. E., W. A. Hauser, J. K. Penry, and C. F. Sings (Eds.), pp. 21–33. New York, Raven Press.
Klingler, D. and P. Wessely. 1977. Influence of alcohol on photomyoclonic and photoconvulsive responses. Electroencephalogr. Clin. Neurophysiol. 43:272–273.
Kohsaka, M., M. Hiramatsu, and A. Mori. 1978. Brain catecholamine concentrations and convulsions in El mice. Adv. Biochem. Psychopharmacol. 19:389–392.
Kollevold, T. 1978. Immediate and early cerebral seizures after head injuries. Part III. J. Oslo City Hosp. 28:77–86.
Kristensen, O. and E. H. Sindrup. 1978. Psychomotor epilepsy and psychosis. Acta Neurol. Scand. 57:361–377.
Kurokawa, M., H. Naruse, and M. Kato. 1966. Metabolic studies on *ep* mouse, a special strain with convulsive predisposition. Prog. Brain Res. 21A:112–130.
Lennox, W. G. and M. A. Lennox. 1960. Epilepsy and Related Disorders. Boston, Little, Brown.
Lennox-Buchthal, M. 1971. Febrile and nocturnal convulsions in monozygotic twins. Epilepsia 12:147–156.
Lennox-Buchthal, M. 1973. Febrile convulsions: A reappraisal. Electroencephalogr. Clin. Neurophysiol. (Suppl. 32):1–138.

Leviton, A. and L. D. Cowan. 1982. Epidemiology of seizure disorders in children. Neuroepidemiology 1:40–83.

Levitt, P. and J. L. Noebels. 1981. Mutant mouse *tottering*: Selective increase of locus ceruleus axons in a defined single-locus mutation. Proc. Natl. Acad. Sci. USA 78:4630–4634.

Lindsay, J., C. Ounsted, and P. Richards. 1980. Long-term outcome in children with temporal lobe seizures. IV. Genetic factors, febrile convulsions and the remission of seizures. Develop. Med. Child. Neurol. 22:429–439.

Loughnan, P. M., H. Gold, and J. C. Vance. 1973. Phenytoin teratogenicity in man. Lancet 1:70–72.

MacGillivray, R. C. 1967. Epilepsy in Down's anomaly. J. Ment. Defic. Res. 11:43–48.

Maraschio, P., O. Zuffardi, F. Bernardi, M. Bozzola, C. DePaoli, C. Fonatsch, S. C. Flatz, L. Ghersini, G. Gimelli, M. Loi, R. Lorini, D. Peretti, L. Poloni, D. Tonetti, R. Vanni, and G. Zamboni. 1981. Preferential maternal derivation in inv dup (15). Analysis of eight new cases. Hum. Genet. 57:345–350.

Mathai, K. V., D. P. Dunn, L. T. Kurland, and F. A. Reeder. 1968. Convulsive disorders in the Mariana Islands. Epilepsia 9:77–85.

Matthes, A. 1969. Genetic studies in epilepsy. In The Physiopathogenesis of the Epilepsies, Gastaut, H., H. Jasper, J. Bancaud, and A. Waltregny (Eds.), pp. 26–35. Springfield, Illinois, Charles C. Thomas.

Maxson, S. C. 1980. Febrile convulsions in inbred strains of mice susceptible and resistant to audiogenic seizures. Epilepsia 21:637–645.

Maxson, S. C., J. S. Cowen, and P. Y. Sze. 1977. Pharmacogenetic differences in audiogenic seizure priming of C57BL/6Bg and DBA/1Bg-*asr* mice. Pharmacol. Biochem. Behav. 7:221–226.

Maxson, S. C., M. D. Fine, B. E. Ginsburg, and D. L. Koniecki. 1983. A mutant for spontaneous seizures in C57BL/10Bg mice. Epilepsia 24:15–24.

McCall, R. D. and D. Frierson, Jr. 1981. Evidence that two loci predominantly determine the difference in susceptibility to the high pressure neurologic syndrome type I seizure in mice. Genetics 99:285–307.

McClearn, G. E., J. R. Wilson, D. R. Petersen, and D. L. Allen. 1982. Selective breeding in mice for severity of the ethanol withdrawal syndrome. Subst. Alcohol Actions Misuse 3:135–143.

McKusick, V. A. 1978. Mendelian Inheritance in Man. Catalogs of Autosomal Dominant, Autosomal Recessive and X-linked Phenotypes. 5th ed. Baltimore, Johns Hopkins University Press.

Metrakos, J. D. and K. Metrakos. 1960. Genetics of convulsive disorders. I. Introduction, problems, methods, and base lines. Neurolology (Minneapolis) 10:288.

Metrakos, J. D. and K. Metrakos. 1970. Genetic factors in epilepsy. Mod. Probl. Pharmacopsychiatry 4:71–86.

Metrakos, K. and J. D. Metrakos. 1961. Genetics of convulsive disorders. II. Genetic and electroencephalographic studies in centrencephalic epilepsy. Neurology (Minneapolis) 11:474–483.

Metrakos, K. and J. D. Metrakos. 1974. Genetics of epilepsy. In Handbook of Clinical Neurology. Vol. 15. The Epilepsies, Magnus, O. and A. M. Lorentz de Haas (Eds.), pp. 429–439. Amsterdam, North-Holland.

Moll, P. P. 1982. Alternative genetic models: An application to epilepsy. In Genetic Basis of the Epilepsies, Anderson, V. E., W. A. Hauser, J. K. Penry, and C. F. Sing (Eds.), pp. 277–289. New York, Raven Press.

Monaghan, H. P., M. O'Sullivan, and N. V. O'Donohoe. 1982. HLA antigens and cryptogenic myoclonic epilepsy. Ir. J. Med. Sci. 151:188–189.

Myrianthopoulos, N. C. (Ed.) 1981. Neurogenetic Directory, Part I. Vol. 42 of Handbook of Clinical Neurology, Vinken, P. J. and G. W. Bruyn (Eds.), pp. 667–719. Amsterdam North-Holland.

Neugebauer, R. and M. Susser. 1979. Epilepsy: Some epidemiological aspects. Psychol. Med. 9:207–215.

Newmark, M. E. and J. K. Penry. 1979. Photosensitivity and Epilepsy: A Review. New York, Raven Press.

Newmark, M. E. and J. K. Penry. 1980. Genetics of Epilepsy: A Review. New York, Raven Press.

Newmark, M. E. and R. J. Porter. 1982. Clinical research trends in the genetics of epilepsy. In Genetic Basis of the Epilepsies, Anderson, V. E., W. A. Hauser, J. K. Penry, and C. F. Sing (Eds.), pp. 161–168. New York, Raven Press.

Noebels, J. 1979. Analysis of inherited epilepsy using single locus mutations in mice. Fed. Proc. 39:2405–2410.
Noebels, J. L. 1982. Defined gene models of the inherited epilepsies. In Genetic Basis of the Epilepsies, Anderson, V. E., W. A. Hauser, J. K. Penry, and C. F. Sing (Eds.), pp. 211–223. New York, Raven Press.
Palayoor, S. T. and T. N. Seyfried. 1983. Association between Ca^{2+}-ATPase activity and audiogenic seizures in mice. Trans. Am. Soc. Neurochem. 14:154.
Penfield, W. and H. H. Jasper. 1954. Epilepsy and the Functional Anatomy of the Human Brain. Boston, Little, Brown.
Phelan, M. C., J. M. Pellock, and W. E. Nance. 1982. Discordant expression of fetal hydantoin syndrome in heteropaternal dizygotic twins. New Engl. J. Med. 307:99–101.
Phillips, G. 1954. Traumatic epilepsy after closed head injury. J. Neurol. Neurosurg. Psychiatry 17:1–10.
Procopis, P. G. and H. D. Jameson. 1974. The photoconvulsive response. Modification by monocular occlusion in man. Arch. Neurol. 31:31–34.
Pueschel, S. M. and J. E. Rynders. 1982. Down Syndrome. Advances in Biomedicine and the Behavioral Sciences. Cambridge. Massachusetts, Ware Press.
Quattlebaum, T. G. 1979. Benign familial convulsions in the neonatal period and early infancy. J. Pediatr. 95:257–259.
Quesney, L. F., F. Andermann, and P. Gloor. 1981. Dopaminergic mechanism in generalized photosensitive epilepsy. Neurology (NY) 31:1542–1544.
Rabending, G. and H. Klepel. 1970. Fotokonvulsivreaktion und Fotomyoklonus: Altersabhängige genetisch determinierte Varianten der gesteigerten Fotosensibilität. Neuropädiatrie 2:164–172.
Reed, S. C. 1980. Counseling in Medical Genetics. 3rd ed. New York, Alan R. Liss.
Rimoin, D. L. and J. D. Metrakos. 1964. The genetics of convulsive disorders in the families of hemiplegics. Proc. 2nd Intl. Cong. Hum. Genet. 3:1655–1658.
Rivas, M. L. 1983. Genetic analyses of petit mal epilepsy. I. Evaluation of HLA, blood groups, serum proteins, and red cell enzymes. Epilepsia 24:115.
Rosenblatt, D. E., C. J. Lauter, and E. G. Trams. 1976. Deficiency of a Ca^{2+}-ATPase in brains of seizure prone mice. J. Neurochem. 27:1299–1304.
Schiøttz-Christensen, E. 1972. Genetic factors in febrile convulsions. Acta Neurol. Scand. 48:538–546.
Schmidt, R., L. Eviatar, H. M. Nitowsky, M. Wong, and S. Miranda. 1981. Ring chromosome 14: A distinct clinical entity. J. Med. Genet. 18:304–307.
Schmitt, F. O. 1982. A protocol for molecular genetic neuroscience. In Molecular Genetic Neuroscience, Schmitt, F. O., S. J. Bird, and F. E. Bloom (Eds.), pp. 1–9. New York, Raven Press.
Schreck, R. R., W. R. Breg, B. F. Erlanger, and O. J. Miller, 1977. Preferential derivation of abnormal human G-group-like chromosomes from chromosome 15. Hum. Genet. 36:1–12.
Schuman, S. H. and L. J. Miller. 1966. Febrile convulsions in families: Findings in an epidemiologic survey. Clin. Pediatr. 5:604–608.
Segal, S. 1979. Anticonvulsants and pregnancy. Pediatrics 63:331–333.
Seyfried, T. N. 1979. Audiogenic seizures in mice. Fed. Proc. 38:2399–2404.
Seyfried, T. N. 1982b. Developmental genetics of audiogenic seizure susceptibility in mice. In Genetic Basis of the Epilepsies, Anderson, V. E., W. A. Hauser, J. K. Penry, and C. F. Sing (Eds.), pp. 199–210. New York, Raven Press.
Seyfried, T. N. 1982a. Convulsive disorders. In The Mouse in Biomedical Research, Vol. IV, Foster, H. L., V. D. Small, and J. G. Fox (Eds.), pp. 97–124. New York, Academic Press.
Seyfried, T. N. 1983. Genetic heterogeneity for the development of audiogenic seizures in mice. Brain Res. 271:325–329.
Seyfried, T. N. Itoh, G. H. Glaser, N. Miyazawa, and R. K. Yu. 1981. Cerebellar gangliosides and phospholipids in mutant mice with ataxia and epilepsy: The Tottering/Leaner syndrome. Brain Res. 216:429–436.
Shamansky, S. L. and G. H. Glaser. 1979. Socioeconomic characteristics of childhood seizure disorders in the New Haven area: An epidemiologic study. Epilepsia 20:457–474.
Sing. C. F., C. L. Hanis, and P. P. Moll. 1982. Questions, measures, and analytical strategies in human genetics. In Genetic Basis of the Epilepsies, Anderson, V. E., W. A. Hauser, J. K. Penry, and C. F. Sing (Eds.), pp. 239–247. New York, Raven Press.
Smeraldi, E., R. Smeraldi, C. L. Cazzullo, A. Cazzullo, G. Fabio, and R. Canger. 1975. Immunogenetics of the Lennox-Gastaut syndrome. Frequency of HLA antigens and haplotypes in patients and first-degree relatives. Epilepsia 16:699–703.

Smith, C. 1970. Heritability of liability and concordance in monozygous twins. Ann. Hum. Genet. 34:85-91.
Suzuki, J. and Y. Nakamoto. 1977. Seizure patterns and electroencephalograms of El Mouse. Electroenceph. Clin. Neurophysiol. 43:299-311.
Takahashi, T. 1976. Photoconvulsive response and disposition. Folia Psychiatr. Neurol. Jpn. 30:349-356.
Tangye, S. R. 1979. The EEG and incidence of epilepsy in Down's syndrome. J. Ment. Defic. Res. 23:17-24.
Thoenen, H. 1982. Strategies for application of molecular genetics to neurobiology. In Molecular Genetic Neuroscience, Schmitt, F. O., S. J. Bird, and F. E. Bloom (Eds.), pp. 459-466. New York, Raven Press.
Trams, E. G. and C. J. Lauter. 1978. Ecto-ATPase deficiency in glia of seizure-prone mice. Nature 271:270-271.
Treiman, D. M., M. M. Chern, J. F. Collins, and the VA Epilepsy Study Group. 1983. Possible genetic effects in partial seizure patients. Epilepsia 24:115.
Tsuboi, T. 1980. Genetic aspects of epilepsy. Folia Psychiatr. Neurol. Jpn. 34:215-225.
Tsuboi, T. 1982. Febrile convulsions. In Genetic Basis of the Epilepsies, Anderson, V. E., W. A. Hauser, J. K. Penry, and C. F. Sing (Eds.), pp. 123-134. New York, Raven Press.
Tsuboi, T. and W. Christian. 1973. On the genetics of the primary generalized epilepsy with sporadic myoclonias of impulsive petit mal type. A clinical and electroencephalographic study of 399 probands. Humangenetik 19:155-182.
Tsuboi, T. and S. Endo. 1977. Incidence of seizures and EEG abnormalities among offspring of epileptic patients. Hum. Genet. 36:173-189.
van den Berg, B. J. 1974. Studies on convulsive disorders in young children. IV: Incidence of convulsions among siblings. Develop. Med. Child Neurol. 16:457-464.
van Gelder, N. M. 1981. The role of taurine and glutamic acid in the epileptic process: A genetic predisposition. Rev. Pure Appl. Pharmacol. Sci. 2:293-316.
Varma, R., G. A. Michos, R. S. Varma, J. Stolar, R. Mesmer, and C. R. Joy. 1982. Serum beta-glucuronidase in epilepsy. Res. Comm. Psychol. Psychiatr. Behav. 7:377-380.
Veall, R. M. 1974. The prevalence of epilepsy among mongols related to age. J. Ment. Defic. Res. 18:99-106.
Vogel, F. 1970. The genetic basis of the normal human electroencephalogram (EEG). Hum. Genet. 10:91-114.
Vogel, F., E. Schalt, J. Krüger, and G. Klarich. 1982. Relationship between behavioral maturation measured by the "Baum" test and EEG frequency. A pilot study on monozygotic and dizygotic twins. Hum. Genet. 62:60-65.
Ward, R. H. 1982. Genetics and epidemiology: A fruitful interface for the study of epilepsy. In Genetic Basis of the Epilepsies, Anderson, V. E., W. A. Hauser, J. K. Penry, and C. F. Sing (Eds.), pp. 317-332. New York, Raven Press.
Watson, C. W. and E. M. Marcus. 1962. The genetics and clinical significance of photogenic cerebral electrical abnormalities, myoclonus, and seizures. Trans. Am. Neurol. Assoc. 87:251-253.
White, R. and M. Skolnick. 1982. DNA sequence polymorphism and the genetics of epilepsy. In Genetic Basis of the Epilepsies, Anderson, V. E., W. A. Hauser, J. K. Penry, and C. F. Sing (Eds.), pp. 311-316. New York, Raven Press.
Whyte, M. P. and V. V. Weldon. 1981. Idiopathic hypoparathyroidism presenting with seizures during infancy: X-linked recessive inheritance in a large Missouri kindred. J. Pediatr. 99:608-611.
Wisniewski, L., T. Hassold, J. Heffelfinger, and J. V. Higgins. 1979. Cytogenetic and clinical studies in five cases of inv dup (15). Hum. Genet. 50:259-270.

3 The Genetics of Muscular Dystrophies

Peter S. Harper

University of Wales, Welsh National School of Medicine,
Department of Medicine, Heath Park, Cardiff, Wales

INTRODUCTION

At the time the muscular dystrophies were last reviewed in this series, (Emery and Walton, 1967), the clinical and genetic classification of the major forms was well established, extensive information existed on prevalence and on formal genetics, and considerable interest was being taken in the special problems of genetic counseling in Duchenne muscular dystrophy (DMD). On the other hand, the primary gene products remained entirely unknown; methods of heterozygote detection, especially in Duchenne dystrophy, were far from satisfactory; and prenatal diagnosis was not feasible.

Looking at the situation today, it has to be admitted that much remains unresolved in these major areas. Nevertheless, there is no doubt that our understanding of the genetics of the muscular dystrophies has increased, and it is likely that the new techniques of molecular genetics will soon place us in the position of being able to identify the genes concerned and the nature of the mutational defects directly, as well as providing reliable methods of carrier detection and prenatal diagnosis. This chapter describes the preliminary steps already made using these techniques.

No attempt is made here to cover the numerous rare myopathies that have been delineated in the past few years and that mostly fall outside the accepted definition of *muscular dystrophy*. They have been fully covered in recent reviews and monographs (Dubowitz 1978; Walton 1981), and as yet our knowledge of their inheritance is mostly fragmentary. Table 3.1 lists the major types of muscular dystrophy considered in this chapter.

DUCHENNE MUSCULAR DYSTROPHY

Clinical Features

DMD presents a well-defined and relatively uniform picture in all populations. The relatively early onset of symptoms (commonly beginning at 2 to 4 years of age), the characteristic gait and proximal muscle weakness, the pseudohypertrophy of the calves, the steady progression with most patients chair-bound by age 11, and the inevitably fatal course with death in the late teens to midtwenties are all well established and distinctive features. It is worth recognizing, however, some of the less appreciated aspects.

1. Central nervous system involvement is frequent (Worden and Vignos, 1962; Dubowitz, 1965). On the average the intelligence quotient (IQ) distinction is shifted 20 points downward; around 30 percent of the patients are significantly mentally retarded, with an IQ level of less than 75 (Dubowitz, 1978). The relationship of this to possible genetic heterogeneity is disputed (see below), but there is certainly correlation for mental retardation within a kindred (Cohen, Molnar, and Taft, 1968).

TABLE 3.1. The Progressive Muscular Dystrophies

	Recessive		Autosomal Dominant
	X-linked	Autosomal	
Duchenne dystrophy	X		
Becker dystrophy	X		
Emery-Dreifuss dystrophy	X		
Limb-girdle dystrophy		X	
Early onset, "Duchenne-like" dystrophy		X	
Facioscapulohumeral dystrophy			X
Oculopharyngeal dystrophy			X
Distal myopathy			X
Myotonic dystrophy			X

2. Cardiac muscle involvement is usual (Perloff et al., 1967; Griggs, Reeves, and Moxley, 1977) and—while often asymptomatic—may cause sudden death in patients whose physical condition is otherwise fair. Electrocardiographic abnormalities are also common in heterozygotes (Emery, 1969) (see the section on carrier detection).
3. Motor developmental landmarks are usually delayed even when presentation is relatively late (Dubowitz, 1978). This may be of importance in consideration of early screening for the disorder.

An interesting observation concerns a boy with DMD and coincidental growth hormone deficiency whose dystrophy ran a very mild course in comparison with other family members (Zatz, Betti, and Way, 1981). His case may prove to be relevant to therapy approaches.

Genetic Heterogeneity

Genetic heterogeneity has been postulated on clinical grounds by the finding that mental retardation may be restricted to particular families (Cohen, Molnar, and Taft, 1968). Emery and coauthors have also suggested that this finding is correlated with a relatively mild course of physical disease and late dependence on a wheelchair (Emery, Skinner, and Holloway, 1979). Others have not found such a clear relationship (Worden and Vignos, 1962; Dubowitz, 1977). Correlation of other characteristics within families has been noted, in particular creatine kinase levels (Sibert, Harper, and Thompson, 1979) and occurrence of the manifesting carrier state (Moser and Emery, 1974), but such findings do not themselves imply heterogeneity. Recent genetic linkage analysis using DNA polymorphisms gives no support for more than one genetic locus. Studies by O'Brien and colleagues (1983) show no significant difference in map distance when families are subdivided according to presence or absence of mental retardation.

Incidence

The incidence of DMD is remarkably constant for those populations where accuracy is possible. Table 3.2 lists some of the estimates. Although the average is around 1 in 5,000 male births, it is likely that the true incidence corresponds to the upper limits—possibly as high as 1 in 3,000 male births. Recent surveys of newborn screening using creatine kinase (ck) have suggested an even higher incidence (DellaMonica, 1978), but it is probable that these studies have also included cases of Becker dystrophy and possibly other myopathies.

In general, the more recent studies—as well as those reflecting many years of continued study—show a higher incidence than the earlier ones, but there is contemporary evidence for some decline in the incidence of new cases in Canada (Hutton and Thompson, 1976), Australia (Hurse and Kakulas, 1974), and Wales (Harper, 1982), possibly owing to genetic counseling and carrier detection. When studying such trends it is essential for data on first cases in a family (currently not preventable) to be analyzed separately from subsequent cases.

Mutation Rate

DMD's mutation rate has been the subject of considerable dispute during recent years, and it is unlikely that it will be resolved fully until a direct test for presence of the gene has been developed. Since DMD is genetically lethal,

TABLE 3.2. Incidence of Duchenne Muscular Dystrophy in Various Populations

Country	Source	Incidence (per 100,000 male births)
United States	Stephens and Tyler, 1951	28.6
England	Blyth and Pugh, 1959	15.1
	Gardner-Medwin, 1970	32.6
Scotland	Brooks and Emery, 1977	25.9
Switzerland	Moser et al., 1966	21.8
Italy	Bertolotto et al., 1981	24.0
Australia	Lawrence et al., 1973	21.9
Japan	Kuriowa and Miyazaki, 1967	19.4
	Kondo, 1980	21.7

one third of the abnormal genes must be lost in each generation and, in an equilibrium, would be expected to be replaced by new mutations, as originally postulated by Haldane (1935). If the mutation rate is equal in the two sexes, one third of all cases should represent new mutations; the proportion of *isolated* cases representing mutation should be somewhat higher and will depend on a number of variables, including family size and structure, early diagnosis, and the effects of the disease and of genetic counseling in limiting further reproduction. Inequality of mutation rate in the sexes would alter this pattern, since mutations occurring in male gametes are transmitted exclusively to females in the next generation.

Most family studies of DMD (Thompson, Murphy, and McAlpine, 1967; Gardner-Medwin, 1970; Sibert, Harper, and Thompson, 1979; Yasuda and Kondo, 1980) fit reasonably well with the concept that one third of cases result from mutation, and Davie and Emery (1978) found good agreement with this when analyzing the data by a number of different procedures. Bundey (1981), studying the disorder in the West Midlands, England, and reviewing previous population studies, concluded that an excess of isolated cases existed, suggesting a relative preponderance of female mutations; certainly indirect studies of mutation rate lend no support to a deficiency of female mutation.

Most criticism of the "classical" view of mutation in DMD has arisen because of a supposed deficiency of noncarrier mothers when various carrier tests have been applied in the direct estimate of mutation, including red cell membrane studies (Roses, Herbstreith, and Metcalf, 1976), lymphocyte capping (Pickard, Gruemer, and Verrill, 1978), lactic dehydrogenase (LDH) isoenzyme measurement (Roses, Roses, and Nicholson, 1977), and combinations of CK with other serum tests such as pyruvate kinase and hamopexin. Other X-linked diseases, such as Lesch-Nyhan syndrome (Francke, Felsenstein, and Gartler, 1976) and hemophilia A (Graham, 1979), where the great majority of mothers of isolated cases have proved to be carriers have also been cited in support of the relative rarity of mutation.

Unfortunately, much of the evidence relating to DMD is unreliable because of the unsatisfactory nature of the carrier tests used, and most of the likely errors are in the direction of overestimation of carriers. Use of CK alone carries the danger of misinterpreting minimal elevations, especially where the prior genetic risk is low, whereas the addition of other tests such as pyruvate kinase or lactate dehydrogenase isoenzymes is likely to increase the proportion of carriers detected only at the expense of introducing more false positives. Finally, many of the hypotheses concerning mutation rate have been prompted by tests that are likely to be completely erroneous or unrelated to the disorder, such as those based on studies of erythrocyte membranes and lymphocyte capping.

Taking those studies based on CK alone, and allowing for the proportion of carriers known to fall within the normal range, most agree reasonably well with the estimated view of one third of cases representing new mutations, with approximately equal mutation rates in the two sexes (Thompson, Murphy, and McAlpine, 1967; Gardner-Medwin, Pennington, and Walton, 1971; Zatz et al., 1976). The study of Caskey et al. (1980) provides a particularly cogent review of the situation.

Grandparental age has been analyzed in relation to mutation (Emery, 1980); in two out of three studies an increased grandparental age was found in the maternal grandfathers where the mother appeared to be a carrier, but with no evidence of carrier state in preceeding generations. Emery found an excess of 3.75 years compared with matched controls.

If mutation is accepted as causing around one third of all cases of DMD, the mutation rate is certainly remarkably high. Estimates of between 7×10^{-5} and 10×10^{-5} seem likely, and this could suggest that the region of genetic material involved is greater than a single locus. No visible deletion or other chromosomal defect has ever been noted, but it will be of great interest to see whether molecular genetic techniques can identify more extensive changes than base substitution.

Gene Mapping

Paucity of gene markers on the X chromosome has hampered the localization of the DMD locus, and only recently has any positive evidence been forthcoming on this. Several thorough studies have shown no linkage between DMD and color blindness, glucose 6-phosphate dehydrogenase (G6PD), or the Xg blood group (Blyth, Carter, and Dubowitz, 1965; Emery, 1966; Zatz et al., 1974); by contrast there was initially suggestive, but not conclusive, evidence for linkage of Becker dystrophy with color blindness and G6PD (see below). The first evidence for localization arose from observation of girls with clinical DMD who showed a balanced X autosome translocation with the breakpoint on the short arm of the X chromosome at p. 21 (Verellen et al., 1978; Lindenbaum, Clarke, and Patel, 1979). Further chromosomally abnormal cases in girls have been reported (Canki, Dutrillaux, and Tivadar, 1979; Greenstein, Reardon, and Chan, 1980; Jacobs et al., 1981); and although the rearrangements have been different, in each case a similar breakpoint at 21 was observed, supporting the view that the DMD locus has been involved and is near this site. It is of interest that clinical reassessment of some of these cases—including the patient of Verellen et al.—has shown a clinical picture closer to Becker than to Duchenne dystrophy.

Independent evidence has now confirmed the short arm localization of DMD and represents an advance of major significance. DNA probes identifying polymorphisms based on restriction fragment length are now available, several of which are located on the short arm of the X chromosome, and have been studied in relation to DMD. The first of these (RC8) was derived from an X chromosome–specific DNA library obtained after separation of X chromosome DNA by a fluorescence-activated cell sorter (Davies et al., 1981). The probe shows a polymorphism using the restriction enzyme Taq 1. Family studies with this polymorphism (Murray et al., 1982), for which around 20 percent of the normal female and DMD carrier population are heterozygous, have shown genetic linkage between the loci, with a distance of 15–20 centimorgans based on the most recent data. A second probe (L128), derived from a total human DNA library and localized to the X chromosome by cell hybrid studies, appears also to be around 20 centimorgans distant from DMD (Davies, Pearson, et al., 1983). The DMD locus appears to be flanked by the probes, and the ordering shown in Figure 3.1 is based on their chromosomal localization as well as on the fact that they show frequent recombination with each other.

The more distally located of the probes, RC_8, shows a suggestion of weak linkage with the Xg blood group (Murray et al., 1982; Sarfarazi et al., 1983); so bearing in mind the likely length of the short arm, it is probable that the DMD locus is only just out of mapping range of Xg, which is now known to be terminally located. In this respect it is of interest that DMD clearly shows X inactivation and that the distal region of the short arm not showing this, including the Xg and steroid sulfatase loci, does not seem to extend as far as the DMD locus.

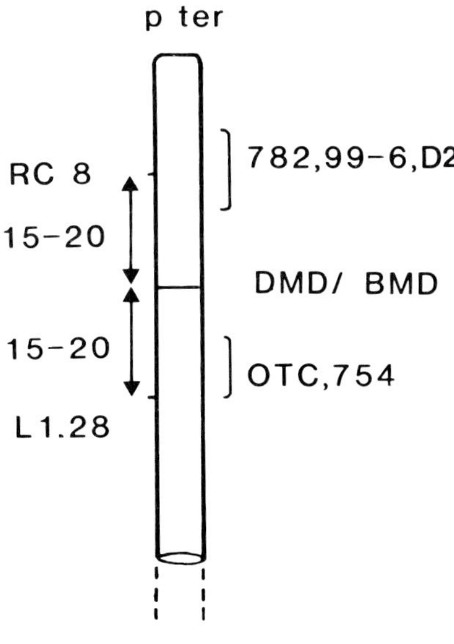

FIGURE 3.1. Schematic map of X chromosome short arm to show location of DMD and BMD in relation to DNA probes. Distances given are still provisional and are based on Davies et al (1983), Kingston et al (1983, 1984) and Brown et al (1985).

Studies of inter-probe genetic distances on large normal families (Drayna et al., 1984) are beginning to produce an accurate framework of genetic markers for the X chromosome into which diseases can be fitted. The availability of a number of further markers in the region of the DMD gene will also allow greater precision in mapping. Some of these are indicated on Figure 3.1. At least four probes distal to the locus and three proximal now exist, though precise ordering remains uncertain. Families heterozygous for more than one marker are particularly valuable, since it may be possible to order markers by seeing at what point crossing over has occurred.

Rapid advances are now likely in the more precise identification of the DMD locus as more polymorphic probes become available. Other molecular approaches are also being used, including analysis of the region corresponding to the p. 21 breakpoint in the translocation. Worton et al. (1984) have studied hybrid cell lines from the patient of Verellen et al. (1978) with an x = 21 translocation and have shown that the autosomal breakpoint involves genes for ribosomal (RNA). Study of gene probes for these could lead to identification of adjacent X chromosome DNA sequences that should be close to the DMD locus.

The use of overlapping DNA sequences may allow "gene walking" to be undertaken from closely linked probes, and a deletion would be particularly amenable to detection if this is indeed the abnormality (or one of them) in DMD. The existing probes are probably too distant to use as a starting point, but they have already helped to establish that there is no obvious genetic heterogeneity within DMD; they have also provided important evidence on the question as to the relationship with Becker dystrophy (see below).

Clinical use of the current linked-gene probes is still limited by their uncertain distances. Those families informative for flanking gene probes (around 60 percent including the newer probes) could be given an accurate prediction regarding inheritance of the DMD gene if neither of the flanking probes shows crossing over, whereas information from one probe alone can be incorporated into odds regarding carrier status (Harper et al., 1983; Pembre et al., 1984; Williams et al., 1983). The lack of interference from X inactivation and the feasibility of very early prenatal diagnosis by direct deoxyribonucleic acid (DNA) analysis of chorionic villi (Williamson et al., 1981; Elles et al., 1983; Williams et al., 1983) suggest that this field will soon be of practical as well as theoretical importance, but we need more precise information on margins of error from crossing over before linked markers can be considered reliable diagnostic tools.

Heterozygote Detection

Reliable detection of the heterozygote for DMD remains the central problem in the genetics of the disorder; current uncertainty is responsible not only for the confusion over mutation rate but also for most of the practical difficulties in genetic counseling. The author has discussed these fully in a recent review (Harper, 1982) in which detailed references can be found.

Table 3.3 lists some of the approaches used in attempting to identify the carrier state; the division into three principal categories is somewhat arbitrary but of practical use. Of the tests in section 1, clinical examination is probably the one of most general use; recognition of "manifesting carriers" is particularly important in the absence of a family member with DMD, since they can easily be misdiagnosed. Minor degrees of weakness (Roses, Nicholson, and Kircher, 1977) are of less value and should probably not be used alone as evidence of the carrier state. Histological studies are currently only helpful when performed in a unit with extensive experience with both DMD and normal muscle; recent quantitative studies and analysis of intracellular and intranuclear calcium (Bodensteiner and Engel, 1978; Maunder-Sewry and Dubowitz, 1979, 1980) suggest that these techniques will help to detect some of those carriers who are normal by conventional CK testing. The widespread use of needle biopsy (Edwards, Young, and Wiles, 1980) has made this a much less traumatic procedure than formerly. Electromyography, by contrast (Gardner-Medwin, 1968; Moosa, Brown, and Dubowitz, 1972), seems to have fallen into disuse, perhaps in part due to its time-consuming nature as well as to difficulties in interpretation.

TABLE 3.3. Approaches to Carrier Detection in Muscular Dystrophy

Studies primarily related to muscle:
Clinical examination
Muscle biopsy
  Light microscopy
  Electron microscopy
  Biochemical changes
Electromyography
Electrocardiography

Studies on serum:
Creatine kinase (+ isoenzymes)
Aldolase
Lactic dehydrogenase
Pyruvate kinase
Myoglobin
Hemopexin

Studies on other cell types:
Red blood cell membrane abnormalities
  Protein phosphorylation
  Morphologic changes
  Biophysical changes
Lymphocyte capping

Studies based on DNA:
Linked X-chromosome markers

Alterations in the electrocardiogram have been shown to be frequent in carriers, and Lane et al. (1980) have shown that they appear to be independent of the results of CK testing, again suggesting that this may be a valuable, as well as simple, adjunct in carrier detection.

Coming to the second group of carrier tests, the use of CK is the current mainstay of carrier detection and will be discussed separately. Most of the other tests in this group are also muscle enzymes or nonenzymatic proteins, and the principal question is whether any of these are superior to CK or whether combined use will increase the proportion of carriers detectable. The situation at present would seem to be that CK isoenzymes (Tzvetanova, 1978) do not give any special advantage, that the claims for LDH isoenzymes (Roses, Roses, and Nicholson, 1977) have not been confirmed (Burt and Emery, 1979), and that pyruvate kinase has given conflicting results (Yamuna, Valmiknathan, and Burt, 1977; Zatz, Shapiro, and Campion, 1978; Percy, Murphy, and Oss, 1979), possibly as a result of technical difficulties (Aston and Stansbie, 1982). Myoglobin (Adornato, Kagen, and Engel, 1978) and hemopexin (Adornato, Kagen, and Engel, 1978) also posed measurement difficulties when first introduced, but a radioimmunoassay for myoglobin (Malvano et al., 1979) has recently been shown to be at least as effective as CK testing and to be free of most of the perplexing variation, which is a serious drawback to the use of CK (Nicholson, 1982). In particular, no change was found in pregnancy, and samples were stable over a long period of time when frozen. If multiple tests are being used, care must be taken to use an appropriate statistical analysis; otherwise overestimation of carriers occurs. It is of interest that a careful recent study using logistic discrimination (Percy, Andrews, and Thompson, 1982) showed the expected proportion of new mutations.

The third group of tests, based on cell types other than muscle (in particular, fibroblasts, red blood cells, and lymphocytes), can currently be said to be of no practical use in carrier detection in DMD. Most of these tests have postulated a systemic disturbance in structure and function of cell membranes, a subject that has stimulated much debate in recent years but that now appears not to be a primary factor in DMD. It seems unlikely that any test that is open to question in the affected male will be of much help in detection of the carrier state. Fuller details of these experimental tests may be found in books by Rowland (1978) and Schotland (1982).

Creatine Kinase in Heterozygote Detection

Although very well established, with its practical use and limitations well described in a number of reviews, several points need to be made regarding the use of CK for the detection of carriers, since such important decisions may stem from the results and their interpretation. A survey of clinical geneticists in the United Kingdom by Bundey (1978) revealed what can only be described as inadequate skills of interpretation. The list that follows highlights some of these problem areas.

1. The sensitivity of samples to light, temperature, hemolysis, and passage of time requires careful attention in taking samples (Thomson, 1969). The laboratory itself must have an adequate normal range of its own.
2. Variation due to exercise, age (Gilboa and Swanson, 1976; Smith, Elton, and Thomson, 1979), pregnancy (Bundey, Crawley, and Edwards, 1979), and other medical conditions must be considered. The rise of CK with exercise is too unpredictable to be utilized in carrier detection, whereas the fact that the 90th percentile in pregnancy corresponds to the mean in nonpregnant women (Bundey, Crawley, and Edwards, 1979; Smith, Elton, and Thomson, 1979) means that testing in pregnancy is hazardous. The normal range in young girls is somewhat higher than adults (Nicholson et al., 1979), which may lead to false identification of carriers if early testing is performed.
3. Even when conditions are controlled carefully, CK results show considerable variation in both normals and carriers; so it is wise to use the mean of three results (Sibert, Harper, and Thompson, 1979; Percy, Andrews, and Thompson, 1982).
4. Since only around 70 percent of carriers show results over the 95th percentile, it is wise to express results in terms of likelihood ratios or probabilities based on the normal and carrier ranges, as in Table 3.4. This mode of expression avoids the errors resulting from results close to the upper limit of the normal range being interpreted as "normal" or "abnormal" when they are in fact almost identical.
5. Any odds must be combined with the appropriate genetic risks, prior and conditional, in a "Bayesian" manner, as clearly described by numerous authors (Murphy and Mutalik, 1969; Emery, 1980). Failure to do this can lead to serious errors in risk estimation, especially when the prior risk is low.

Table 3.5 provides practical examples of these situations.

Future Developments in Carrier Detection

The current far-from-satisfactory situation results partly from our ignorance of the primary defect in DMD, partly from the variability in gene expression produced by X chromosome inactivation. For both these reasons a test based directly on DNA will totally change the situation; and, as discussed above, there is a real likelihood of this being the case within a few years. The currently available linked DNA probes are still too distant for satisfactory use. Even closely linked ones will require family studies and be

TABLE 3.4. Likelihood of Being a Carrier for Different Levels of Serum Creatine Kinase

CK (IU/l)	Likelihood Ratio	Probability
<40	0.12	0.11
40–	0.12	0.11
50–	0.16	0.14
60–	0.27	0.22
70–	0.46	0.32
80–	0.86	0.46
90–	1.67	0.63
100–	3.28	0.77
110–	6.49	0.87
120–	10.79	0.93
130–	25.12	0.96
140–	49.02	0.98
150–	94.34	0.99
160–	180.9	0.995
170–	342.5	0.997
180–	641.0	0.998
190+	>1000	>0.999

CK = creatine kinase
Source: Based on data from Sibert, Harper, and Thompson, 1979.

of limited help to relatives of an isolated case. So direct identification of a change in the DNA of the DMD locus itself is the goal toward which a number of groups are now working.

Genetic Counseling

The difficulties of carrier detection are not the only problems that confront those providing genetic counseling for DMD. The uncertainty as to what prior risk to assign the mother of an isolated case, the need for incorporation of data on normal sons, and the importance of integrating carrier testing and genetic information on the whole kindred can produce an extremely complex situation on occasions. For such large and extended kindreds it is helpful to have a computerized risk estimation; that provided by the program (PEDIG) (Heuch and Li, 1972; Conneally and Heuch, 1974) is extremely helpful in this respect. It will often be found when using this program that low-risk estimates are made much lower by PEDIG than when calculated by hand. This is not due to error but to the fact that the program can take into consideration a number of pieces of normal information that are usually ignored in simple manual risk estimates. The author has found this program particularly useful in constructing risk estimates for a systematic genetic register of carriers (Harper, 1982).

The counseling of extended families constitutes an essential factor in any attempt to prevent DMD, and a number of centers have made a systematic attempt to do this. As a result a decline in incidence has been noted (as discussed above), although no change has been evidenced in Scotland (Brooks and Emery, 1977); since small families (one or two children) are currently the norm in many countries, the remaining first cases in a family will prevent any considerable further decline. It is especially important that early diagnosis is achieved in an initial case, so that genetic counseling and carrier detection can be given in time to prevent the birth of subsequent affected males in the family and in other branches. The subject of newborn screening is discussed later.

TABLE 3.5. Potential Pitfalls in Using Creatine Kinase to Determine Carrier Risk

Subject	Prior Genetic Risk	CK Level (IU/l)	Odds of Being a Carrier from CK Result	Final Risk Estimate (in percent)
NOT ALL "NORMAL" RESULTS ARE THE SAME				
A	1/2	94	1.7 : 1	63
B	1/2	35	0.12 : 1	11

Subjects A and B are unmarried daughters of an obligatory carrier for DMD. Both have CK levels within the normal range, but using specific odds, the final risk for each is very different. Had their results simply been regarded as "normal," the two would have been given identical risks.

Subject	Prior Genetic Risk	CK Level (IU/l)	Odds of Being a Carrier from CK Result	Final Risk Estimate (in percent)
THE GENETIC RISK AS WELL AS THE CK LEVEL MUST DETERMINE THE FINAL RISK				
A	1/2	120	10 : 1	91
B	1/20	120	10 : 1	31
C	1/2,000	120	10 : 1	0.5

All three subjects have a CK level outside the normal range. For subject A with a high prior risk, the test provided very strong evidence of being a carrier. For subject B, who might be a moderately close relative, the test has increased the chances of being a carrier, but it is still more likely that she is not. For subject C—detected through a population screening study—the final chance is still overwhelmingly against her being a carrier despite the high CK level.

CK = creatine kinase
DMD = Duchenne muscular dystrophy
Source: After Harper, 1982.

Prenatal Diagnosis

The absence of any clear knowledge regarding the primary defect means that only secondary changes can be evaluated for prenatal diagnosis of DMD. Since there is no obvious change in constituents of amniotic fluid, the use of CK in fetal blood samples has received most attention. Techniques now exist for direct fetoscopic blood sampling (Rodeck and Campbell, 1978), and several centers have obtained data on the normal range for this, along with samples from pregnancies at risk (Emery, 1980). Unfortunately, some pregnancies giving results in the normal range have later proved to be affected (Ionasescu, Zellweger, and Cancilla, 1978; Golbus et al., 1979; Emery, 1980), removing confidence in its use as a clinical procedure. A further problem is the difficulty on account of postmortem artefact in interpreting fetal muscle abnormalities in those cases where termination was undertaken (whether because of abnormal CK results or for other reasons). Quantitative fiber analysis has shown some changes (Toop and Emery, 1974; Emery, Burt, and Dubowitz, 1979), but no definitive test exists yet.

At present the only reliable prenatal procedure for those at high risk is fetal sexing with termination of a male pregnancy. This practice is unacceptable to many families and traumatic to almost all (Bundey and Ebdy, 1982)—something that emphasizes the need for careful risk evaluation in possible carriers before a pregnancy is undertaken. A further problem is the dysgenic effect of fetal sexing, since a proportion of couples who would otherwise not have had children will transmit the gene to daughters; this effect would be less were there to be a direct prenatal diagnostic test.

Two factors, already mentioned, are likely to change the situation soon regarding prenatal diagnosis. First, the development of gene probes closely linked to and (hopefully before long) directly related to the DMD locus will dramatically improve diagnosis. Second, the use of first trimester diagnostic techniques based on chorionic villi has already been shown to be capable of application to DNA polymorphisms (Williamson et al., 1981; Elles et al., 1983); these techniques are already in clinical use in the hemoglobinopathies (Old et al., 1982). Fetal sexing using X chromosome- and Y chromosome-specific DNA probes (Gosden et al., 1982; Page et al., 1982) is also now becoming possible in the first trimester. Combined use of three such probes (a Y chromosome-specific repetitive sequence; an X-Y homologous probe; and an X chomosome probe detecting a polymorphism) in a pregnancy at risk for DMD has been reported by Williams et al. (1983). Rapid direct cytogenetic techniques should also be useful (Simoni et al., 1983).

Newborn Screening for DMD

If an effective treatment for DMD existed, there would be a strong case for the presymptomatic detection of all cases to allow therapy before irreversible muscle damage occurred, as is currently feasible for disorders such as phenylketonuria and hypothyroidism. The absence of satisfactory treatment makes the case much weaker, but it has nevertheless been advocated in order to help prevent the birth of further affected male infants, which may occur in the interval between birth and diagnosis of the first case in a family.

Two principal questions need to be answered about newborn screening for DMD. First, Is it feasible? Second, Is it desirable?

Zellweger and Antonik (1975) were the first to evaluate the use of CK on dried filter paper blood spots for screening; subsequently, the work of Scheuerbrandt (Beckmann and Scheuerbrandt, 1976; Scheuerbrandt, 1980), using a sensitive chemiluminescent method, has provided considerable experience in Germany. Other centers in France (Dellamonica, 1978; Dellamonica et al., 1979) and Scotland (Skinner et al., 1982) have used different methods of analysis but have had technically satisfactory results.

These studies leave no doubt that the affected male with DMD can be detected by a sensitive CK assay on a filter paper dried blood sample, such as is commonly taken for phenylketonuria screening between 6 and 10 days of age. The extremely high levels of CK seen in DMD at this age give a wide separation from the normal range; and few false positives are encountered, provided a high cutoff point (such as 400 IU/l) is taken. It is too early to be certain whether some cases are missed using this method, but the high incidence of confirmed DMD from screening suggests that this will be uncommon. Careful investigation and follow-up to exclude other myopathies is important; the distinction of Becker dystrophy may be particularly difficult at this early stage.

Screening of female births to detect the carrier state is generally agreed to be impossible (Skinner et al., 1982). This finding was entirely predictable from the overlap between carrier and normal CK ranges already discussed, and it dramatizes the need for considering the prior risk (in this situation exceedingly low) in the interpretation of a result.

Assuming the feasibility of newborn screening and turning to its desirability, the issues become less clear. Screening itself would not prevent the first affected birth in a kindred but could prevent subsequent births. However, with small families the proportion of second births occurring before diagnosis of the first, which are both male and affected, is low, certainly less than 15 percent and probably considerably less in societies where two-child families are the norm (Grimm, 1981). Against this must be set the trauma caused to the family by diagnosis of an incurable disorder in an apparently healthy infant, as well as the worry for those families where an abnormality proves not to be due to DMD. A compromise has been suggested (Gardner-Medwin, Bundey, and Green, 1978) of screening all boys with developmental delay, such as those not walking by 18 months. A program to evaluate this approach is being developed in Wales by the author and colleagues.

A newborn screening for DMD should not be undertaken until the more basic steps of carrier testing of extended families and setting up a register of carriers have been achieved. Centers already in this situation may find it helpful to evaluate screening as a research project, but widespread screening as a service should probably await the development of some more satisfactory form of treatment.

DMD in Girls

The rare but well-documented occurrence of clinical Duchenne dystrophy in a girl has a number of possible causes and always requires careful investigation.

Chromosome analysis is essential; the occurrence of the DMD phenotype in a number of girls carrying a balanced X autosome translocation has already been mentioned and is of great theoretical interest for the localization of the gene. All recorded cases so far have shared a breakpoint at p. 21, have arisen de novo, and have exhibited no detectable deletion of chromosomal material. These girls have not shown the multiple abnormalities characteristic of unbalanced translocations, but there seems little doubt that they represent a true abnormality at the site of the DMD locus.

The reasons why such individuals should show clinical DMD have been discussed by Jacobs et al. (1981). It seems likely that the normal X chromosome is preferentially inactivated and that one does not have to postulate that these girls also carry the DMD mutation, a view supported by the absence of the disease in male relatives.

The Turner (XO) chromosome constitution is a further cytogenetic cause of DMD and should especially be suspected when an affected girl occurs in a well-established X-linked kindred.

The existence of an autosomal recessive disorder mimicking DMD has been the subject of much debate in the past (Walton and Gardner-Medwin, 1981), but reinvestigation of such cases both by cytogenetic studies and by muscle biopsy now suggests that they are extremely rare, certainly too rare to affect genetic counseling for the isolated male case. Lower CK levels, less active degenerative changes on biopsy, a more benign course, and lack of electrocardiogram (ECG) changes all help to distinguish the severe end of the autosomal recessive limb-girdle dystrophy group.

A final cause of DMD in girls that has to be considered is the manifesting carrier state, although whether this can really be considered responsible for typical DMD in a girl is doubtful. Mild clinical manifestations in girls or moderate progressive disease in adults is a more usual result. The principal danger is that such individuals will be confused with autosomal recessive limb-girdle dystrophy and given falsely reassuring genetic counseling (Moser and Emery, 1974; Emery, 1980).

The Primary Defect in DMD

Were there any clear indication as to the nature of the primary defect in DMD, it would be of prime importance in relation to the genetic basis of the disorder, as well as to practical aspects of carrier detection and prenatal diagnosis. Unfortunately, this is not the case. Since the literature is voluminous, no attempt will be made here to discuss the various defects that have at various times been claimed to be of major importance. Full references to this work can be found in books edited by Rowland (1978), Walton (1981), Schotland (1982), and Engel and Banker (1985).

Animal models are likewise not discussed here, since they appear to bear little relationship genetically to Duchenne or other human dystrophies. Mention should be made, however, of an X-linked dystrophic condition in the mouse, reported by Bulfield et al. (1984). In view of the evolutionary conservation of the X chromosome, this disorder could indeed be homologous to human X-linked dystrophies, and it is being studied in detail.

BECKER MUSCULAR DYSTROPHY

Becker muscular dystrophy—inherited as an X-linked recessive like Duchenne dystrophy but of later onset and more benign course—was first clearly delineated by Becker and Kiener (1955). Considerable reluctance existed initially to accept that the two conditions were distinct; despite the occurrence of some phenotypic overlap, the early genetic evidence suggested that they were separate conditions (Blyth and Pugh, 1959), the two forms running separately in families.

Clinically, the distribution of muscle weakness is similar to that in DMD, with proximal muscles mainly involved, particularly of the pelvic girdle, and sparing of facial muscles. Difficulty in walking or running is the usual presentation; age at onset is later than in DMD and is more variable, some individuals being asymptomatic into late adolescence. The extensive study of Emery and Skinner (1976) showed that the most reliable criterion of separation from DMD was age of entry into a wheelchair; by 12 years of age 90 percent of DMD boys were chair-bound, but only very few with Becker dystrophy. Most patients remain ambulant well into adult life, but disability is severe by early middle age. Cardiac and central nervous system involvement occur but are less common then in DMD (Kingston, 1983). Despite these major differences, the distinction of Becker dystrophy from DMD may not be easy at first presentation; the course of the disease in a previous generation may prove the most important guide in this situation.

Genetics

There is no doubt about the X-linked recessive nature of Becker dystrophy. Not only is the disorder transmitted through unaffected females, but the lack of male-to-male transmission and the carrier status of daughters can be shown directly, in contrast to DMD. Genetic fitness has been estimated at two thirds of normal (Emery and Skinner, 1976). The proportion of new mutations might correspondingly be expected to be less than in DMD, according to the formula:

$$\mu = \tfrac{1}{3}\, I(1-f),$$

where (I = incidence and f = genetic fitness of affected males).

In practice the estimation of mutation rate is hampered by the same problems as for DMD—notably the lack of any definitive test for excluding the carrier state in the mother of a supposed new mutation. In fact the situation is made more difficult by the lower proportion of carriers detectable (see below); so it is probably wise to assume a prior risk of around 90 percent for the mother of an isolated case being a carrier.

The incidence of Becker dystrophy is less well established than for DMD, and the value of 2.8 per 100,000 given by Kloepfer and Emery (1974) is probably an underestimate. The author's impression is that many of the more mildly affected patients escape ascertainment for a considerable time, especially single cases; the difficulty of confidently separating these from the autosomal recessive limb-girdle dystrophy adds another variable.

Gene Mapping

A critical test of distinctiveness of Becker and Duchenne dystrophies would have been their assignment to separate regions of the X chromosome (see Figure 3.1), and there was in fact some early evidence to support this. Emery, Smith, and Sanger (1969) and Skinner, Smith, and Emery (1974) found a suggestion of genetic linkage between Becker dystrophy and color blindness, the peak lod score being 1.25 at a recombination fraction of 23 cM. There was no evidence of linkage with the Xg blood group. Zatz et al. (1974), in Brazil, also found some evidence for loose linkage with G6PD, already known to be closely linked to color blindness. However, recent evidence has come from the study of Becker families with the short arm DNA probes shown to be linked to DMD, as discussed above (Kingston et al., 1983, 1984).

This study shows conclusive evidence for linkage between Becker dystrophy and probe L1.28, at a distance (around 20 cM) comparable with that for DMD. The data for the distal probe RC8 also show similar linkage (though with a less conclusive lod score), whereas probe DXYS1, located close to the centromere on the long arm (Page et al., 1982), shows loose linkage. No evidence was found for linkage with color blindness in this study, nor was there anything to suggest heterogeneity. Subsequent studies using a total of seven DNA probes located on the short arm of the X chromosome shows comparable distances for DMD and BMD in all cases (Brown et al., 1985, in press).

These data do not prove allelism, but they are compatible with it and indicate at the least a very similar localization; taking clinical and other features into consideration, allelism now seems a reasonable working hypothesis. The possibility of allelism bears important consequences for the way in which we view the relationship between the two disorders, since *allelism* implies a common gene product. It also has the practical consequence that diagnostic and predictive tests based on DNA are likely to be equally applicable to both disorders.

Genetic Counseling and Carrier Detection

Although the principles of genetic counseling are the same as for DMD, the risks for offspring of the affected male are frequently misunderstood, both by patients and clinicians. Daughters are not infrequently reassured that they are unlikely to be carriers on the basis of normal CK levels when they are of course obligatory carriers. The disorder can probably always be confirmed or excluded in males at risk within the first six months of life by the extremely high CK in affected inplants (usually over 500 IU, although not always as high as in DMD). This probably also means that cases of Becker dystrophy will be detected by newborn screening using CK (see above), an important point to be considered in explaining the prognosis in such a situation. The variability in course is something that also needs careful discussion in genetic counseling, although it is important to stress that there is no chance of typical DMD occurring in a Becker dystrophy family.

Carrier detection in Becker dystrophy has not received such detailed study as in DMD, but it appears that the proportion of carriers falling outside the normal range is around 50 percent (Kingston, 1983), compared with 70 percent for DMD. Skinner et al. (1975) found that CK levels in obligatory carriers decreased with age and have published a curve that allows this to be taken into account. However, one advantage over DMD is the feasibility of classifying young girls in BMD families as obligatory carriers, although numbers studied in this age group are still insufficient to determine whether testing in childhood has advantages over later life. Manifesting carriers do not seem to occur in Becker dystrophy.

Some families with X-linked muscular dystrophy do not appear to fit clinically into either the Becker or Duchenne categories, and it seems likely that there is at least one further distinct type following X-linkage. This form was first recognized by Emery and Dreifuss (1966) and is characterized by variable (usually mild) disability, by early contractures, and by prominent and often fatal cardiomyopathy. Other kindreds have been recorded (Cammann, Vehreschild, and Ernst, 1974; Hopkins, Jackson, and Elsas, 1981), and it is possible that the large family of Mabry et al. (1965) may have had this form. The linkage relationships of the Emery-Dreifuss type are currently being studied by DNA probes to assess whether the localization is similar to Duchenne and Becker dystrophies.

AUTOSOMAL RECESSIVE LIMB-GIRDLE MUSCULAR DYSTROPHY

Autosomal recessive limb-girdle muscular dystrophy is undoubtedly the most poorly defined group within the major dystrophies. It will probably prove to contain further heterogeneity. Because of this and clinical variability our understanding of the genetics is limited, and there are serious pitfalls in genetic counseling.

The typical clinical picture (Walton and Nattrass, 1954) is one of late childhood or adolescent onset of difficulty in walking, with the pelvic girdle predominantly affected, the upper limbs and shoulder girdle less so, and the face spared. The distribution of muscle involvement is similar to Becker and (at a different age) Duchenne dystrophies, but calf pseudohypertrophy is uncommon. The lack of facial involvement distinguishes the disorder from facioscapulohumeral dystrophy. The pelvic girdle involvement results in relatively early disability in walking, so that many patients enter a wheelchair in their twenties or before, even though active in other respects.

In the past many patients were confused with the late onset (Kugelberg-Welander) forms of spinal atrophy; it still remains difficult to distinguish the isolated male case from Becker dystrophy. Another clinical problem with profound genetic consequences is the ease of confusion of female cases with the manifesting Duchenne carriers (Moser and Emery, 1974; Emery, 1980). Early onset cases in girls, that resemble Duchenne dystrophy, should have chromosome studies. Despite these major difficulties, there is little doubt that most cases follow autosomal recessive inheritance (Morton and Chung, 1959; Morton, Chung, and Peters, 1963). The prevalence appears to vary considerably among populations. In most of Europe and the United States it would seem to be much less common than Becker and Duchenne dystrophies; Morton and Chung estmated around 3 per 100,000 incidence from pooled data. In other populations, including some of the U.S. Amish isolates, it is the most common adult dystrophy.

There is no adequate information on localization of the gene (Stevenson, Cheeseman, and Huth, 1955), which is not surprising considering the problems of delineation. An early study suggesting linkage with the Pelger-Huët anomaly—a variation in leucocyte morphology (Schneiderman et al., 1969) that was based with an unusual kindred following autosomal dominant inheritance—has not been confirmed. Similarly uncertain is whether the heterozygotes show any changes in CK or in muscle histology. This is of practical importance in the distinction of the isolated male case from Becker dystrophy, and at present it is probably wise to assume that a marked elevation of CK in the mother of an isolated case favors the diagnosis of the X-linked Becker form. No biochemical abnormalities suggestive of the primary defect have so far been discovered.

In summary, it is usually only possible to be certain about the inheritance of this group when the individual pedigree is typical. Counseling for the numerous isolated cases often remains very provisional, even after the most careful clinical and genetic investigations have been made.

FACIOSCAPULOHUMERAL DYSTROPHY

Facioscapulohumeral dystrophy, the most benign of the major progressive muscular dystrophies, is characterized by a combination of facial and neck muscle weakness with predominant involvement of the shoulder girdle. The relative sparing of the pelvic girdle and thigh muscles keeps patients mobile to an advanced stage, although distal lower limb involvement, especially of the peroneal muscles, is frequent.

The disorder has been the subject of a number of reports, but a recent extensive study by Padberg (1982) has clarified the genetic aspects, as well as contributed some important practical points regarding genetic counseling. The following account is largely based on this study, which also provides a full account of the older literature.

Inheritance is clearly autosomal dominant, and penetrance is nearly complete by middle age. The sex incidence is equal, and there is no clear sex difference in severity, although Padberg found most of his propositi to be male, a finding also seen with myotonic disorders. Fitness is not significantly reduced, although there may be bias toward ascertainment of more fertile individuals, and life span is normal. Prevalence estimates vary greatly from 1 in 20,000 for West Germany (Becker, 1953) to around 1 in 400,000 in Switzerland and Wisconsin (Morton and Chung, 1959; Moser et al., 1966). Padberg (1982) found a prevalence of 1 in 46,000 in Holland and, allowing for lack of ascertainment, estimated a true prevalence of 1 in 21,000. Since many cases do not seek medical advice until a late stage, if at all, it would not be surprising if a frequency of around 1 in 20,000 is usual in most populations.

There is no convincing evidence to suggest heterogeneity in facioscapulohumeral dystrophy, although care must be taken to exclude neurogenic scapuloperoneal syndromes and rare mitochondrial myopathies. So far, genetic linkage data have not provided any conclusive localization, but Padberg obtained a suggestion of linkage with the serum protein polymorphism GM (maximum lod score of 1.4 at $\theta = 0.20$).

The principal genetic counseling problem in this disorder is the usual one with variable dominantly inherited conditions: risk for the apparently normal but relatively young family member transmitting the condition. The chances that a young adult who is entirely normal on clinical examination has the gene would seem to be extremely small, but the importance of careful examination is shown by the finding of Padberg that one third of his cases were asymptomatic at the time of study. CK is of little value except that an elevated level frequently confirms the clinical findings in asymptomatic individuals. Relatives often need careful explanation of why their genetic risks are different from those in Duchenne dystrophy.

OCULOPHARYNGEAL MUSCULAR DYSTROPHY

Since recognition of the mitochondrial myopathies, a proportion of cases of myopathy with major ocular involvement have been separately classified, especially those with congenital onset and those with central nervous system involvement. There remains, however, a distinct disorder following autosomal dominant inheritance and combining progressive ophthalmoplegia with progressive dysphagia from pharyngeal and esophageal muscle dysfunction. Onset is commonly late, and general motor disability is slight in comparison with the severe problems caused by the visual and swallowing difficulty. Most of the large families have been of French-Canadian origin (Hayes et al., 1963; Barbeau, 1966) and may have resulted from a single mutation in a founding member. In other populations the disorder is extremely rare.

DISTAL MYOPATHY

A diagnosis of distal myopathy should only be accepted with caution since most patients with progressive distal weakness and wasting prove to have neuropathic or anterior horn cell disorders. However, the existence of a true myopathy with a benign course, affecting small muscles of hands and feet, has been well documented by Welander (1951, 1957) in a large inbred Swedish kindred. Inheritance was clearly autosomal dominant, but some individuals were symptomless until after 60 years of age. More severely affected individuals with two affected parents were noted who probably represented homozygotes. Another autosomal dominant family was described by Sumner, Crawfurd, and Harriman (1971), but the condition appears to be very rare. Walton (1963) described isolated cases of distal muscular dystrophy, but it is not certain whether all belonged to the autosomal dominant form.

CONGENITAL AND METABOLIC MYOPATHIES

Our knowledge of the genetics of a group of largely nonprogressive congenital and metabolic muscle disorders has grown less rapidly than the subdivision that has occurred, so that for the most part we are able to say relatively little unless a clear inheritance pattern is shown by a particular family. Table 3.6 summarizes the main facts known for congenital myopathies. Particular attention should be drawn to the X-linked recessive lethal form of centronuclear myopathy and to the marked variation in severity within a family seen in thc dominantly inherited forms of central core disease and nemaline myopathy.

The small but growing group of myopathies for which a clear primary metabolic cause is known is shown in Table 3.7. All are probably autosomal recessive.

TABLE 3.6. Congenital Myopathies

Type	Inheritance	Comment
Central core disease	Commonly AD, occasionally AR	Clinical expression variable and often mild
Nemaline myopathy	AR or AD	Most cases isolated
Centronuclear (myotubular) myopathy		
Severe infantile form	XR	Usually fatal in neonatal period
Milder form	Usually Ad	
Multicore myopathy	Uncertain	
Fiber-type disproportion	Uncertain	
Congenital muscular dystrophy	Usually AR	Frequent in Japan
Mitochondrial myopathies		
Cytochromic oxidase deficiency	AR	Usually fatal in infancy
With ophthalmoplegia (Kearns Sayre type)	Usually sporadic	Probably heterogeneous
Other infantile types	Variable	
Late onset with ophthalmoplegia	Often AD	

AD = autosomal dominant
AR = autosomal recessive
XR = X-linked recessive

TABLE 3.7. Rare Inherited Myopathies Associated with Specific Metabolic Defects

Carnitine palmityl transferase deficiency
Glycogenoses
Type II (Pompe)
Type III (Debrancher)
Type V (McArdle)
Cytochrome oxidase deficiency
Muscle phosphorylase deficiency

MYOTONIC DYSTROPHY AND OTHER MYOTONIC DISORDERS

Table 3.8 lists the major myotonic disorders and their inheritance. All are nonprogressive except for myotonic dystrophy; so only this disorder will be considered in detail. Becker's (1977) monograph on the nonprogressive myotonias and a separate work on paramyotonia (Becker, 1970) give full accounts of these other conditions.

Myotonic dystrophy is one of the most variable genetic disorders known and, despite extensive clinical and genetic studies (Harper, 1979), still presents a number of puzzling features that defy easy explanation. Clinically, the main feature is the presence of myotonia, in conjunction with a rather characteristic distribution of muscle weakness and wasting—the face, jaw, anterior neck, and distal limb muscles being principally involved. The phenomenon of myotonia—a delay in relaxation of voluntary muscles or percussion-induced contraction—has been fully studied (Bryant, 1973; 1977; Rudel, 1985) and shown to be the result of a number of possible abnormalities of sodium and potassium ion flux across the muscle membrane. Apart from voluntary muscle, there is extensive involvement of smooth and caridac muscle and of other systems. Despite much experimental work, no primary biochemical defect has been identified (Harper, 1979, 1984).

Genetically myotonic dystrophy behaves as an autosomal dominant, with penetrance close to completion by old age, but with some individuals minimally, even subclinically, affected. Transmission by the sexes is equal when all ages are considered (Harper, 1972, 1979), and the incidence in the sexes is also equal, although propositi are predominantly male (Thomasen, 1948).

Estimates of prevalence are shown in Table 3.9. The variation is likely

TABLE 3.8. The Myotonic Disorders

	Inheritance
Myotonic dystrophy	Autosomal dominant
Myotonia congenita	
Thomsen's disease	Autosomal dominant
Recessive type	Autosomal recessive
With painful cramps	Autosomal dominant
Paramyotonia congenita	Autosomal dominant
Periodic paralysis	
Hypokalemic	Autosomal dominant
Normo/hyperkalemic (adynamia episodica)	Autosomal dominant
Chondrodystrophic myotonia (Schwartz-Jampel syndrome)	Autosomal recessive
Acquired myotonia	
Drug induced	
Associated with malignancy	

TABLE 3.9. Prevalence of Myotonic Dystrophy

Source	Place	Frequency per 100,000
Klein, 1958	Switzerland	4.9
Lynas, 1957	Northern Ireland	2.4
Kurland, 1958	Rochester, Minnesota, United States	3.3
Grimm, 1975	West Germany	5.5
Todorov et al., 1970	Switzerland	13.5*

*Heterozygote frequency, not disease frequency.

to result from incomplete ascertainment more than from true geographical factors, and the true frequency is probably at least as great as the highest estimates. Particular foci in some regions (for example, parts of Canada) may reflect operation of the "founder principle" in these populations.

Genetic fitness is moderately reduced in both sexes, being around 70 percent in most series (Bell, 1948; Klein, 1958; Harper, 1979), but it is possible that lack of ascertainment of reproductively normal mild cases may have biased such estimates. Similar omission of fatal neonatal cases could result in the opposite, but parents may continue reproducing to replace such infants; so the true overall fitness of the gene is hard to assess accurately. New mutations are extremely unusual; almost all isolated cases where both parents are available for careful study prove to have been transmitted. And the connection of apparently separate kindreds by individuals showing only cataract and no muscle abnormalities has been recognized for many years (Fleischer, 1918).

Genetic heterogeneity in myotonic dystrophy has been much debated (Bundey and Carter, 1972; Harper, 1979). Originally, there was argument as to its distinctiveness from the nonprogressive myotonia congenita (Maas and Paterson, 1939); this is now beyond doubt, but whether heterogeneity exists within myotonic dystrophy itself is doubtful. Undoubtedly some kindreds show atypical features, such as that described from Labrador (Webb et al., 1978; Pryse-Phillips, Johnson, and Lassen, 1979); Bundey (1981) has claimed that at least two distinct clinical types can be recognized, but the experience of most workers, including the author, is that all degrees of severity and variations in the form of expression may be seen within a single kindred.

This remarkable variation has prompted a number of explanations, notably the phenomenon of "anticipation"; anticipation holds that a progressive worsening with reduced age at onset occurs in successive generations. This is undoubtedly a striking feature of many families, but whether it requires an unorthodox biological explanation is a different matter. Penrose (1948) elegantly pointed out the biases that would favor the milder, older generation reproducing more and being observed more frequently, whereas mild or late onset disease in the younger generation would more often escape observation unless this was continued over a prolonged period. The converse would apply to severe and early onset cases. However, Penrose and other early workers had not appreciated one important factor favoring severe childhood disease: transmission specifically through an affected mother, often herself mildly affected.

Congenital Myotonic Dystrophy

This remarkable form of myotonic dystrophy, first recorded clearly by Vanier (1960) and subsequently studied by Dyken and the author (Harper and Dyken, 1972; Dyken and Harper, 1973), shows an entirely different clinical picture to that seen in adult life. Table 3.10 summarizes the main abnormalities. Frequently, intrauterine onset occurs, and death is common in the first hours of days or life; the underlying pathology is of extreme muscle hypoplasia accompanied by histological changes of immaturity rather than of degeneration (Sarnat and Silbert, 1976; Young et al., 1981). Myotonia is insignificant, and extreme hypotonia is the predominant neuromuscular problem. In survivors, features of the classical disease gradually appear during childhood, and the disease is never truly transient, as is the case in congenital myasthenia gravis. Mental retardation of varying degrees occurs in the majority of congenital cases (Harper, 1975a).

The common feature of all congenital cases is their exclusively maternal transmission (Harper, 1975b); although some childhood, even a few infantile onset, cases have an affected father, this has not been recorded where there is clear intrauterine onset, nor have such cases been recorded as new mutations. This maternal transmission initially prompted speculation as to a biochemical or immunological circulating maternal factor (Harper and Dyken, 1972). But no such factor has been identified, and it seems likely that whatever the mechanism the immediate cause of the congenital disease is a widespread failure of intrauterine fetal muscle development. Genetic risks for the congenital disease are discussed below.

Gene Mapping

Myotonic dystrophy is the best understood of all autosomal muscular dystrophies in terms of its linkage relationships. It has in many ways been a prototype for the study of autosomal linkage. In 1954 Mohr showed that the Lutheran and Lewis blood groups were linked and that a possibility also existed of linkage between them and myotonic dystrophy. Subsequently, it was shown that the secretor locus was the primary determinant of the Lewis type of red cells, and linkage of myotonic dystrophy with secretor was confirmed (Renwick et al., 1971; Harper et al., 1972). This triple linkage remained unassigned for 10 years until the demonstration of linkage of secretor with the complement component C3 (Eiberg et al., 1981), already localized to chromosome 19 by somatic cell hybrid studies (Whitehead et al., 1982),

TABLE 3.10. Congenital Myotonic Dystrophy: Major Clinical Features

Bilateral facial weakness
Hypotonia
Delayed motor development
Mental retardation
Neonatal respiratory distress
Feeding difficulties
Talipes
Hydramnios in later pregnancy
Reduced fetal movements

allowed the whole group to be placed on this chromosome. Linkage data between myotonic dystrophy and C3 support this (Eiberg et al., 1981), including data from a C3 DNA polymorphism (Davies, Jackson, and Williamson, 1983). Further confirmation has come from the finding of very close linkage between another chromosome 19 marker, the enzyme peptidase D, to myotonic dystrophy (O'Brien, Ball, and Sarforazi, 1983). Figure 3.2 shows one of the families in which this marker is segregating and demonstrates the lack of recombination. Unfortunately, the polymorphism is too infrequent to be of more than occasional use in prediction.

A new and highly polymorphic DNA marker which appears to be closely linked to myotonic dystrophy is the gene probe for the apolipoprotein ApoCII. This probe, which is closely linked to ApoE (Humphries et al., 1984),

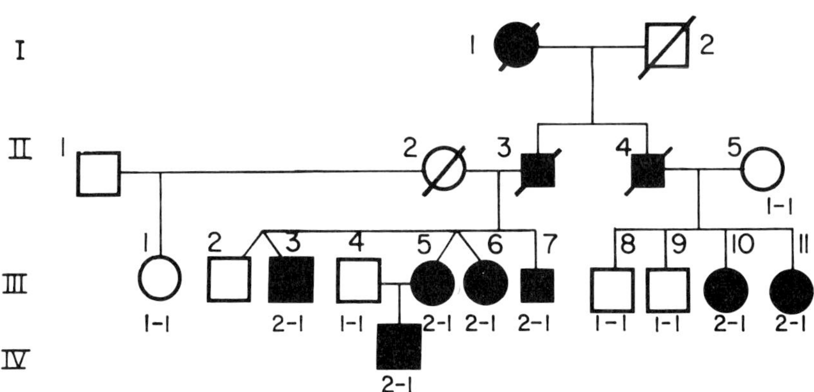

FIGURE 3.2. Linkage between myotonic dystrophy and the enzyme peptidasc D. One of the families reported by O'Brien, Ball, and Sarfarazi (1983). The myotonic dystrophy gene is in coupling with the variant "2" peptidase D allele.

and thus probably on the opposite side of myotonic dystrophy to C3, has been found to be closely linked to myotonic dystrophy by Shaw et al. (1985). The most likely distance, based on 12 families was 4 cM, with a peak lod score of 5.

No other serious diseases have so far been shown to be closely linked to myotonic dystrophy, but an interesting observation on neurofibromatosis and myotonic dystrophy in the same large kindred suggests the possiblility of linkage between the two disorders (Ichikawa et al., 1981).

The current state of the chromosome 19 gene map is shown in Figure 3.3. At present the linkage map cannot be matched up with the structural map, but this will undoubtedly be resolved in the near future. Other genes of interest recently localized to chromosome 19 include those for familial hypercholesterolemia, apolipoprotein E, one subunit of ferritin, and β human chorionic gonadotrophin, making 19 a relatively well mapped chromosome (Human Gene Mapping, 1984).

Localization of genes to specific regions of chromosome 19 is being helped by the use of rodent-human hybrid cell lines, derived from translocations involving 19, in which particular parts of the chromosome are missing. Using this approach, most of the markers closely linked to myotonic dystrophy are located in the central region of the chromosome, apart from C3, which is placed more distally on the short arm. This observation also allows orientation of the linkage group (Brook et al., 1984), the most likely site for myotonic dystrophy being on the short arm close to the centromere.

The precise localization of the myotonic dystrophy gene is now being studied using DNA probes derived from libraries relatively specific for chromosome 19; one such sequence has been derived from a flow-sorted library (Brook et al., 1984), but the most encouraging source is a hybrid cell line in which chromosome 19 is the only recognizable human chromosome; this library has already yielded a number of DNA sequences specific for chromosome 19 and should prove useful for studying not only myotonic dystrophy but also other disorders on the chromosome. It should produce correspondingly valuable results for prediction and prenatal detection, as well as help to resolve the question of genetic heterogeneity and eventually to identify the nature of the mutational defect. So far linkage data have not produced evidence for involvement of more than one locus.

Detection and Prediction

The remarkable variation in age at onset and severity of myotonic dystrophy has already been mentioned. The practical consequence of this is the difficulty in predicting whether an asymptomatic individual at risk actually has the gene. Curves of age at onset have been produced (Harper, 1977); however, since these are based on age at diagnosis, they are likely to underestimate the proportion that can be detected at a particular age. Study

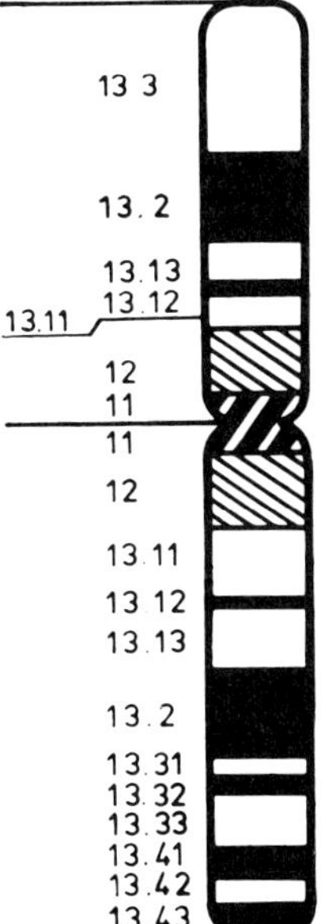

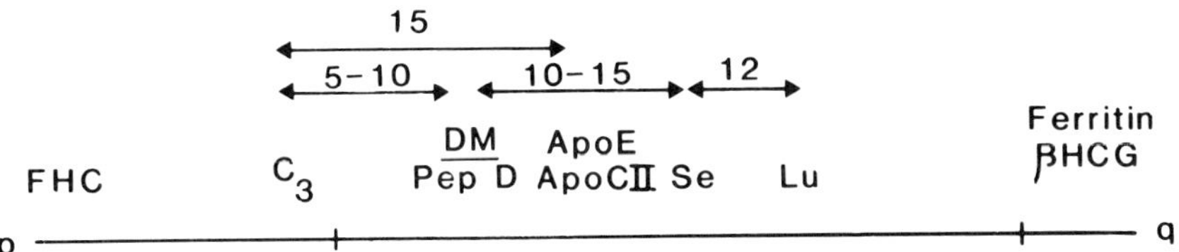

FIGURE 3.3. **(A)** Structural and **(B)** linkage map of chomosome 19, showing some of the distances estimated from published data. The order of loci and relationship to the chromosome structure remain uncertain.

of the offspring of affected individuals has suggested that in many families the expected 50 percent affected is seen by age 14 (Harper, 1973), but in some families age at first detection may be later (Bundey, 1974).

The main practical aids to detection include careful clinical examination (the author found 14 percent of asymptomatic relatives to be clinically affected [Harper, 1973]), slit-lamp examination of the eyes for the characteristic early lens opacities, and electromyography for myotonia (Bundey, Carter et al., 1979; Polgar et al., 1972). Numerous other investigations have been claimed to be of additional help. However, none are convincing, and many of the reservations already discussed in relation to Duchenne dystrophy also apply here. Table 3.11 lists some of the tests that have been applied. Application of genetic linkage data is limited by the distance of the available markers; although the C3 locus is more polymorphic than the secretor locus, neither is sufficiently close to allow accurate prediction. It is possible that the peptidase D locus will prove close enough to be of use in the small number of informative families, but closely linked DNA markers will hopefully be available in the near future.

Prenatal diagnosis of myotonic dystrophy is not feasible directly at present, although prenatal prediction by secretor typing has been attempted (Schrott, Karp, and Omenn, 1973; Insley et al., 1976), since this test can be performed on amniotic fluid (Harper et al., 1971; Teichler-Zallen and Doherty, 1980). Again, the use of DNA markers in chorionic villi should prove valuable in the future.

In addition to prediction of risk, prediction of severity is important in a disorder as variable as myotonic dystrophy. In particular, the likelihood of congenital or severe childhood disease in the offspring of affected women is of major concern. The risks are summarized in Table 3.12. The table shows that for a woman who has already had a congenitally affected child the risks are exceedingly high and that they are considerable for the offspring of affected women overall.

TABLE 3.11. Preclinical Tests for Detection of Myotonic Dystrophy

Established
Electromyography
Slit-lamp examination
Limited value
Serum creatine kinase
Serum immunoglobulins
Serum FSH
Electroretinography
Genetic linkage markers (secretor, C_3)

FSH = follicle-stimulating hormone

TABLE 3.12. Approximate Risks for Offspring of Women with Myotonic Dystrophy

	Risk (in percent)
Previous congenitally affected child	
Unaffected	50
Severe childhood disease	25
Neonatal death	25
No previous congenitally affected child	
Unaffected	50
Severe childhood disease	10
Neonatal death	10
Later affected	30

Source: Based on Harper, 1972, 1979, and unpublished data.

REFERENCES

Adornato B. T., L. J. Kagen and W. K. Engel. 1978. Myoglobinaemia in Duchenne muscular dystrophy patients and carriers: a new adjunct to carrier detection. Lancet 2 : 499–501.

Aston J. P. and D. Stansbie. 1982. Plasma pyruvate kinase activity: disruption of platelets and leucocytes results in increased plasma activity. Clin. Chim. Acta 123 : 199–202.

Barbeau A. 1966. The syndrome of hereditary late onset ptosis and dysphagia in French Canada. In: Progressive muskeldystrophie, myotonie, myasthenie. Ed. E. Kuhn. Springer, New York.

Becker P. E. 1977. Myotonia congenita and syndromes associated with myotonia. Thieme, Stuttgart.

Becker P. E. 1970. Paramyotonia congenita (Eulenberg). Thieme, Stuttgart.

Becker P. E. 1953. Dystrophia musculorum progressiva. Eine genetische und klinische Untersuchung der Muskeldystrophien. Thieme, Stuttgart.

Becker, P. E. and F. Kiener. 1955. Eine neue X chromosomale Muskeldystrophie. Zeitschrift für Neurologie. 193 : 427.

Beckmann R. and G. Scheuerbrandt. 1976. Screening auf erhöhte CK-Aktiväten. Kinderarzt 7 : 1267–1272.

Bell J. 1948. Dystrophia myotonica and allied diseases. In: Treasury of human inheritance. Vol. 4. Cambridge University Press, Cambridge.

Bertolotto A., M. De Marchi, C. Doriguzzi, T. Mongini, C. Monnier, L. Palmucci, D. Schiffer and L. Verze. 1981. Epidemiology of Duchenne muscular dystrophy in the province of Turin. Ital. J. Neurol. Sci. 2 : 81–84.

Blyth H., C. O. Carter and V. Dubowitz. 1965. Duchenne's muscular dystrophy and the Xg blood groups: a search for linkage. J. Med. Genet. 2 : 157–160.

Blyth H. and R. J. Pugh. 1959. Muscular dystrophy in childhood. The genetic aspect. Ann. Hum. Genet. 23 : 127.

Bodensteiner J. B. and A. G. Engel. 1978. Intracellular calcium accumulation in Duchenne dystrophy and other myopathies: a study of 567,000 muscle fibres in 114 biopsies. Neurology (NY) 28: 439–446.

Brook J. D., D. J. Shaw, A. L. Meredith, Brums, Gap and P. S. Harper. 1984. Localization of genetic markers and orientation of the linkage group and chromosome 19. Human Genetics 68 : 282–285.

Brooks A. P. and A. E. H. Emery. 1977. The incidence of Duchenne muscular dystrophy in the South East of Scotland. Clin. Genet. 2 : 290–294.

Brown C. S., N. S. T. Thomas, M. Sarfarazi, K. E. Davies, L. Kunkel, P. L. Pearson, H. M. Kingston, D. J. Shaw and P. S . Harper. 1985. Genetic linkage relationships of seven DNA probes with Duchenne and Becker muscular dystrophy. Human Genetics. submitted.

Bryant S. H. 1977. The physiological basis of myotonia. In: Pathogenesis of human muscular dystrophies. Ed. L. P. Rowland, pp. 715–728. Excerpta Medica, Amsterdam.

Bryant S. H. 1973. The electrophysiology of myotonia, with a review of congenital myotonia of the goat. In: New developments in electromyography and clinical neurophysiology. Vol. 1. Ed. J. E. Desmedt, pp. 420–450. Karger, Basel.

Bulfield G., W. G. Siller, P. A. L. Wight and K. J. Moore. 1984. X-chromosome-linked muscular dystrophy (mdX) in the mouse. PNAS. 81 : 1189–1192.

Bundey S. 1981. A genetic study of Duchenne muscular dystrophy in the West Midlands. J. Med. Genet. 18 : 1–7.

Bundey S. 1978. Calculation of genetic risks in Duchenne muscular dystrophy by geneticists in the United Kingdom. J. Med. Genet. 15 : 249–253.

Bundey S. 1974. Detection of heterozygotes for myotonic dystrophy. Clin. Genet. 5 : 107–109.

Bundey S. and C. O. Carter. 1972. Genetic heterogeneity for dystrophia myotonica. J. Med. Genet. 9 : 311–315.

Bundey S., C. O. Carter and J. F. Soothill. 1979. Early recognition of heterozygotes for the gene of dystrophia myotonica. J. Neurol. Neurosurg. Psychiat. 33 : 279–293.

Bundey S., J. M. Crawley and J. H. Edwards. 1979. Serum creatine kinase levels in pubertal, mature, pregnant and postmenopausal women. J. Med. Genet. 16 : 117–121.

Bundey S., and J. Ebdy. 1982. Fetal sexing in possible carriers for Duchenne Muscular Dystrophy. Prenatal Diagnosis *2*, 1–6.

Burt D. and A. E. H. Emery. 1979. Serum LDH-S in carriers of Duchenne muscular dystrophy. Neurology (NY) 29 : 239–241.

Cammann R., T. Vehreschild and K. Ernst. 1974. Eine neue Sippe von X-chromosomealer benigner Muskeldystrophie mit Fruhkontrakturen (Emery-Dreifuss). Psychiatrie Neurologie und medizinische Psychologie (Leipzig) 26 : 431–438.

Canki N., B. Dutrillaux and I. Tivadar. 1979. Dystrophie musculaire de Duchenne chez une petite fille porteuse d'une translocation t (X;3) (p21:q13) de novo. Ann. Genet. 22 : 35–39.

Caskey C. T., R. L. Nussbaum, L. C. Cohan and L. Pollack. 1980. Sporadic occurrence of Duchenne muscular dystrophy: evidence for new mutation. Clin. Genet. 18 : 329–341.

Cohen H. J., G. E. Molnar and L. T. Taft. 1968. Dev. Med. Child Neurol. 10 : 754.

Conneally P. M. and I. Heuch. 1974. A computer program to determine genetic risks: a simplified version of PEDIG. Am. J. Hum. Genet. 26 : 773–775.

Davie A. M. and A. E. H. Emery. 1978. Estimation of proportion of new mutants among cases of Duchenne muscular dystrophy. J. Med. Genet. 15 : 339–345.

Davies K. E., J. Jackson, R. Williamson, P. S. Harper, S. Ball, M. Sarfarazi, L. Meredith and G. Fey. 1983. J. Med. Genet.

Davies K. E., P. L. Pearson, P. S. Harper, J. M. Murray, T. O'Brien, M. Sarfarazi and R. Williamson. 1983. Genetic analysis of the short arm of the X-chromosome defined by two random cloned DNA sequences flanking Duchenne muscular dystrophy. Nucleic Acids Research 11 : 2303–2312.

Davies K. E., B. D. Young, R. G. Elles, M. E. Hill and R. Williamson. 1981. Cloning of a representative genomic library of the human X chromosome after sorting by flow cytometry. Nature 293 : 374–376.

Dellamonica C. 1978. Thesis. University of Lyon.

Dellamonica C., J. M. Robert, J. Cotte, H. Plauchu and C. Dorche. 1979. Dépistage néonatal systematique de la myopathie de Duchenne de Boulogne. Nouv. Presse Med. 8 : 1491–1493.

Drayna D., K. Davies, D. Hartley, J.-L. Mandel, G. Camerino, R. Williamson and R. White. 1984. Genetic mapping of the human X chromosome by using restriction fragment length polymorphisms. Proc. Natl. Acad. Sci. USA. 81 : 2836–2839.

Dubowitz V. 1978. Muscle disorders in childhood. W. B. Saunders, Philadelphia.

Dubowitz V. 1977. Mental retardation in Duchenne muscular dystrophy. In: Pathogenesis of human muscular dystrophies. Ed. L. P. Rowland, pp. 688–694. Excerpta Medica, Amsterdam.

Dubowitz V. 1965. Intellectual impairment in muscular dystrophy. Arch. Dis. Child. 40 : 296–301.

Dyken P. R. and P. S. Harper. 1973. Congenital dystrophia myotonica. Neurology (Minneapolis) 23 : 465–473.

Edwards R., A. Young and M. Wiles. 1980. Needle biopsy of skeletal muscle in the diagnosis of myopathy and the clinical study of muscle function and repair. N. Eng. J. Med. 302 : 261–271.

Edwards R. J., C. H. Rodeck, and D. C. Watts. 1984. Errors in plasma CK estimation on fetal blood samples resulting from contamination with amniotic fluid and maternal blood. Prenatal Diagnosis 4 : 119–134.

Eiberg H., J. Mohr, L. Staub-Nielsen and N. Simonsen. 1981. Linkage relationship between the locus for C3 and 50 polymorphic systems: assignment of C3 to the DM-SE-LU linkage group: confirmation of C3-LES linkage; support of LES-DM synteny. Proc. VI Inter. Cong. Hum. Genet. (Jerusalem). Alan Liss, New York.

Elles R. G., R. Williamson, M. Niazi, D. V. Coleman and D. Horwell. 1983. Absence of maternal contamination of chorionic villi for fetal-gene analysis. New Eng. J. Med. 308 : 1433– 1435.

Emery A. E. H. 1980. Duchenne muscular dystrophy. Genetic aspects, carrier detection and antenatal diagnosis. Br. Med. Bull. 36 : 117–122.

Emery A. E. H. 1969. Abnormalities of the electrocardiograms in female carriers of Duchenne muscular dystrophy. Br. Med. J. 2 : 418–420.

Emery A. E. H. 1966. Genetic linkage between the loci for colour blindness and Duchenne type muscular dystrophy. J. Med. Genet. 3 : 92–95.

Emery A. E. H., D. Burt and V. Dubowitz. 1979. Antenatal diagnosis of Duchenne muscular dystrophy. Lancet 1 : 847–849.

Emery A. E. H. and F. E. Dreifuss. 1966. Unusual type of benign X-linked muscular dystrophy. J. Neurol. Neurosurg. Psychiat. 29 : 338.

Emery A. E. H. and R. Skinner. 1976. Clinical studies in benign (Becker type) X-linked muscular dystrophy. Clin. Genet. 10 : 189.

Emery A. E. H., R. Skinner and S. Holloway. 1979. A study of possible heterogeneity in Duchenne muscular dystrophy. Clin. Genet. 15 : 444–449.

Emery A. E. H., C. A. B. Smith and R. Sanger. 1969. The linkage relations of the loci for benign (Becker type) X-borne muscular dystrophy, colour blindness and the Xg blood groups. Ann. Hum. Genet. 32 : 261.

Emery A. E. H. and J. N. Walton. 1967. Prog. Med. Genet. 5 : 116–145.

Engel A. and B. Q. Banker. Eds. 1985. Myology. McGraw Hill, New York.

Fleischer B. 1918. Über myotonischer dystrophie mit katarakt. Albrecht von Graefes. Arch. Ophthalmol. 96 : 91–133.

Francke U., J. Felsenstein and S. M. Gartler. 1976. The occurrence of new mutants in the X-linked recessive Lesch-Nyhan disease. Am. J. Hum. Genet. 28 : 123–136.

Gardner-Medwin D. 1970. Mutation rate in Duchenne type of muscular dystrophy. J. Med. Genet. 7 : 334.

Gardner-Medwin D. 1968. Studies of the carrier state in the Duchenne type of muscular dystrophy. 2. Quantitative electromyography as a method of carrier detection. J. Neurol. Neurosurg. Psychiat. 31 : 124–134.

Gardner-Medwin D., S. Bundey and S. Green. 1978. Early diagnosis of Duchenne muscular dystrophy. Lancet 1 : 1102.

Gardner-Medwin D., R. J. Pennington and J. N. Walton. 1971. The detection of carriers of X-linked muscular dystrophy genes: a review of some methods studied in Newcastle upon Tyne. J. Neurol. Sci. 13 : 459–474.

Gilboa H. and J. B. Swanson. 1976. Serum creatine phosphokinase in normal newborns. Arch. Dis. Child. 51 : 283–285.

Golbus M. S., J. D. Stephens, M. J. Mahoney, J. C. Hobbins, F. P. Haseltine, C. T. Caskey and B. Q. Banker. 1979. Failure of fetal creatine phosphokinase as a diagnostic indicator of Duchenne muscular dystrophy. New Eng. J. Med. 300 : 860.

Gosden J. R., A. R. Mitchell, C. M. Gosden, C. H. Rodeck and J. M. Morsman. 1982. Direct vision chorion biopsy and chromosome-specific DNA probes for determination of fetal sex in first-trimester prenatal diagnosis. Lancet 2 : 1416–1418.

Graham J. B. 1979. Genotype assignment (carrier detection) in the haemophilias. Clinics in Haematology 8 : 115.

Greenstein R. M., M. P. Reardon and T. S. Chan. 1980. An (X:11) translocation in a girl with Duchenne muscular dystrophy. Cytogenet. Cell Cenet. 27 : 268.

Griggs R. C., W. Reeves and R. T. Moxley. 1977. The heart in Duchenne dystrophy. In: Pathogenesis of human muscular dystrophies. Ed. L. P. Rowland, pp. 661–671. Excerpta Medica, Amsterdam.

Grimm T. 1981. Newborn screening for Duchenne muscular dystrophy. Monatsschr. Kinderheilkd. 129 : 414–417.

Grimm T. 1975. Thesis. University of Gottingen.
Haldane J. B. S. 1935. The rate of spontaneous mutation of a human gene. J. Genet. 31 : 317-326.
Harper P. S. 1984. Myotonic disorders. In: Myology. Ed. A. Engel, in press.
Harper P. S. 1982. Carrier detection in Duchenne muscular dystrophy. A critical assessment. In: Disorders of the motor unit. Ed. D. L. Schotland. John Wiley & Sons, New York.
Harper P. S. 1979. Myotonic dystrophy. Saunders, Philadelphia.
Harper P. S. 1977. Phenotypic variation in myotonic dystrophy: causes and consequences. In: Pathogenesis of human muscular dystrophies. Ed. L. P. Rowland, pp. 705-712. Excerpta Medica, Amsterdam.
Harper, P. S. 1975a. Congenital myotonic dystrophy in Britain 1. Clinical aspects. Arch. Dis. Child. 50 : 505-513.
Harper, P. S. 1975b. Congenital myotonic dystrophy in Britain 2. Genetic basis. Arch. Dis. Child. 50 : 514-521.
Harper P. S. 1973. Presymptomatic detection and genetic counselling in myotonic dystrophy. Clin. Genet. 4 : 134-140.
Harper P. S. 1972. Genetic studies in myotonic dystrophy. Thesis. University of Oxford.
Harper P. S., W. B. Bias, J. R. Hutchinson and V. A. McKusick. 1971. ABH secretor status of the fetus: a genetic marker identifiable by amniocentesis. J. Med. Genet. 8 : 438-440.
Harper P. S. and P. R. Dyken. 1972. Early onset dystrophia myotonia – evidence supporting a maternal environmental factor. Lancet 2 : 53-55.
Harper P. S., T. O'Brien, J. M. Murray, K. E. Davies, P. Pearson and R. Williamson. 1983. The use of linked DNA polymorphisms for genotype prediction in families with Duchenne muscular dystrophy. J. Med. Genet. 20 : 252-254.
Harper P. S., M. L. Rivas, W. B. M. Bias, J. R. Hutchinson, P. R. Dyken and V. A. McKusick. 1972. Genetic linkage confirmed between the loci for myotonic dystrophy, ABH secretion and Lutheran Blood Group. Amer. J. Hum. Genet. 24: 310-316.
Hayes R., W. London, J. Seidman and L. Embree. 1963. Oculopharyngeal muscular dystrophy. New Eng. J. Med. 268 : 163.
Heuch I. and F. H. F. Li. 1972. A computer program for calculation of genotype probabilities using phenotype information. Clin. Genet. 3 : 501-504.
Hopkins L. C., J. A. Jackson and L. J. Elsas. 1981. Emery-Dreifuss humeroperoneal muscular dystrophy: an X-linked myopathy with unusual contractures and bradycardia. Ann. Neurol. 10 : 230-237.
Human Gene Mapping 7. Birth Defects Original Articles Series 20 No 2. 1984.
Humphries S. E., K. Berg, L. Gill, A. M. Cumming, F. W. Robertson, A. F. H. Stalenhoef, R. Williamson, and A-L. Borresen. 1984. The gene for apolipoprotein C-II is closely linked to the gene for apolipoprotein E on chromosome 19. Clin. Genet. 26 : 389-396.
Hurse P. V. and B. A. Kakulas. 1974. Genetic counselling in inherited muscle disease in Western Australia. Excerpta Medica Int. Congr., Amsterdam Series 334 : 88.
Hutton E. M. and M. W. Thompson. 1976. Carrier detection and genetic counselling in Duchenne muscular dystrophy: a follow-up study. Can. Med. Assoc. J. 115 : 749-752.
Ichikawa K., C. Croseley, A. Culebras and L. Weitkamp. 1981. Coincidence of neurofibromatosis and myotonic dystrophy in a kindred. J. Med. Genet. 18 : 134-138.
Insley J., G. W. G. Bird, P. S. Harper and G. W. Pearce. 1976. Prenatal prediction of myotonic dystrophy. Lancet 1 : 806.
Ionasescu V., H. Zellweger and P. Cancilla. 1978. Fetal serum-creatine-phosphokinase not a valid predictor of Duchenne muscular dystrophy. Lancet 2 : 1251.
Jacobs P. A., P. A. Hunt, M. Mayer and R. D. Bart. 1981. Duchenne muscular dystrophy (DMD) in a female with an X/autosome translocation: further evidence that the DMD locus is at Xp21. Am. J. Hum. Genet. 33 : 513-518.
Kingston H. M. 1983. Clinical and genetic studies of Becker muscular dystrophy. M.D. thesis. University of Manchester.
Kingston H. M., M. Sarfarazi, N. S. T. Thomas, P. S. Harper. 1984. Localization of the Becker muscular dystrophy gene on the short arm of the X chromosome by linkage to cloned DNA sequences. Human Genetics 67 : 6-17.
Kingston H. M., N. S. T. Thomas, P. L. Pearson, M. Sarfarazi and P. S. Harper. 1983. Genetic linkage between Becker muscular dystrophy and a polymorphic DNA sequence on the short arm of the X chromosome. J. Med. Genet. 20 : 255-258.
Klein D. 1958. La dystrophie myotonique (Steinert) et la myotonie congénitale (Thomsen) en Suisse. J. Genet. Humaine. Suppl. 1 : 1-328.

Kloepfer H. W. and A. E. H. Emery. 1974. Genetic aspects of neuromuscular disease. In: Disorders of voluntary muscle. Ed. J. N. Walton. Churchill Livingston, Edinburgh.
Kondo K. 1980. Genetics of Duchenne muscular dystrophy and care program in Japan. Adv. Neurol. Sci. 24 : 701.
Kuriowa Y. and T. Miyazaki. 1967. Epidemiological study of myopathy in Japan. In: Exploratory concepts in muscular dystrophy. Ed. A. T. Milhorat, p. 98. Excerpta Medica, Amsterdam.
Kurland L. T. 1958. Descriptive epidemiology of selected neurologic and myopathic disorders with particular reference to a survey in Rochester, Minnesota. J. Chronic Dis. 8 : 378–418.
Lane R. J. M., D. M. Gardner-Medwin, A. D. Roses. 1980. Electrocardiographic abnormalities in carriers of Duchenne muscular dystrophy. Neurology 30 : 497–501.
Lindenbaum R. H., G. Clarke and C. Patel. 1979. Muscular dystrophy in an X;1 translocation female suggests that Duchenne locus is on X chromosome short arm. J. Med. Genet. 16 : 389–392.
Lynas M. A. 1957. Dystrophia myotonica, with special reference to Northern Ireland. Ann. Hum. Genet. 21 : 318–351.
Maas O. and A. S. Paterson. 1939. The identity of myotonia congenita (Thomsen's disease), dystrophia myotonica (myotonia atrophica) and paramyotonia. Brain 62 : 198–212.
Mabry C. C., I. E. Roeckel, R. L. Munich and D. Robertson. 1965. X-linked pseudohypertrophic muscular dystrophy with a late onset and slow progression. New Eng. J. Med. 273 : 1062.
Malvano R., A. Massaglia, M. Marchisio, P. A. Giacesa, V. Dona, G. C. Zuccelli, A. Clerico and M. G. Del Chicca. 1979. Radioimmunoassay of circulating human myoglobin. J. Nucl. Med. All. Sci. 22 : 169–175.
Maunder-Sewry C. A. and V. Dubowitz. 1980. Intranuclear calcium in carriers of Duchenne muscular dystrophy. In: Muscular dystrophy research. Advances and new trends. Ed. C. Angelini, G. A. Darieli and D. Fontanari. Excerpta Medica Int. Cong. Series 527 : 14–22.
Maunder-Sewry C. A. and V. Dubowitz. 1979. Myonuclear calcium in carriers of Duchenne muscular dystrophy: an X-ray microanalysis study. J. Neurol. Sci. 42 : 337–347.
Mohr J. 1954. A study of linkage in man. Munskgaard, Copenhagen.
Moosa A., B. H. Brown and V. Dubowitz. 1972. Quantitative electromyography: carrier detection in Duchenne type muscular dystrophy using a new automatic technique. J. Neurol. Neurosurg. Psychiat. 35 : 841–844.
Morton N. E. and C. S. Chung. 1959. Formal Genetics of muscular dystrophy. Am. J. Hum. Genet. 11 : 360–369.
Morton N. E., C. S. Chung and H. A. Peters. 1963. Genetics of muscular dystrophy. In: Muscular dystrophy in man and animals. Ed. G. H. Bourne and N. Golarz. Karger, Basel.
Moser H. and A. E. H. Emery. 1974. The manifesting carrier in Duchenne muscular dystrophy. Clin. Genet. 5 : 271–284.
Moser H., V. Weismann, R. Richterich and E. Rossi. 1966. Progressive muskeldystrophie VIII. Haufigeit, Klinik und Genetik der typen I and III. Schweizer Med. Wschr. 96 : 169–174.
Murphy E. A. and C. S. Mutalik. 1969. Hum. Hered. 19 : 126–151.
Murray J. M., K. E. Davies, P. S. Harper, L. Meredith, C. R. Mueller, P. N. Goodfellow and R. Williamson. 1982. A cloned DNA sequence on the short arm of the X-chromosome: linkage relationship to Duchenne muscular dystrophy. Nature 300 : 69–71.
Nicholson G. A., D. Gardner-Medwin, R. J. T. Pennington and J. N. Walton. 1979. Carrier detection in Duchenne muscular dystrophy: assessment of the effect of age on detection rate with serum creatine-kinase activity. Lancet 1 : 692–694.
Nicholson L. V. B. 1982. Myoglobin levels in human serum and their usefulness in detecting carriers of X-linked muscular dystrophy genes. Thesis. University of Newcastle.
O'Brien T. A., S. A. Ball, M. Sarfarazi, P. S. Harper and E. B. Robson. 1983. Genetic linkage between the loci for myotonic dystrophy and peptidase D. Ann. Hum. Genet. 47 : 117–121.
O'Brien T. A., P. S. Harper, K. Davies, J. M. Murray, M. Sarfarazi and R. Williamson. 1983. Absence of genetic heterogeneity in Duchenne muscular dystrophy shown by a linkage study using two cloned DNA sequences. J. Med. Genet. 20 : 249–251.
Old J. M., R. H. T. Ward, M. Petrov, F. Karogozlu, B. Modell and D. J. Weatherall. 1982. First trimester fetal diagnosis for haemoglobinopathies: three cases. Lancet 2 : 1413–1416.
Padberg G. 1982. Facioscapulohumeral disease. Thesis. University of Leiden.
Page D., B. DeMartinville, D. Barker, A. Wyman, R. White, U. Francke and D. Botstein. 1982. Single-copy sequence hybridizes to polymorphic and homologous loci on human X and Y chromosomes. Proc. Natl. Acad. Sci. U.S.A. 79 : 5352–5356.
Pembrey M. E., K. E. Davies, R. M. Winter, R. G. Elles, R. Williamson, J. A. Fazzone, C.

Walker. 1984. Clinical use of DNA markers linked to the gene for Duchenne muscular dystrophy. Arch. Dis. Child. 59 : 208-216.

Penrose L. S. 1948. The problem of anticipation in pedigrees of dystrophia myotonica. Ann. Eugen. (London) 14 : 125-132.

Percy M. E., D. F. Andrews and M. W. Thompson. 1982. Duchenne muscular dystrophy carrier detection using logistic discrimination: serum creatine kinase, hemopexin, pyruvate kinase and lactate dehydrogenase in combination. Am. J. Med. Genet. 13 : 27-38.

Percy M. E., I. Murphy and C. Oss. 1979. Serum creatine kinase and pyruvate kinase in Duchenne muscular dystrophy carrier detection. Muscle and Nerve. 2 : 329-339.

Perloff J. K., W. C. Roberts, A. C. de Leon and D. O'Doherty. 1967. The distinctive electrocardiogram of Duchenne's progressive muscular dystrophy. Am. J. Med. 42 : 179.

Pickard N. A., H. D. Gruemer and H. L. Verrill. 1978. Systemic membrane defect in the proximal muscular dystrophies. N. Eng. J. Med. 299 : 841-846.

Polgar J. G., W. G. Bradley, A. R. M. Upton, J. Anderson, J. M. L. Howat, F. Petito, D. F. Roberts and J. Scopa. 1972. The early detection of dystrophia myotonica. Brain 95 : 761-776.

Pryse-Phillips W., G. J. Johnson and B. Lassen. 1979. A partial syndrome of myotonic dystrophy. Canad. J. Neurol. 6 : 388.

Renwick J. H., S. E. Bundey, M. A. Ferguson-Smith and M. M. Izatt. 1971. Confirmation of the linkage of the loci for myotonic dystrophy and ABH secretion. J. Med. Genet. 8 : 407-416.

Rodeck C. H. and S. Campbell. 1978. Sampling pure fetal blood by fetoscopy in the second trimester of pregnancy. Br. Med. J. 2 : 728-730.

Rodeck C. H. and J. M. Morsman. 1983. First-trimester chorion biopsy. Br. Med. Bull. 39 : 338-342.

Roses A. D., M. J. Roses, B. S. Metcalf, K. L. Hull, G. A. Nicholson, G. B. Hartwig and C. R. Roe. 1977. Pedigree testing in Duchenne muscular dystrophy. Ann. Neurol. 2 : 271.

Roses A. D., M. J. Roses and G. A. Nicholson. 1977. Lactate dehydrogenase isoenzyme 5 in detecting carriers of Duchenne muscular dystrophy. Neurology (Minneapolis) 27 : 414-421.

Roses, A. D., M. Herbstreith and B. Metcalf. 1976. Increased phosphorylated components of erythrocyte membrane spectrin band II with reference to Duchenne muscular dystrophy. J. Neurol. Sci. 30 : 167-178.

Roses M. J., M. T. Nicholson and C. S. Kircher. 1977. Evaluation and detection of Duchenne's and Becker's muscular dystrophy carriers by manual muscle testing. Neurology (Minneapolis) 27 : 20-25.

Rowland P. 1978. Pathogenesis of human muscular dystrophies. Elsevier.

Rudel R. 1985. The pathophysiologic basis of the myotonias and of the periodic paralyses. In: Myology. Ed. A. G. Engel and B. Q. Banker. In Press.

Sarfarazi M., P. S. Harper, H. M. Kingston, J. M. Murray, T. O'Brien, K. E. Davies, R. Williamson, P. Tippett and R. Sanger. 1983. Genetic linkage relationships between the Xg blood group system and two X chromosome DNA polymorphisms in families with Duchenne and Becker muscular dystrophy. Hum. Genet. 65 : 169-171.

Sarnat H. B. and S. W. Silbert. 1976. Maturational arrest of fetal muscle in neonatal myotonic dystrophy. Arch. Neurol. 33 : 466-475.

Scheuerbrandt G. 1980. Recent advances and new trends. Proceedings of international symposium on muscular dystrophy, Venice. Excerpta Medica, Amsterdam, pp. 157-166.

Schneidermann L. J., W. I. Sampson, W. C. Schoene and G. B. Haydon. 1969. Genetic studies of a family with two unusual autosomal dominant conditions: muscular dystrophy and Pelger-Huet anomaly. Am. J. Med. 46 : 380.

Schotland D. L. 1982. Disorders of the motor unit. Ed. D. L. Schotland. John Wiley & Sons, New York.

Schrott H. G., L. Karp and G. S. Omenn. 1973. Prenatal diagnosis of myotonic dystrophy. Clin. Genet. 4 : 38-45.

Shaw D. J., A. L. Meredith, M. Sarfarazi, S. M. Huson, J. D. Brook, O. Mykleboft and P. S. Harper. 1985. The apolipoprotein CII gene: Subchromosomal localization and linkage to the myotonic dystrophy locus. Human Genetics (in press).

Sibert J. R., P. S. Harper and R. J. Thompson. 1979. Carrier detection in Duchenne muscular dystrophy: evidence from a study of obligatory carriers and mother of isolated cases. Arch. Dis. Child. 54 : 534-537.

Simoni G., B. Brambati, C. Danesino, F. Rosella, G. L. Terzoli and M. Fraccaro. 1983. Hum. Genet. 63 : 349-357.
Skinner R., A. E. H. Emery, A. J. B. Anderson and C. Foxall. 1975. The detection of carriers of benign (Becker type) X-linked muscular dystrophy. J. Med. Genet. 12 : 131.
Skinner R., A. E. H. Emery, G. Scheuerbrandt and J. Syme. 1982. Feasibility of neonatal screening for Duchenne muscular dystrophy. J. Med. Genet. 19 : 1-3.
Skinner R., C. Smith and A. E. H. Emery. 1974. Linkage between the loci for benign Becker-type X-borne muscular dystrophy and deutan colour blindness. J. Med. Genet. 11 : 317.
Smith I., R. A. Elton and W. H. S. Thomson. 1979. Carrier detection in X-linked recessive (Duchenne) muscular dystrophy serum creatine phosphokinase values in premenarchal, menstruating, postmenopausal and pregnant normal women. Clin. Chim. Acta 98 : 207-216.
Stephens F. E. and F. H. Tyler. 1951. Studies in disorders of muscle. V. The inheritance of childhood progressive muscular dystrophy in 33 kindreds. Am. J. Hum. Genet. 3 : 111.
Stevenson A. C., E. A. Cheeseman and M. C. Huth. 1955. Muscular dystrophy in Northern Ireland. III. Linkage data with particular references to autosomal limb-girdle muscular dystrophy. Ann. Hum. Genet. 19 : 165.
Sumner D., M. D. A., Crawfurd and D. G. F. Harriman. 1971. Distal muscular dystrophy in an English family. Brain 94 : 51.
Teichler-Zallen D. and R. A. Doherty. 1980. Amniotic fluid secretor typing: validation for use in prenatal prediction of myotonic dystrophy. Clin. Genet. 18 : 257-267.
Thomasen E. 1948. Myotonia. Universitets forlaget, Aarhus.
Thompson M. W., E. G. Murphy and P. J. McAlpine. 1967. An assessment of the creatine kinase test in the detection of carriers of Duchenne muscular dystrophy. J. Pediat. 71 : 82-93.
Thomson W. H. S. 1969. The biochemical identification of the carrier state in X-linked recessive (Duchenne) muscular dystrophy. Clin. Chim. Acta 26 : 207-221.
Todorov A., M. Jéquier, D. Klein and N. E. Morton. 1970. Analyse de la segregation dans la dystrophie myotonique. J. Genet. Hum. 18 : 387-406.
Toop A. and A. E. H. Emery. 1974. Muscle histology in fetuses at risk for Duchenne muscular dystrophy. Clin. Genet. 5 : 230.
Tzvetanova E. 1978. Serum creatine kinase isoenzymes in progressive muscular dystrophy. Enzyme 23 : 238-245.
Vanier T. M. 1960. Dystrophia myotonica in childhood. Br. Med. J. 2 : 1284-1288.
Verellen C., V. Markovic, R. Demeyer, M. Freund, C. Laterre and R. Worton. 1978. Expression of an X-linked recessive disease in a female due to non-random inactivation of the X chromosome. Am. J. Hum. Genet. 30 : 97A.
Vogel F. 1977. A probable sex difference in some mutation rates. Am. J. Hum. Genet. 29 : 312-319.
Walton J. N. Ed. 1981. Disorders of voluntary muscle. Churchill Livingstone, Edinburgh.
Walton J. N. 1963. Clinical aspects of human muscular dystrophy. In: Muscular dystrophy in man and animals. Ed. G. H. Bourne and M. N. Golarz, p. 263. Hafner, New York.
Walton J. N. and D. Gardner-Medwin. 1981. Progressive muscular dystrophy and the myotonic disorders. In: Disorders of voluntary muscle. Ed. J. Walton. Churchill Livingstone, Edinburgh.
Walton J. N. and F. J. Nattrass. 1954. On the classification, natural history and treatment of the myopathies. Brain 77 : 169.
Webb D., A. Matthews, M. Harris, I. Muir, J. Hostetter, W. Marshall, L. Salimonu, J. Gray, J. Faulkner and G. Johnson. 1978. Myotonia dystrophica: unusual features in a Labrador family. Can. Med. Assoc. J. 118 : 497-500.
Welander L. 1957. Homozygous appearance of distal myopathy. Acta Genetica et Statistica Medica (Basel) 7 : 231.
Welander L. 1951. Myopathia distalis tarda heriditaria; 249 examined cases in 72 pedigrees. Acta Med. Scand. (Suppl.) 141 : 265, 1-124.
Whitehead A. S., E. Solomon, S. Chambers, W. F. Bodmer, S. Povey and G. Fey. 1982. Assignment of the structural gene for the third component of human complement to chromosome 19. Proc. Natl. Acad. Sci. U.S.A. 79 : 5021-5025.
Williams H., A. Roberts, M. Upadhyaya, C. S. Brown, N. S. T., Thomas, P. S. Harper, J. R. Gosden. 1983. First trimester fetal sexing in a pregnancy at risk for Duchenne muscular dystrophy. Lancet 2 : 568-569.

Williamson R., J. Eskdale, D. V. Coleman, N. Niazi, F. E. Loeffler and B. Modell. 1981. Direct gene analysis of chorionic villi: a possible technique for first-trimester antenatal diagnosis of haemoglobinopathies. Lancet 2 : 1125–1127.
Worden D. K. and P. J. Vignos. 1962. Intellectual function in childhood progressive muscular dystrophy. Pediatrics 29 : 968–977.
Worton R. G., C. Duff, J. E. Sylvester, R. D. Schmickel, F. W. Huntington. 1984. Duchenne muscular dystrophy involving translocation of the *dmd* gene next to ribosomal RNA genes. Science 224 : 1447–1449.
Yamuna S., K. Valmiknathan and D. Burt. 1977. Serum pyruvate kinase in carriers of Duchenne muscular dystrophy. Clin. Chim. Acta 79 : 277–279.
Yasuda N. and K. Kondo. 1980. No sex difference in mutation rates of Duchenne muscular dystrophy. J. Med. Genet. 17 : 106–111.
Young R. S. K., D. L. Gang, E. L. Zalnaitis and K. S. Krishnamoorthy. 1981. Dysmaturation in infants of mothers with myotonic dystrophy. Arch. Neurol. 38 : 716–719.
Zatz M., R. T. B. Betti and J. A. Way. 1981. Benign Duchenne muscular dystrophy in a patient with growth hormone deficiency. Am. J. Med. Genet. 10 : 301–304.
Zatz M., O. Frota-Pessoa, J. A. Levy and C. A. Peres. 1976. Creatine-phospho-kinase (CPK) activity in relatives of patients with X-linked muscular dystrophies: A Brazilian Study. J. Genet. Hum. 24 : 153–168.
Zatz M., S. B. Itskan, R. Sanger, O. Frota-Pessoa and P. H. Saldanha. 1974. New linkage data for the X-linked types of muscular dystrophy and G6PD variants, color blindness and Xg blood groups. J. Med. Genet. 11 : 321.
Zatz M., L. J. Shapiro and D. S. Campion. 1978. Serum pyruvate kinase (PK) and creatine-phospho-kinase (CPK) in progressive muscular dystrophies. J. Neurol. Sci. 36 : 349–362.
Zellweger H. and A. Antonik. 1975. Newborn screening for Duchenne muscular dystrophy. Pediatrics 55 : 30–34.

4 Huntington's Disease: Genetics, Chemical Pathology, and Management

Richard H. Myers

Department of Neurology, Massachusetts General Hospital, Harvard Medical School, and Boston University Medical School, Boston, Massachusetts

Miriam Schoenfeld

Department of Human Genetics, Yale University Medical School, New Haven, Connecticut

Edward D. Bird

Department of Neurology, Massachusetts General Hospital and Ralph Lowell Laboratories, McLean Hospital, Harvard Medical School, Boston, Massachusetts

INTRODUCTION

Huntington's disease (HD) is a mid-life onset neurodegenerative disorder with serious medical and social consequences. This dominantly inherited disorder has a prevalence of between 4 and 8 per 100,000 people, with at least twice as many still unaffected gene carriers among the offspring of those affected (Reed et al., 1958; Walker et al., 1981). As part of a New England research program on HD established in 1980, 140 persons diagnosed with HD have been examined. We will review studies attempting to elucidate the primary defect in HD, offer suggestions for managing HD patients, and outline procedures for counseling families who are at risk for this tragic illness.

George Sumner Huntington, while still a child, became aware of an involuntary movement disorder that occurred in certain families in his father's medical practice in East Hampton, Long Island. His later account "On Chorea" described the clinical and genetic characteristics so accurately and vividly that his name has since been associated with this disease (Huntington, 1872).

Some 30 years before Huntington's description, Charles Waters reported hereditary involuntary movements among individuals in the southeastern region of New York State (Dunglison, 1842). In another early publication, Irving Lyon (1863) noted that choreiform movements were confined to certain families. Vessie (1932) traced the ancestors of several persons with chorea in the northeastern United States to immigrants from Bures, England, who arrived in 1630. Many of these families moved to Connecticut and Long Island where Huntington made his original observations. Similarly, HD in Tasmania has been traced to an English woman from Somerset County (Brothers, 1949), and cases in South Africa have been traced to Dutch immigrants who settled there in 1658 (Hayden, 1981).

GENETICS

Mode of Transmission

The primary genetic defect in HD has not been identified. The autosomal dominant transmission places each son or daughter of a gene carrier at a 50 percent risk of inheriting HD. The *penetrance* of the HD gene, defined as the number of gene carriers who actually express the illness, appears to be 100 percent, assuming the individual does not die of unrelated causes prior to the onset of symptoms. The gene has recently been linked to chromosome 4 by a DNA restriction fragment length polymorphism (Gusella et al., 1983), and this will lead to a procedure for the detection of gene carriers prior to the onset of symptoms for some persons at risk for HD.

The overwhelming majority of cases of HD can be confirmed by family history, but the verification process may be impeded by the early death of the affected parent, by adoption, or by uncertain paternity. It is often difficult to document a history when the pattern of age at onset in a family is late and members of prior generations are more likely to have died of other causes before the onset of symptoms.

The mutation rate to HD may be the lowest recognized in any inherited disorder (Shaw and Caro, 1982). Although rare instances of suspected new mutation have been reported (Stevens and Parsonage, 1969), none has met the requirements of (1) serologic confirmation of unaffected parents, (2) autopsy confirmation of HD in the proband, and (3) genetic transmission of the gene to offspring. A family history of HD was obtained in more than 99 percent of the individuals diagnosed with HD in our New England cohort. The absence of the illness among black Africans (Hayden, 1981) and the lower prevalence among black Americans (Wright, Still, and Abramson, 1981), where the gene has presumably been introduced, suggests that mutation to HD is exceedingly rare. The infrequency of mutation may indicate that the mutational event entails more than a change of a single nucleic acid base pair or may occur only when a specific and perhaps uncommon allele is present.

Factors Influencing Onset Age

Age of Onset in HD

The initial symptoms of HD are insidious, and it is difficult to designate an exact age of onset. Mean onset ages ranging from 35 to 42 years (Bell, 1934; Brothers, 1964; Bolt, 1970; Newcombe, 1981; Myers et al., 1982) have been reported. Within the New England sample, the onset age was estimated for 243 persons with a mean of 41 years (Figure 4.1); 6 percent of these cases exhibited initial symptoms before 21 years of age and 23 percent after age 50. The variation in clinical features and in age of onset could suggest that HD represents more than one genetic defect. While it is not possible to disprove genetic heterogeneity, occasionally different members of the same family display substantial differences in clinical features and in onset age of symptoms. For one family in the New England study, a 40-year difference in onset age was recorded for a man and his niece. These variations within families suggest that other environmental or genetic factors may modify the expression of a single HD gene.

Twin Studies

The HD literature contains reports of 13 sets of monozygotic (MZ) twins (Hayden, 1981). The concordance in age of onset and clinical features in these pairs is consistently very high. The difference in onset age within MZ twin pairs is, on the average, less than one year. The difference in onset age within four sets of dizygotic (DZ) twins (Adams and Myers, 1982), averages four years. The differences between these fraternal twins are similar to differences in age of onset found between full siblings (Myers et al., 1982). While the

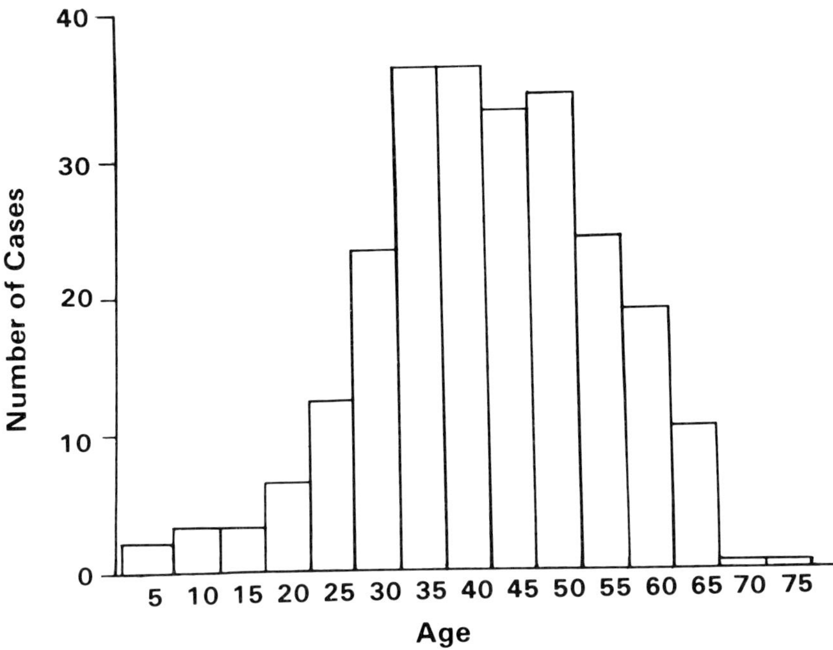

FIGURE 4.1. Onset age of symptoms for 243 persons with Huntington's disease from New England. The average onset age is 41 years.

number of twin pairs is small, the differences between DZ twins are greater than those found between MZ twins. These data suggest that factors modifying the expression of HD are genetically transmitted.

Consistency within Families

The similarity in age of onset among members of the same family has been evaluated for several populations of HD patients (Bell, 1942; Brackenridge, 1972; Myers et al., 1982). Significant correlation coefficients have been reported in the age at onset of siblings, and for parents and their offspring (Table 4.1). The variability in onset age is significantly greater between extended families than within extended families (Newcombe, Walker, and Harper, 1981; Myers et al., 1982). Thus a substantial amount of variability in onset age of HD can be attributed to differences between families. However, the considerable age differences that occasionally exist within families indicate that prediction of onset age from one sibling to the next or from parent to child must be made with reservation.

Sex of Affected Parent

A preponderance of paternal transmission is found in cases of early onset HD and has been noted by investigators in England (Bird, Caro, and Pilling, 1974; Newcombe, Walker, and Harper, 1981), Belgium (Vegter-van

TABLE 4.1. Correlation Coefficients

Father-Child		Mother-Child		Parent-Child		Sib-Sib		
Coefficient	Number[a]	Coefficient	Number	Coefficient	Number	Coefficient	Number	Reference
				0.593	153	0.465	442	Bell, 1942[b]
0.582	340	0.533	282	0.531	622	0.587	831	Brackenridge, 1972[c]
				0.78	51			Myers et al., 1982[d]
0.70	51	0.61	49			0.78[e]	83	Myers et al., 1983[f]

[a]Number of pairs.
[b]Literature review.
[c]Literature review.
[d]Southeastern United States.
[e]Unpublished data.
[f]New England, United States.

der Vlis, Volkers, and Went, 1976), South Africa (Hayden et al., 1982), and the United States (Myers et al., 1982). In the New England cohort, 90 percent of cases with onset before age 19 inherited the HD gene from an affected father (Myers et al., 1983); the remaining 10 percent inherited HD from a mother who also had early onset of the disorder.

Among 66 patients with late onset HD in whom initial symptoms appeared at age 50 or later, significantly more inherited the gene from an affected mother than from an affected father (Myers et al., 1983). Furthermore, affected offspring of late onset females also had late ages at onset of symptoms, whereas offspring of late onset males had significantly earlier onset ages. In the rare instances that late onset individuals inherited the gene from a father, the father was of mid-life onset.

The unusual influence of the sex of the affected parent on the age of onset in the offspring supports a hypothesis in which extrachromosomal or cytoplasmic factors of exclusively maternal origin modify the expression of the HD gene (Boehnke et al., 1983; Myers et al., 1983). According to this model the juvenile onset cases will inherit from their unaffected mothers extrachromosomal factors that precipitate the onset of HD. Juvenile onset would seldom be inherited from an affected mother because affected females with early onset producing extrachromosomal factors will be early onset themselves and therefore less likely to reproduce.

In summary, our understanding of the genetics in HD has been enhanced by three observations. First, the high concordance in onset among MZ twins compared with DZ twins and full siblings suggests that onset age is substantially under the control of genetically transmitted mechanisms. Second, significant consistency in age of onset exists within families, suggesting that a discrete number of additional genes modify the expression of the HD gene. Third, the role of a maternally transmitted factor is suggested by the unusual influence of the affected parent's sex on the expression of the HD gene in offspring.

PATHOLOGY

A general loss of tissue occurs throughout the HD brain, with the greatest atrophy being in the basal ganglia. Alzheimer (1911) was the first to identify the loss of neurons in the caudate and putamen in HD where up to 60 percent loss of tissue can occur. Marked neuronal loss is also found in the globus pallidus. The small neurons (Golgi type II) are lost preferentially. The substantia nigra often appears darker than normal and shows a greater loss of volume in the ventral pars reticulata than in the dorsal pars compacta (Bird, 1980). In advanced cases the whole brain weight may be decreased by 20 to 30 percent. Juvenile cases show greater atrophy than the adult form (Byers, Gilles, and Fung, 1973). Increased concentrations of glial cells are seen particularly in the caudate nucleus and putamen; however, Lange et al. (1976)

found the total glial cell number in each nucleus of the basal ganglia to be the same as in normal brain. Brain biopsies of the frontal cortex reveal proliferation and hypertrophy of astrocytes containing large amounts of lipofuscin and a high activity of acidphosphatase (Tellez-Nagel et al., 1970). Electronmicroscopy on frontal cortex biopsy tissue reveals increased lipofuscin granules. Other ultrastructural features were increased amounts of smooth endoplasmic reticulum and vesicles associated with Golgi complexes (Tellez-Nagel, Johnson, and Terry, 1973).

Approximately 7 percent of histopathologically examined cases diagnosed clinically as HD actually have other neurological disorders (Bird, 1980). Alzheimer's disease is the condition most commonly misdiagnosed as HD; other conditions include inherited autosomal dominant cerebellar ataxia and, rarely, Jakob-Creutzfeldt disease. In view of the implication for "at-risk" members of choreic families, it is imperative that every case with the diagnosis of Huntington's disease be confirmed eventually by autopsy. The neuropathological report should subsequently be filed in the patient's clinical notes for future reference by succeeding generations.

CLINICAL FEATURES

Neurological and Psychiatric Manifestations

Huntington's disease results in progressive involuntary choreiform (dance-like) movements and dementia (Bird, 1980; Bruyn, 1968; Folstein and Folstein, 1981). In an estimated 50 percent of all patients, emotional disturbance may either precede or accompany the onset of chorea. The initial signs of involuntary movement are subtle. In most cases chorea begins in the extremities with intermittent "piano playing" movements of the fingers or with slight facial twitching. As the disease progresses, the movements become more pronounced and eventually impair the ability to stand and walk. Difficulty in speaking and swallowing interferes with communication and proper nutritional balance. Patients sometimes die choking on food or develop aspiration pneumonia in the terminal phase of the disease. The average age of death is 55 years, usually 15 to 25 years after the onset of chorea.

The cognitive impairment and emotional disturbances seen in HD may initially be the most troubling symptoms to the patient and family. Changes in work performance or in the ability to manage household responsibilities may precipitate seeking medical attention. Impaired memory, inability to concentrate, and decreased expressive language are hallmarks of HD.

Although the disorder has been considered in the past to be heralded by abnormal movements, many patients will manifest behavioral changes before the onset of chorea. Emotional disturbance may be present for up to a decade prior to the onset of movement disorder, particularly for those patients who exhibit initial signs of chorea in their late twenties or early thirties (Brackenridge, 1971). When this occurs, patients may be admitted to a psychiatric hospital with the diagnosis of schizophrenia, which is later clarified with the advent of abnormal movements. Depression is frequently a feature of HD early in the illness and may also occur prior to the onset of involuntary movement. In addition, there may be changes in sleep patterns, such as an inability to sleep, frequent awakenings during the night, or the adoption of daytime sleep periods.

In the early stages of HD, extreme forms of violence, shoplifting, or arson may result in criminal prosecution. The judicial system and the general community need to be aware of the various ways in which this disorder may present and that certain HD patients may require psychiatric commitment rather than incarceration.

Age of Onset

Juvenile Onset

In approximately 3 percent of HD cases, onset of symptoms occurs before the age of 15 and usually presents with intellectual decline, seizure disorder, and myoclonic or dystonic movements rather than chorea. The course of the illness is rapid, so that muscle rigidity occurs early and is present throughout the terminal stages of the disease (Westphal, 1905). The diagnosis is made more complicated when the parent, usually the father, does not develop choreiform movement until some years after the manifestation of disease in the child. Since many early onset cases present without involuntary movement, the nosology has changed from Huntington's chorea to Huntington's disease.

Late Onset

The onset of choreiform movement in some persons occurs in the sixth or seventh decade of life (Myers et al., in press). In these cases the progress of the disease is generally slower than with mid-life onset. Movements may last as long as 20 to 25 years, with death resulting from causes other than those usually associated with HD. Cognitive impairment is less pronounced when the onset of HD is late in life. The diagnosis of HD in the elderly is sometimes difficult, since frequently there is no family history of chorea. This is understandable since in the general population the age of death in previous generations was younger and ancestors would have died before manifestation of the disorder.

Physical Examination and Differential Diagnosis

The neurological examination reveals abnormalities in the motor system; the sensorium is intact. Lingual instability is revealed by the difficulty patients may have in maintaining a protruded tongue. Muscle strength is generally good, but tonicity may vary from hypotonia early in the disease to hypertonia in the later stages. Abnormalities of saccadic eye movements and pursuit occur, and optokinetic nystagmus is commonly lost within five years of onset. Reflexes are exaggerated even in the early stages of the disease. Involuntary movement may be elicited when the patient extends both hands above his or her head with eyes pressed shut and tongue protruded. The choreoathetotic nature of the movement disorder gives the gait a dancelike appearance, which is sometimes mistaken for inebriation. Generalized rigidity often occurs in the terminal stages.

In the early stages of the disease, the diagnosis can be difficult to make owing to the insidious nature of onset (Folstein and Folstein, 1981). The cerebellar ataxias, which may be inherited either as autosomal dominant, recessive, or sex-linked, may be difficult to differentiate from HD. However, as HD progresses, choreiform movements and the onset of dementia will differentiate it from the ataxias.

The presence of myoclonic movements in a demented patient with mid-life familial Alzheimer's disease may suggest HD. However, the rapid progress of the dementia in patients with Alzheimer's disease is far more severe than that found in HD. Similarly, the abnormal movements that occur with Jakob-Creutzfeldt disease may be mistaken for HD; but again, the rapid downhill course of this illness will usually differentiate it from the slower progress of HD. Wilson's disease, which usually occurs at a young age (Bearn, 1957), and choreoacanthocytosis (Sakai et al., 1981), a mid-life onset illness, are both autosomal recessive disorders and can be differentiated from HD by appropriate blood tests. Tardive dyskinesias and basal ganglia infarcts may mimic HD, but the nonprogressive nature of these illnesses, as well as the different antecedents and nature of onset, will distinguish them from HD.

Management

Care of the HD Patient

The burden of managing a patient with HD usually falls upon the spouse and at-risk offspring who often become physically and mentally exhausted. The primary physician, therefore, has to care for the whole family and will often need to seek help from consultant neurologists, psychiatrists, genetic counselors, and allied health professionals.

Medical. When movements begin to appear and the diagnosis is almost certain, the patient should be seen frequently, at least every four months. Development of rapport with the patient during this period allows exploration of the impact of the illness and assists the patient to accept the diagnosis. The confirmation of the diagnosis permits the patient to discuss the concerns involved in having HD, which he or she may have suspected for some time.

The patient should be encouraged to remain active in employment, which will require some understanding on the part of employers, welfare, and rehabilitative services.

Although no specific treatment will halt the progressive degeneration of neurons, a great deal can be done to make the HD patient's life more comfortable. Prior to the onset of abnormal movements, a number of psychiatric problems may arise, the most common being depression. The incidence of suicide among at-risk individuals and patients in the early stages of the disorder is six times the incidence in the general population. Whether earlier detection and treatment of the depression will alter this incidence remains to be seen. Nevertheless, the tricyclic antidepressants, such as Elavil (amitriptyline), are the most popular agents used for depression in HD.

Although not as common as depression, the patient may manifest a psychosis indistinguishable from severe schizophrenic psychoses, and this occurs more often in the second or third decade of life, commonly in HD families where the onset of movement is before age 35. Neuroleptic agents are usually beneficial; however, these behavioral conditions may be resistant to treatment and the patient may require psychiatric care.

Agents that block dopamine at the postsynaptic receptor, such as the phenothiazines or butyrophenones, will reduce the prominent choreiform movements to some degree, but unless the patient becomes a danger to himself by falling, these agents probably should be withheld until absolutely necessary. Some patients do well on small doses of a butyrophenone, whereas others seem to do better with a phenothiazine; there is no evidence that one phenothiazine is superior to another. A number of new agents are under trial, but a word of caution regarding clinical trials in HD is warranted. A number of positive effects by various agents in HD have been reported prior to being evaluated in a double-blind protocol, but most of these early reports have not been substantiated with time. Lithium was thought to reduce the movements (Dalen, 1973); although lithium may be helpful to some HD patients with depression, it has not found a place in the treatment of HD (Carmen, Shoulson, and Chase, 1974; Vestergaard, Baastrup, and Petersson, 1977). Isoniazid (INH) was considered by Perry et al. (1979) to be a useful agent in HD, but this has not been supported in follow-up (McLean, 1982).

Patients appear to develop tolerance to a number of neuropharmacologic agents with time, and when medication is either stopped or switched to another agent, there can be some relief. Benzodiazepines such as Valium® or Librium® may be effective in reducing anxiety and the movements. Unfortunately, tolerance to benzodiazepines occurs, and while an increased dosage may be temporarily effective, a change in medication is usually required.

Health-Related Services. Eventually, the patient may require chronic hospitalization to relieve the burden on the family—a distressing event for the patient who will have previously admitted their parent in the final stages of HD. This can also be frightening to the hospital staff who may have never encountered an HD patient and who may therefore be reluctant to admit such patients. In Massachusetts there is a chronic facility that is well staffed with experienced nurses and health-related services where it has been possible to admit a few HD patients that were able to return home again after rehabilitation.

Almost all patients on admission to hospital are underweight. The increased caloric intake required by HD patients is difficult to provide in a home setting. It can take up to 5,000 calories per day to bring the patient back to a normal weight. The increased caloric utilization is assumed to be due to the excessive movement, and patients maintain this need in the later stages of the disorder when they become rigid. We have noted a general improvement in function accompanying weight gain.

As the disease progresses, communication becomes one of the most frustrating problems of the HD patient. Speech may become unintelligible, but speech therapists help patients make their needs understood by having them speak slowly or use communication tools such as wordboards. Swallowing is disturbed, and the use of breathing techniques during eating can help prevent aspiration.

Occupational therapists can design special feeding utensils and teach the patient proper eating positions. In addition, they can help patients with any adaptive devices required for activities of daily living such as dressing or bathing. The physical therapist can help the patient gain confidence walking with various assisting apparatuses. Exercise sessions on stationary bicycles and aerobic exercises to music are popular. Group recreation sessions involving singing are effective as HD patients often respond to old tunes learned years previously. Other sessions making craft objects give the patient a sense of accomplishment.

HD patients require a considerable amount of nursing care. Attention to protective lounge chairs and padded beds is important to reduce the multiple injuries these patients sustain. The problems of aspiration of food or incontinence of urine and stool require regular monitoring and special care of the skin.

Genetic Counseling in HD

Counseling in HD will be changed dramatically for persons at risk when procedures for presymptomatic and prenatal testing become available. Nevertheless, many of the issues described here will continue to be relevant. Genetic counseling requires an unusual combination of listening to the concerns of the counselee and presenting important factual information (Wexler, 1979). As a result of the role the counselor plays in imparting information, the counselee may also expect to receive advice. While it is certainly quicker and easier to be directive, it is more effective for the counselor to assist the counselee in making his or her own decision.

Some of the primary goals of genetic counseling with HD family members are:

1. To present information in an understandable manner;
2. To impart information at a rate at which it can be assimilated without overwhelming the counselee;
3. To listen to the impact new information has on the counselee;
4. To present a realistic prognosis of HD but also to acknowledge the important contributions that at-risk individuals make to their families and to society;
5. To present alternatives for childbearing;
6. To discuss the manner by which the counselee shares information with other relatives and prospective mates;
7. To appreciate the unique concerns of each counselee;
8. To present suggestions for preparing for the possibility of HD; and
9. To remain a nondirective source of information.

Families living with HD experience stress over a period of many years because of its highly debilitating nature and mode of transmission. Although a test to identify gene carriers prior to the onset of symptoms may be available in the future, at present persons at risk do not know if they will develop the illness. The intensity of the stress waxes and wanes in conjunction with important life decisions such as marriage and childbearing and with disease-related events including the diagnosis, institutionalization, or death of an affected family member. Genetic counselors working with families having diagnosed members may expect an initial consultation with the family during a period of acute stress, followed by intermittent meetings over many years as difficult situations arise.

Family History. The first step in genetic counseling is collection of the family history. For the overwhelming majority of HD cases, a family history of movement disorder can be confirmed by aggressive investigation. If possible, the presence of symptoms consistent with HD should be documented in three generations to verify an autosomal dominant pattern of inheritance. Recessive disorders, such as Wilson's disease or choreoacanthocytosis, whose clinical features may be mistaken for HD, may appear in two consecutive generations if pseudodominance occurs. It is imperative to establish the diagnosis of HD by postmortem examination of the brain in at least one member of each family, as some at-risk persons have been counseled for HD and subsequently this diagnosis was found to be incorrect. Collection of the family history also gives the counselor insight into the counselee's experience and perceptions of HD.

Medical records of deceased relatives are an essential source of information and may be obtained from hospitals that maintain records over many years. Unfortunately, state law does not mandate the indefinite preservation of medical records, and they may be destroyed after the death of the patient. Frequently, however, medical records can be obtained with proper written consent of living relatives. Affected members of prior generations are sometimes misdiagnosed with Parkinson's disease, organic brain syndrome, or other neuropsychiatric disorders. Information about the symptoms displayed and the age at which they first appeared should be documented for as many family members as possible. The variability in onset age among families is significantly greater than within family variability. By ascertaining a pattern in onset age in some families, it is possible to reduce some of the uncertainty experienced by persons at risk for HD.

Death certificates also provide valuable information about cause of death and names of the decedent's ancestors. In cases where a history of HD seems obscure, medical records and death certificates should be obtained for every potential gene carrier, regardless of reports that the individual was not affected.

Imparting Genetic Information. The nature of the counseling session depends upon the counselee's relationship to the patient and to his or her previous experiences. Ideally, information about HD should be presented in a series of meetings. In the first meeting, fundamental information on HD should be imparted to all counselees, ensuring that they are adequately informed about the illness should they choose not to return for follow-up counseling. Thus a first meeting should include a discussion of the genetic risks, the variability in age at onset of symptoms, and the prognosis of HD. Visual aids, such as a chromosome karyotype and a frequency histogram of onset ages (Figure 4.1), may aid the counselee's comprehension of the concept of autosomal dominant inheritance as well as other facts about HD. The preliminary family history is collected, and permission to request medical records of other family members is also obtained during the first meeting.

In a second meeting, basic factual information on HD can be reinforced. In addition, the counseling process may involve a discussion of the impact of the illness on the family, concerns about marriage and reproduction, and implications for other family members. This discussion as well as the presentation of the genetic facts about HD may be facilitated through the updating of the family history with a review of medical records on other affected family members. This enables the counselee to gain an appreciation of the effects of HD in his or her family.

A third meeting, if indicated, may include a standard neurological evaluation to determine whether any symptoms of HD are present. Subsequent counseling will depend upon the outcome of that evaluation. During the entire counseling process, time should be set aside so that questions can be answered and much factual information can be imparted in response to the questions posed by the counselee.

Counseling with HD-Diagnosed Patients and Their Families. Concerns of the diagnosed patient and his or her family will vary depending on the stage of the illness but frequently focus on four main areas:

1. Expectations for the future as the disease progresses;
2. Management of the patient at home;

3. Whether and when to consider institutionalization; and
4. How to impart information to children or other relatives who may not be aware of the hereditary implications of the illness.

Following the diagnosis of HD, family members often ask questions about the course of the illness and what they can expect to happen to their affected relative. Because of the variability in symptoms expressed in HD, there is no typical course of the illness and thus no typical family experience. It is not possible to provide family members with a precise timetable of the disease process. On the other hand, it is important to convey realistic information that enables families to prepare for the inevitable deterioration. Information about the course of HD can be imparted by discussing the progression of the illness in other affected family members. In addition, families can be helped to evaluate the rate of progression in HD by assessing the rapidity of changes in the years since symptoms first appeared.

As the HD patient becomes increasingly disabled, he or she may be left at home during the day as other family members go to work. Concerns may focus on the patient's safety and establishing meaningful daytime activities. Unfortunately, day-care facilities are rarely equipped for patients with HD, and there are relatively few organized activities for patients disabled by neurological disease. Whenever possible, home health aides or the Visiting Nurse Association should be engaged to provide some daytime care. Eventually, the HD patient can no longer be left at home alone for fear of injury.

Despite the often overwhelming burden of care imposed on the family by an affected relative, the decision to institutionalize is very difficult and fraught with guilt and uncertainty. The counselor may relieve some of the guilt involved in this decision by acknowledging that home care is difficult, that this option is necessary for many families, and that institutionalization is not an abrogation of family responsibilities.

Some families erroneously believe that only "mental" institutions will accept their ill relative for care. While this may have been true in the past, many HD patients are now receiving expert care in veterans' hospitals, private nursing homes, and state chronic care facilities. A small percentage of patients with severe psychiatric problems are difficult to manage in facilities equipped exclusively to deal with physical illness; finding placements for these patients can be very difficult.

Whenever a parent is diagnosed with HD, a new generation of children becomes at risk. In some families this possible threat was well known prior to the diagnosis and at-risk children have an understanding of the genetic implications of their parent's illness. In other families, especially those for whom a history of HD was obscure, children may have no knowledge of their possible inheritance, and their at-risk status must be explained. This task often falls on the unaffected parent, who is also dealing with the recent and perhaps unexpected diagnosis of the spouse. Parents are often afraid to share information about the hereditary aspects of the illness with their children lest the children reject or blame them for their at-risk status. The task is so difficult and painful that some parents chose not to tell their children of the genetic implications of the illness. This may be done with the hope that the children will not have inherited the disease and will never need to know. In these instances the realization that the family history was concealed can engender strong feelings of resentment and anger toward the parents. The genetic counselor may be of assistance in imparting information about the genetic risks, thus lessening the burden of the unaffected parent at this difficult time.

Counseling Individuals at Risk for HD and Their Spouses. Some individuals at risk for HD seek counseling when making plans about marriage and childbearing. While many at-risk persons marry and have children, such decisions are often not made easily. With respect to reproductive choices made by individuals at risk for HD, the role of the genetic counselor has been controversial. While a direct stance against reproduction has been advocated by some counselors (Wallace, 1972), evidence is available that a nondirective approach may be effective in reducing the birthrate in HD families (Harper et al., 1981) and thus the incidence of the disease. Decisions to have children usually involve factors other than the advice imparted by the counselor. Furthermore, if a decision is made that is contrary to the suggestions made during genetic counseling, the couple or individual may be deterred from returning in the future because advice from the counselor was not heeded. On the other hand, a neutral counselor will be more likely to be called upon should problems arise in the future. Finally, one half of those persons at risk for HD are not gene carriers, and there are ethical considerations in directive counseling of persons who will not transmit the disease.

The counseling setting provides a forum for discussion of the options available to at-risk individuals contemplating children. When the male is at risk, artificial insemination by donor (AID) will allow a couple to experience pregnancy yet ensure that the HD gene will not be transmitted to a subsequent generation. In vitro fertilization with a donor egg will, no doubt, be available to at-risk women in the future. Adoption is not usually an alternative for at-risk couples owing to the reluctance of adoption agencies to place children with families in which one parent is likely to develop a disabling disease. Sterilization of at-risk individuals who have not had children is not recommended because this option may be irreversible. With the identification of a DNA RFLP (restriction fragment length polymorphism) linked to the HD gene (Gusella et al., 1983), a test to recognize gene carriers prior to onset and prenatally will be available in the near future. This will alter the options for procreation available to at-risk persons dramatically.

Because of their heightened sensitivity to the early signs of HD, another frequent concern of at-risk individuals is whether they are manifesting early symptoms. Normal benign occurrences such as myoclonic jerks, muscle fasciculations, occasional clumsiness, or changes of mood may be perceived as signs of incipient disease. Preoccupation with one's neurological status may create anxiety, depression, and loss of sleep, which in turn are also mistakenly interpreted as pathological. At-risk individuals with such concerns are advised to have a neurological examination by a neurologist experienced with HD. While the possibility of diagnosis exists, many at-risk individuals can have their fears at least temporarily allayed. An annual neurological examination is encouraged for at-risk individuals so that unfounded fears do not intensify, and this enables the neurologist to gain a sense of the premorbid personality and neurological status.

Given the high probability of developing HD, ways of planning for this eventuality should also be discussed with at-risk individuals. Securing good disability and health insurance policies or selecting employment that offers such fringe benefits is encouraged. Spouses of those at risk may want to obtain skills for employment, should the need arise. "Hoping for the best but preparing for the worst" is a philosophy that helps to minimize the effects of HD on family functioning, should a member become disabled.

Impact on the Family. The consequences of HD extend beyond the physical disability resulting from the illness and include a far-reaching and serious impact on the social and psychological well-being of the patient and his or her family. Despite the absence of a cure, clinicians who are aware of and acknowledge the wide-ranging implications of HD can help families to manage more effectively with the illness.

The first symptoms of HD may appear at any age, but their usual occurrence at mid-life corresponds to a time when responsibilities at home or at work are great and the young family has not yet accumulated financial resources on which to fall back. When the husband is affected, the family may be confronted by the onset of HD when involuntary movements or intellectual impairment result in a change in job performance. While employment may continue after the diagnosis has been made, the patient may have to assume positions of increasingly less responsibility and financial remuneration as the disease progresses. Eventually, when the patient becomes unable to work, severe financial hardship and disruption of the usual family routine may result. The financial impact of HD may be greater in families where an affected husband provides the only source of income and the wife must find employment to support the family, as well as maintain her usual household responsibilities.

When a mother with young children becomes affected by HD, her ability to care effectively for her children may become severely impaired as the disease progresses. If outside family or community support is unavailable, the family unit may dissolve, necessitating that children be placed in adoptive or foster care. In less extreme situations, the husband will begin to assume the roles of both parents, providing financial support as well as taking over child care and other household responsibilities.

The increasing responsibility placed on the unaffected spouse can engender anger and resentment as the struggle to provide basic necessities intensifies and as time and money for leisure become scarce. Community organizations such as the Visiting Nurse Association and home health services may be effective in reducing the responsibilities of the unaffected spouse in the care of the affected individual, but these resources may be expensive and beyond the financial means of the young couple.

When the onset of HD is at a later age, the impact of the illness on day-to-day family functioning is less pronounced. Children are usually older, already independent, and more able to assume responsibility. Eligibility for Social Security disability benefits is also easier to obtain, and community services are better prepared to care for disabled persons in an older age bracket.

While some spouses divorce or desert their affected partner, many marriages do not fail, and the unaffected spouse serves as an important source of support throughout the illness. Patients with concerned families are more likely to stay at home longer, and institutionalization in a state hospital becomes less likely. The tragedy of HD, however, is that regardless of the care and treatment for the affected partner, the spouse will not escape the legacy of the illness. Most likely, he or she will relive the experience of HD with affected children and grandchildren. It is not uncommon when working with families with HD to meet a grandparent who nursed an affected spouse as well as affected children and is now observing the first signs of HD in a grandchild. The unaffected spouse may experience profound feelings of guilt over the role he or she played in possibly having transmitted the gene to at-risk children (Hans and Koeppen, 1980).

The unaffected spouse plays a critical role in maintaining family stability as the disease progresses. The spouse's reaction to the illness may be in large part influenced by the way in which he or she first learns of the illness. In families where knowledge of a family history of HD was concealed prior to marriage and childbearing, the unaffected spouse has no opportunity to prepare for the possibility that HD may develop. He or she may feel anger and resentment toward the affected partner and his family at this deception and at having been deprived of the opportunity to make important life decisions with knowledge of this unwanted inheritance. In situations in which the unaffected spouse is fully aware of all the implications of the illness, his or her decision to enter into marriage implies, in part, a willingness to deal with and accept the illness. Of course, not all individuals are aware of the presence of the HD gene in their families, especially when it is obscured by early death of a gene-carrying parent or adoption of an at-risk individual. In these cases there is no one to "blame" for the concealed family history, and the shock of its discovery is borne equally by all family members.

Thus the manner in which an unaffected spouse learns about HD may affect the integrity of the marriage and the willingness of the spouse to care for the affected mate. A candid approach to the subject by individuals at risk for HD with prospective mates prior to marriage can circumvent the later dire consequences of concealed information.

The Development of Presymptomatic and Prenatal Testing for HD. The gene for HD has recently been localized to chromosome 4 by a linked DNA restriction fragment length polymorphism (RFLP). The initial report of the linked RFLP examined DNA obtained from blood samples (DNA can be obtained from any unfixed tissue) of two large HD families, one from the United States, and one from Venezuela (Gusella et al., 1983). The form of the linked RFLP was found to be different for these two families. This indicates that in different HD families, different forms of the linked RFLP are associated with the gene for HD. Therefore, although the linked RFLP will likely prove useful as a test procedure for presymptomatic detection or even prenatal detection of probable HD gene carriers, the form of the linked marker will need to be determined in each family with members wishing to be tested.

The method of determining the form of the linked marker is to ascertain which DNA marker is common to family members affected by HD and absent from those old enough to be known non-carriers. Thus, tissue from at least one and perhaps more affected individuals will usually be required for linkage testing (Conneally et al., 1984) in each family. For some families, brain tissue of affected members may have been saved frozen through one of the Brain Tissue Banks for HD and DNA may be obtained from this source. For other families, however, no HD affected relative is living, and for these individuals testing will usually not be possible under current conditions. Some families may have members in late stages of HD or elderly unaffected relatives who will be informative for future testing. A DNA blood storage facility under the direction of Dr. P. Michael Conneally has been established at the University of Indiana to preserve samples for future testing.

At the time of the preparation of this chapter, presymptomatic and prenatal testing for linkage of the HD gene is not expected to become available before 1986. Two additional questions must be resolved before this linkage marker can be used diagnostically. First, it must be determined that all cases of HD are caused by a gene carried at the identified locus on chromosome 4. This requires linkage testing among many additional large HD families. Second, the frequency of genetic recombination (cross-over events between linked RFLP and HD gene) must be determined. The frequency of recombination will determine the accuracy of the test procedure and the frequency of false positive and false negative results.

Many difficult ethical issues are raised in the use of a presymptomatic test procedure for an illness which has no effective treatment. Considerable pre and post test counseling will be necessary. In the final analysis, the decision rests with the at-risk individual. Our inquiries suggest that perhaps 75 percent of persons at risk for HD would seek testing (Schoenfeld et al., 1984) to assist them in making decisions for marriage, career, and procreation.

NEUROCHEMISTRY

The measurement of certain neurotransmitter substances and their related biosynthetic enzymes in postmortem brain tissue from patients dying with extrapyramidal disorders has provided a better understanding of the clinical manifestations and neuropharmacological basis of such conditions. The decreased concentration of dopamine in the basal ganglia in Parkinson's disease (Ehringer and Hornykiewicz, 1960) and the decreased concentration of gamma-aminobutyric acid (GABA) (Perry, Hansen, and Kloster, 1973) and its biosynthetic enzyme, glutamic acid decarboxylase (GAD) (Bird et al., 1973), in the same nuclei in Huntington's disease may reflect the essential neurochemical differences between these two clinical syndromes.

Several chemical transmitters and related enzymes are fairly stable in postmortem brain. Brain cooling has been shown to start immediately after death; and provided the cadaver has been placed in a refrigerator promptly, this rate of cooling progresses until the center of the brain reaches 4°C, usually about 24 hours after death (Spokes and Koch, 1978). Enzyme activities decline to a stable level during this first 24 hours. Control brain tissues used for comparison have been taken from nonneurological cases dying from natural causes including bronchopneumonia, a common cause of death in HD. Patients dying in mental hospitals who have received neuroleptic agents may serve as another group for comparison with the choreic group (Bird, Spokes, and Iversen, 1979).

Gamma-aminobutyric Acid (GABA)

Gamma-aminobutyric acid (GABA) is widely distributed throughout the mammalian central nervous system (CNS) and is considered to be an important inhibitory neurotransmitter at virtually all levels of nervous system function (Roberts, 1976). Studies on the regional distribution of GABA in the mammalian CNS, including that of man, have shown highest levels in the pallidum, substantia nigra (SN), and the striatum (Fahn and Cote, 1968).

Bird and Iversen (1974) found that the activity of GAD was decreased in the striatum and SN but not in the cortex of the brain from choreics. This selective loss of enzyme activity does not appear to be related to chronic hospitalization or to neuroleptic drug administration, as normal GAD values are observed in chronically hospitalized psychotic patients (Bird, Spokes, and Iversen, 1979). Bowen et al. (1976) noted low postmortem brain GAD values from patients dying with conditions likely to lead to cerebral hypoxia. Iversen et al. (1978) compared the GAD activities in postmortem brain from both controls and choreics that suffered either sudden death or a prolonged death with bronchopneumonia. Although a significant decrease in GAD activity was found in the controls that died from bronchopneumonia, the respective GAD activity for choreics in both the acute and prolonged death groups was significantly lower.

It was hoped that replacement of the GABA deficiency in HD might at least alleviate the chorea of the illness. In practice, however, all therapeutic attempts at increasing brain GABA activity have proved fruitless. The use of potent postsynaptic GABA agonists such as muscimol have not improved the chorea in HD, nor has the administration of other drugs thought to manipulate this transmitter system, such as baclofen. There is a considerable loss of GABA receptor sites in the brains of patients with HD, a loss that makes it most unlikely that therapy aimed at increasing function of this neurotransmitter will be successful.

Glutamate

Glutamate, the substrate for GABA production, is of special interest in HD since it appears to be an excitatory neurotransmitter that can destroy nerve cells when administered in large doses. Intracerebral injections of glutamate—and more potent excitatory analogues, such as kainic acid—produce neuronal degeneration of the striatum in experimental animals that is similar to that found in HD (Coyle and Schwarcz, 1976; McGeer and McGeer, 1976). GABA neurons are highly sensitive to the toxic effects of kainic acid. Glutamate is reported to stimulate dopamine-containing neurons (Roberts and Anderson, 1979), and this effect might account for the apparent increase in dopaminergic activity in HD. The findings in HD have been postulated to be related to increased levels of glutamate or to an endogenous analogue of kainic acid that might have neurotoxic effects (Sandberg and Johnston, 1981). Recent studies indicate that kainic acid is only a weak agonist at glutamate receptor sites (Constanti and Nistri, 1976) and that it may act on specific kainate receptors (Evans, Francis, and Watkins, 1977). This is further supported by the observation that glutamate-induced excitation of neurons in rat striatum can be blocked by agents that are ineffective for blocking kainate effects (Hall, Hicks, and McLennan, 1978). These findings have led to the proposal that an endogenous substance that interacts with the kainate receptor is responsible for the neurodegeneration found in HD (London et al., 1981).

Dopamine

Klawans (1970) pointed out that chorea in HD is aggravated by drugs that enhance dopaminergic action. Chorea is the most common type of dyskinesia provoked by L-dopa therapy in patients with Parkinson's disease. It thus seems likely that either relative or absolute dopaminergic overactivity in the nigrostriatal pathway might be implicated in the pathophysiology of chorea. Medications that block dopamine (DA) receptors (phenothiazines and butyrophenones) or agents that deplete DA (reserpine and tetrabenazine) are recognized as being effective in reducing the choreiform movements in HD.

The activity of tyrosine hydroxylase (T-OH), the biosynthetic enzyme for DA, was found to be increased in the SN of HD brain (Bird and Iversen, 1977), and more recently significant increases in DA concentrations were found in the corpus striatum, SN, and nucleus accumbens (Spokes, 1980). These increases could not be attributed to chronic neuroleptic drug administration, which is common clinical practice in the treatment of chorea, as normal striatal DA values were found in similarly treated patients dying with a hospital diagnosis of schizophrenia (Bird, Spokes, and Iversen, 1979). Moreover, drug-treated choreic subjects had DA values similar to their drug-free counterparts (Spokes, 1980). Animal studies support the view that acute and long-term treatment with neuroleptic drugs increases striatal DA levels (O'Keefe, Sharman, and Vogt, 1970; Lloyd et al., 1977). However, there is evidence that chronic neuroleptic administration to primates significantly decreases the DA concentration in the striatum (Bird and Anton, 1982). The most likely explanation for the increases in DA concentrations in HD brain is that there is an increased density of dopaminergic terminals in the striatum and SN owing to loss of neurons other than monoamine-containing neurons.

Since there is approximately a 50 percent loss of tissue in the corpus striatum, a 200 percent increase in the DA concentration of the remaining tissue would be expected if the net content of DA were unchanged. However, the increases found were only about 70 percent in the putamen and pallidum, and only 32 percent in the caudate nucleus. The total content of DA in the choreic corpus striatum is, therefore, likely to be considerably reduced. Note that the greatest increases in DA were found in the nucleus accumbens, a region that exhibits less atrophy than the body of the caudate.

It would appear that there may be at least three factors affecting the absolute concentrations of DA found in the HD brain. First, there may be functional increases due to the loss of neuroinhibitory control by GABA. Second, there may be some DA decrease due to a retrograde loss of DA terminals in the body of the caudate where the greatest loss of neurons is occurring.

A useful method for estimating the functional activity of DA-containing neurons is to measure brain and cerebrospinal fluid (CSF) concentrations of the major DA metabolite, homovanillic acid (HVA), formed from the metabolism of DA released at synapses. Reduced CSF/HVA concentrations have been found in Huntington's disease (Curzon, Gumpert, and Sharper, 1972; Chase, 1973), suggesting that the turnover of the remaining DA in the corpus striatum is reduced. However, HVA concentration in CSF may also depend on ventricular volume, and the decreased HVA concentrations in CSF from choreic subjects may, to some extent, be related to ventricular enlargement, which is invariably present in HD (Curzon, 1975). CSF levels of HVA have been shown to be particularly low in rigid choreic patients (Caraceni et al., 1977), and it has been shown that rigid cases have marked atrophy of the corpus striatum (Liss, Paulson, and Sommer, 1973). The estimate of DA turnover in postmortem brain, as measured by the ratio of HVA/DA, is decreased (Walsh, Bird, and Stevens, 1982), supporting the preterminal CSF data above.

The neuropharmacological evidence suggesting increased DA function as the cause of chorea in HD cannot, therefore, be attributed to an absolute overactivity of the nigrostriatal DA system. It is possible, however, that there is a relative overactivity due to the impact of a normal DA release onto a grossly reduced population of neurons bearing DA receptors in the striatum. Indeed, there is a considerable decrease in DA receptor density measured in striatal tissue in HD (Reisine et al., 1979).

Acetylcholine (ACh)

The enzyme choline acetyltransferase (CAT), responsible for the biosynthesis of acetylcholine (ACh), is considered to be exclusively localized in cholinergic neurons and is present in high activity in basal ganglia tissue, especially the putamen. Although there is an average 50 percent loss of CAT activity per unit weight in putamen and caudate nucleus from choreic brains, a substantial portion of the choreic cases show CAT activities in basal ganglia that are within the normal range (Bird and Iversen, 1974). The remaining cases have severely reduced basal ganglia CAT activities (approximately 85 percent reduction). The finding in some choreic cases that CAT activity was only slightly reduced in caudate nucleus may support the suggestion of McGeer, McGeer, and Fibiger (1973) that the degeneration of the cholinergic innervation in choreic basal ganglia occurs in a "patchy" manner. This has also been noted by Aquilonius, Eckernas, and Sundwall (1975). The findings of only slightly reduced CAT activity in the choreic nucleus accumbens is consistent with other reports (Aquilonius, Eckernas, and Sundwall, 1975; McGeer and McGeer, 1976) and, again, emphasizes the relative sparing of this region compared with the striatum.

Normal CAT values were obtained in the striatum and other brain regions from long-term psychotic inpatients, indicating that neither hospitalization nor neuroleptic drug treatment is responsible for these changes in HD (Bird, Spokes, and Iversen, 1979). No significant differences in CSF levels of ACh or of choline—the precursor of ACh—has been found in early HD (Welsh et al., 1976), but in late stages of the disease choline is reduced (Aquilonius et al., 1972).

There is a considerable reduction in the density of muscarinic receptors in the caudate nucleus and putamen (Hiley and Bird, 1974; Enna et al., 1976). This loss of muscarinic receptors probably reflects the fact that many lie upon GABA neurons in the striatum, which themselves are grossly depleted.

Serotonin (5-Hydroxytryptophan)

Serotonin-containing nerve fibers arise from cell bodies in the midbrain raphe nuclei and end in a variety of brain regions. Serotonin concentrations in various brain regions from six HD cases were examined by Bernheimer and Hornykiewicz (1973); although there were increased concentrations in the putamen, globus pallidus, and central gray of the midbrain, these were not considered to be significantly different from normal. Significant increases in the concentrations of 5-hydroxytryptophan (5-HT) and 5-hydroxyindoleacetic acid (5-HIAA)—a metabolite of 5-HT—in nucleus accumbens, caudate, and putamen were found by Walsh, Bird, and Stevens (1982). This increase may be relative to the loss of other neurons; however, the nucleus accumbens exhibits the largest increase in 5-HT and 5-HIAA concentrations even though this nucleus is not nearly as atrophic as the caudate or putamen in HD. The function of 5-HT in the striatum is not clear, but lesions that interrupt 5-HT pathways increase DA turnover in the nucleus accumbens (Herve et al., 1979). The increased 5-HT in HD may, therefore, be a contributing factor in the decrease in DA turnover noted in the striatum and nucleus accumbens in this disorder. However, the ratio of 5-HIAA/5-HT, an estimate of serotonin turnover, was not significantly different in HD and controls (Walsh, Bird, and Stevens, 1982). CSF/5-HT concentration is reportedly within the normal range in choreic subjects, both at steady state and after probenecid loading (Chase, 1973). It would appear that 5-HT is not a neurotransmitter that is primarily involved in HD.

Neuropeptides

A number of neuropeptides in human brain are now being measured. Angiotensin-converting enzyme (ACE) converts the inactive decapeptide angiotensin I to the active octapeptide angiotensin II and is ubiquitous in brain tissue, with high activities found in the corpus striatum. Selective depletion of ACE activity has been demonstrated in the corpus striatum of HD brains (Arregui et al., 1979). The decrease was 83 to 92 percent in globus pallidus, and 62 to 69 percent in caudate and putamen. Normal values were obtained for two cortical areas.

The neuropeptide gonadotropin-releasing hormone (GRH) has been measured in the human hypothalamus (Bird, Chiappa, and Fink, 1976; Okon and Koch, 1976). Increased GRH was observed in the median eminence of the female choreic brain but not in the male (Bird, Chiappa, and Fink, 1976). Increased fertility has been noted in female patients with HD. Whether this is due to alterations in GRH function remains to be determined.

Substance P (SP) is an undecapeptide that is unevenly distributed in brain, with the highest concentrations reported in the SN (Brownstein et al., 1977). SP terminals in the SN have been seen in close association with the DA dendrites (Ljungdahl et al., 1978), and it has been thought that SP is modulating the inhibitory stimulus of GABA on the dopaminergic cell. Markedly decreased concentrations of SP have been found in the SN and the inner segment of the globus pallidus (Kanazawa et al., 1979) and the striatum (Buck et al., 1981; Aronin et al., 1983) of HD brain. SP concentrations in CSF are not altered in HD (Cramer et al., 1981).

Cholecystokinin (CCK) has been found within the terminals of DA neurons (Hokfelt et al., 1980). Measurements of CCK-8 in HD brain reveal that the only areas that show significant differences are decreased concentrations in the globus pallidus and SN (Emson et al., 1980). There are no significant differences found in the cortex or caudate and putamen of HD brain. Receptors for CCK are reduced 70 percent in basal ganglia and 40 percent of cerebral cortex (Hayes, Goodwin, and Paul, 1981).

Vasoactive intestinal polypeptide is also one of the intestinal peptides found in high concentrations in cerebral cortex. This peptide was measured in the cerebral cortex and caudate of control and choreic brain, and no differences were noted (Emson, Fahrenkrug, and Spokes, 1979).

In contrast to the decreased concentration of certain peptides in HD brain, thyrotropin-releasing hormone (TRH) and somatostatin (SS) have been noted to be increased in HD brain. TRH levels were elevated nearly threefold over those seen in caudate and putamen from control subjects, whereas hippocampal TRH levels were unaffected (Spindel, Wurtman, and Bird, 1980). The increase in TRH may reflect changes in TRH turnover in the basal ganglia or may simply reflect a relative sparing of TRH neurons in the highly atrophied basal ganglia present in HD. The increase seen in basal ganglia TRH in patients with this disorder resembles the threefold increase previously noted in striata of rats subjected to partial striatal deafferentiation by cerebral hemitransection at midhypothalamic level (Spindel, Wurtman, and Bird, 1981).

SS is increased threefold in HD caudate nucleus and nearly fourfold in putamen and globus pallidus but is not increased in the amygdala, hippocampus, or SN (Aronin et al., 1983). Although these increases may reflect a selective sparing of SS-containing neurons, the magnitude of the increase—the largest yet reported for HD—also suggests that there may be an alteration in the inhibitory influences of SS in HD. The levels of SS in CSF are reported to be reduced (Cramer et al., 1981).

The significance of altered levels of peptides in HD remains to be determined. Definitive pharmacologic agents that alter brain neuropeptide concentrations and functions have not yet been identified, and further research will be necessary to determine the relation of neuropeptides to symptoms or to the underlying abnormality in HD.

Glucose Metabolism

Enzymes involved in glucose metabolism have been studied in the postmortem HD brain to determine whether some fundamental defect in glucose utilization might cause the cell death in the choreic basal ganglia (Bird, Gale, and Spokes, 1977). The enzymes hexokinase, glucose 6-phosphate dehydrogenase, and lactic acid dehydrogenase were normal throughout the choreic postmortem brain. Glutamic dehydrogenase (GDH) was significantly increased in the globus pallidus of choreic brain, but this may be secondary to the increased density of glial cells, which possess GDH at high activity. There was, however, in HD brain, decreased activity in the rate-limiting glycolytic enzyme phosphofructokinase (PFK). Increased concentrations of gamma hydroxybutyrate have been found in both the cortex and striatum of HD brain (Ando et al., 1979).

Recent studies on in vivo desoxyglucose uptake in at-risk patients in the early stages of their disease as measured by positron emission tomography (PET) reveal that there is a significant decrease in labeled glucose uptake in the basal ganglia (Kuhl et al., 1982). This finding in patients that have no demonstrable atrophy of the caudate on computerized axial tomography (CAT) suggests that an early defect in HD brain may be a membrane defect altering the uptake of glucose or a failure to utilize glucose.

Membrane Studies

Initial indications of altered electron spin resonance of erythrocyte membranes (Butterfield, Oeswein, and Markesbery, 1977; Butterfield, Purdy, and Markesbery, 1979; Butterfield and Markesbery, 1981) have not been replicated in subsequent reports (Fung and Ostrowski, 1981; Comings, Pekkala, Schuh, Kuo, and Chan, 1981). Similarly, reports of altered fibroblast membrane structure, measured by fluorescent probe binding (Pettegrew, Nichols, and Stewart, 1979), have not been confirmed (Beverstock and Pearson, 1981; Lakowicz and Sheppard, 1981). Early reports of altered Na^{+}-K^{+} adenosine triphosphatase in erythrocyte membranes (Butterfield et al., 1978) have not been replicated (McLean and Nihei, 1980; Dubbelman et al., 1981). Other negative results include no change in glutamate uptake by fibroblasts (Comings, Goetz, Holden, and Holtz, 1981) and no difference in growth of HD fibroblasts (Goetz et al., 1981).

In spite of a concerted systematic examination of the neurochemical properties in postmorten tissue, the underlying pathogenetic process in HD has eluded discovery. Although a primary gene defect may be among the numerous changes identified in the HD brain, considerably more research of this kind will be required to establish the cause of this disorder.

CONCLUSION

During the past 15 years, a great deal of activity in HD research has taken place. More neurochemical data on HD is available than for any other neurological disorder. While the cause of HD has not yet been identified, research has provided insight into the nature of the chemical dysfunction and early cell loss.

One area of particular interest is the search for the HD gene using recombinant DNA techniques (Housman and Gusella, 1982; Gusella et al., 1983). The recent localization of the HD gene to chromosome 4 will lead to a presymptomatic and prenatal test in the near future. This finding may eventually also lend additional clues in the pathology of the disease with implications for treatment. While these studies provide hope for increasing the understanding of HD, it is impossible to predict how long it will be before these advances will occur, and it is important that false expectations for cure in the foreseeable future not be raised among HD families.

We are encouraged by the recognition and diagnosis of HD at an earlier stage in the community. The more frequent use of the health-related services will be of considerable assistance in providing care for the HD patient and family.

Acknowledgment: The authors are supported by the Huntington's Disease Center, funded by National Institute of Neurological Communicative Diseases and Stroke Grant 16367, and the Brain Tissue Resource Center, MH/NS 31862.

REFERENCES

Adams, P., and R. H. Myers. 1982. Unpublished data.

Alzheimer, A. 1911. Über die anatomische Grundlage der Huntingtonischen Chorea und der choreatischen Bewegungen überthaupt. Z. Neurol. Zbl. 3:566-567.

Ando, N., B. I. Gold, E. D. Bird, and R. H. Roth. 1979. Regional brain levels of gamma-hydroxybutyrate in Huntington's disease. J. Neurochem. 32:617-622.

Aquilonius, S. M., S. A. Eckernas, and A. Sundwall. 1975. Regional distribution of choline acetyltransferase in human brain: changes in Huntington's chorea. J. Neurol. Neurosurg. Psychiatry 38:669-677.

Aquilonius, S. M., B. Nystrom, J. Schuberth, and A. Sundwall. 1972. Cerebrospinal fluid choline in extrapyramidal disorders. J. Neurol. Neurosurg. Psychiatry 35:720-725.

Aronin, N., P. E. Cooper, L. J. Lorenz, E. D. Bird, S. M. Sagar, and J. B. Martin. 1983. Somatostatin is increased in the basal ganglia in Huntington's disease. Ann. Neurol. 13: 519-526.

Arregui, A., L. L. Iversen, E. G. S. Spokes, and P. C. Emson. 1979. Alterations in postmortem brain angiotensin-converting enzyme activity and some neuropeptides in Huntington's disease. In Advances in Neurology, Chase, T. N., N. S. Wexler, and A. Barbeau (Eds.), Vol. 23, pp. 517-525, New York, Raven Press.

Avila-Giron, R. 1973. Medical and social aspects of Huntington's chorea in the state of Zulia, Venezuela. In Advances in Neurology, Barbeau, A., T. N. Chase, and G. W. Paulson (Eds.), Vol. 1, pp. 261-266, New York, Raven Press.

Bearn, A. G. 1957. Wilson's disease. An inborn error of metabolism with multiple manifestations. Am. J. Med. 22:747-757.

Bell, J. 1934. Huntington's chorea. In The Treasury of Human Inheritance, Fisher, R. A. (Ed.), Vol. 4 (1), pp. 1-67, London, Cambridge University Press.

Bell, J. 1942. On the age of onset and age at death in hereditary muscular dystrophy with some observations bearing on the question of antedating. Ann. Eugen. 11:272-289.

Bernheimer, H., and O. Hornykiewicz. 1973. Brain amines in Huntington's chorea. In Advances in Neurology, Barbeau, A., T. N. Chase, and G. W. Paulson (Eds.), Vol. 1, pp. 525-531, New York, Raven Press.

Beverstock, G. C., and P. L. Pearson. 1981. Membrane fluidity measurements in peripheral cells from Huntington's disease patients. J. Neurol. Neurosurg. Psychiatry 44:684-689.

Bird, E. D. 1980. Chemical pathology of Huntington's disease. Ann. Rev. Pharmacol. Tox. 20: 533-551.

Bird, E. D., and A. H. Anton. 1982. Phenothiazine biphasic effect on dopamine concentrations in the basal ganglia of sub-human primates. Psychiatry Research 6:1-6.

Bird, E. D., A. J. Caro, and J. B. Pilling. 1974. A sex related factor in the inheritance of Huntington's chorea. Ann. Hum. Genet. 37:255-260.

Bird, E. D., S. A. Chiappa, and G. Fink. 1976. Brain immunoreactive gonadotropin-releasing hormone in Huntington's chorea and in non-choreic subjects. Nature 260:536-538.

Bird, E. D., J. S. Gale, and E. G. Spokes. 1977. Huntington's chorea: post-mortem activity of enzymes involved in cerebral glucose metabolism. J. Neurochem. 29:539-545.

Bird, E. D., and L. L. Iversen. 1974. Huntington's chorea: post-mortem measurement of glutamic acid decarboxylase, choline acetyltransferase and dopamine in basal ganglia. Brain 97:457-472.

Bird, E. D., and L. L. Iversen. 1977. Neurochemical findings in Huntington's chorea. In Essays in Neurochemistry and Neuropharmacology, Youdim, M. B. H., W. Lovenberg, D. F. Sharman, and J. R. Lagnado (Eds.), Vol. 1, pp. 177-195, London, John Wiley & Sons.

Bird, E. D., A. V. P. Mackay, C. N. Rayner, and L. L. Iversen. 1973. Reduced glutamic acid decarboxylase activity of post-mortem brain in Huntington's chorea. Lancet i:1090-1092.

Bird, E. D., E. G. Spokes, and L. L. Iversen. 1979. Increased dopamine concentration in limbic areas of brain from patients dying with schizophrenia. Brain 102:347-360.

Boehnke, M., P. M. Conneally, and K. Lange. 1983. Two models for a maternal factor in the inheritance of Huntington disease. Am. J. Hum. Genet. 35:845-860.

Bolt, J. M. W. 1970. Huntington's chorea in the West of Scotland. Br. J. Psychiatry 116:259-270.

Bowen, D. M., C. B. Smith, P. White, and A. N. Davison. 1976. Neurotransmitter-related enzymes and indices of hypoxia in senile dementia and other abiotrophies. Brain 99:459-496.

Brackenridge, C. J. 1971. The relation of type of initial symptoms and line of transmission to ages at onset and death in Huntington's disease. Clin. Genet. 2:287–297.

Brackenridge, C. J. 1972. Familial correlations for age at onset and age at death in Huntington's disease. J. Med. Genet. 9:23–32.

Brothers, C. R. D. 1949. The history and incidence of Huntington's chorea in Tasmania. Proc. R. Australian College Physicians 4:48–50.

Brothers, C. R. D. 1964. Huntington's chorea in Victoria and Tasmania. J. Neurol. Sci. 1:405–420.

Brownstein, M. J., E. A. Mroz, M. L. Tappaz, and S. E. Leeman. 1977. On the origin of substance P and glutamic acid decarboxylase (GAD) in the substantia nigra. Brain Res. 135: 315–323.

Bruyn, G. W. 1968. Huntington's chorea: historical, clinical and laboratory synopsis. In Handbook of Clinical Neurology, Vinken, P. J., and G. W. Bruyn (Eds.), Vol. 6, pp. 298–378, Amsterdam, Elsevier.

Buck, S. H., T. F. Burks, M. R. Brown, and H. I. Yamamura. 1981. Reduction in basal ganglia and substantia nigra substance P levels in Huntington's disease. Brain Res. 209:464–469.

Butterfield, D. A., and W. R. Markesbery. 1981. Huntington's disease. A generalized membrane defect. Life Sci. 28:1117–1131.

Butterfield, D. A., J. Q. Oeswein, and W. R. Markesbery. 1977. Electron spin resonance study of membrane protein alterations in erythrocytes in Huntington's disease. Nature 267:453–455.

Butterfield, D. A., J. Q. Oeswein, M. E. Prunty, K. C. Hisle, and W. R. Markesbery. 1978. Increased sodium plus potassium adenosine triphosphatase activity in erythrocyte membranes in Huntington's disease. Ann. Neurol. 4:60–62.

Butterfield, D. A., M. J. Purdy, and W. R. Markesbery. 1979. Electron spin resonance, hematological and deformability studies of erythrocytes from patients with Huntington's disease. Biochim. Biophys. Acta 551:452–458.

Byers, R. K., F. H. Gilles, and C. Fung. 1973. Huntington's disease in children. Neuropathologic study of four cases. Neurology 23:561–569.

Caraceni, T., G. Calderini, A. Consolazione, E. Riva, S. Algeri, F. Girotti, R. Spreafico, A. Branciforti, A. Dall'Olio, and P. L. Morselli . 1977. Biochemical aspects of Huntington's chorea. J. Neurol. Neurosurg. Psychiatry 40:581–587.

Carmen, J. S., I. Shoulson, and T. N. Chase. 1974. Huntington's chorea treated with lithium carbonate. Lancet i:811.

Chase, T. N. 1973. Biochemical and pharmacological studies of monoamines in Huntington's chorea. In Advances in Neurology, Barbeau, A., T. N. Chase, and G. W. Paulson (Eds.), Vol. 1, pp. 533–542, New York, Raven Press.

Comings, D. E., I. E. Goetz, J. Holden, and J. Holtz. 1981. Huntington disease and Tourette syndrome. II. Uptake of glutamic acid and other amino acids by fibroblasts. Am. J. Hum. Genet. 33:175–186.

Comings, D. E., A. Pekkala, J. R. Schuh, P. C. Kuo, and S. I. Chan. 1981. Huntington disease and Tourette syndrome. I. Electron spin resonance of red blood cell ghosts. Am. J. Hum. Genet. 33:166–174.

Constanti, A., and A. Nistri. 1976. A comparative study of the effects of glutamate and kainate on the lobster muscle fiber and the frog spinal cord. Br. J. Pharmacol. 57:359–368.

Conneally, P. M., M. R. Wallace, J. F. Gusella, and N. S. Wexler. 1984. Huntington disease: Estimation of heterozygote status using linked genetic markers. Genet. Epi. 1:81–88.

Coyle, J. T., and R. Schwarcz. 1976. Lesions of striatal neurones with kainic acid provides a model for Huntington's chorea. Nature 263:244–246.

Cramer, H., J. Kohler, G. Oepen, G. Schomburg, and E. Schroter. 1981. Huntington's chorea: measurements of somatostatin, substance P and cyclic nucleotides in cerebrospinal fluid. J. Neurol. 225:183–187.

Curzon, G. 1975. CSF homovanillic acid: an index of dopaminergic activity. In Advances in Neurology, Calne, D., T. N. Chase, and A. Barbeau (Eds.), Vol. 9, pp. 349–357, New York, Raven Press.

Curzon, G., J. Gumpert, and D. Sharpe. 1972. Amine metabolites in the cerebral spinal fluid in Huntington's chorea. J. Neurol. Neurosurg. Pyschiatry 35:514–519.

Dalen, P. 1973. Lithium therapy in Huntington's chorea and tardive dyskinesia. Lancet i:107–108.

Davis, K. L., L. E. Hollister, J. D. Barchas, and P. A. Berger. 1976. Choline in tardive dyskinesias and Huntington's disease. Life Sci. 19:1507–1516.

Dubbleman, T. M., A. W. D.Bruijne, J. V. Steveninck, and G. W. Bruyn. 1981. Studies on erythrocyte membranes of patients with Huntington's disease. J. Neurol. Neurosurg. Psychiatry 44:570–573.

Dunglison, R. 1842. The Practice of Medicine. Ed. 1, pp. 312–313, Philadelphia, Lea and Blanchard.

Ehringer, H., and O. Hornykiewicz. 1960. Distribution of noradrenaline and dopamine (3-hydroxytyramine) in the human brain and their behavior in diseases of the extrapyramidal system. Klin. Wschr. (Berlin) 38:1236–1239.

Emson, P. C., J. Fahrenkrug, and E. G. S. Spokes. 1979. Vasoactive intestinal polypeptide (VIP): distribution in normal human brain and in Huntington's disease. Brain Res. 173:174–178.

Emson, P. C., J. F. Rehfeld, H. Langevin, and M. Rossor. 1980. Reduction in cholecystokinin-like immunoreactivity in the basal ganglia in Huntington's disease. Brain Res. 198:497–500.

Enna, S. J., E. D. Bird, J. P. Bennett, D. B. Bylund, H. I. Yamamura, L. L. Iversen, and S. H. Snyder. 1976. Huntington's chorea: changes in neurotransmitter receptors in the brain. N. Engl. J. Med. 294:1305–1309.

Evans, R. H., A. A. Francis, and J. C. Watkins. 1977. Effects of monovalent cations on the responses of motoneurones to different groups of amino acid excitants in frog and rat spinal cord. Experientia 33:246–248.

Fahn, S., and L. J. Cote. 1968. Regional distribution of gamma-aminobutyric acid (GABA) in brain of the rhesus monkey. J. Neurochem. 15:209–213.

Folstein, S., and M. Folstein. 1981. Diagnosis and treatment of Huntington's disease. Comprehensive Therapy 7:60–66.

Fung, L. W. M., and M. S. Ostrowski. 1981. Spin label electron paramagnetic resonance (EPR) studies of Huntington's disease erythrocyte membranes. Am. J. Hum. Genet. 34:469–480.

Giles, R. E., H. Blanc, H. M. Cann, and D. C. Wallace. 1980. Maternal inheritance of human mitochondrial DNA. Proc. Natl. Acad. Sci. 77:6715–6719.

Goetz, I. E., Roberts, and J. Warren. 1981. Skin fibroblasts in Huntington disease. Am. J. Hum. Genet. 33:187–196.

Growdon, J. H., and R. J. Wurtman. 1979. Oral choline administration to patients with Huntington's disease. In Advances in Neurology, Barbeau, A., T. N. Chase, and N. Wexler (Eds.), Vol. 23, pp. 765–776, New York, Raven Press.

Gusella, J. F., N. S. Wexler, P. M. Conneally, S. L. Naylor, M. A. Anderson, R. E. Tanzi, P. C. Watkins, K. Ottina, M. R. Wallace, A. Y. Sakaguchi, A. B. Young, I. Shoulson, E. Bonilla, and I. B. Martin. 1983. A polymorphic DNA marker genetically linked to Huntington's disease. Nature 306:234–238.

Hall, J. G., T. P. Hicks, and H. McLennan. 1978. Kainic acid and the glutamate receptor. Neurosci. Lett. 8:171–175.

Hans, M. B., and A. H. Koeppen. 1980. Huntington's chorea: its impact on the spouse. J. Nerv. Ment. Dis. 168:209–213.

Harper, P. S., A. Tyler, S. Smith, P. Jones, R. G. Newcombe, and V. McBroom. 1981. Decline in the predicted incidence of Huntington's chorea associated with systematic genetic counselling and family support. Lancet ii:411–413.

Hayden, M. R. 1981. Huntington's Chorea. New York, Springer-Verlag.

Hayden, M. R., J. M. MacGregor, D. S. Saffer, and P. H. Beighton. 1982. The high frequency of juvenile Huntington's chorea in South Africa. J. Med. Genet. 19:94–97.

Hayes, S. E., F. K. Goodwin, and S. M. Paul. 1981. Cholecystokinin receptors are decreased in basal ganglia and cerebral cortex of Huntington's disease. Brain Res. 225:452–456.

Herve, D., H. Simon, G. Blanc, A. Lisoprawski, M. Le Moal, J. Glowinski, and J. P. Tassin. Increased utilization of dopamine in the nucleus accumbens but not in the cerebral cortex after dorsal raphe lesion in the rat. Neurosci. Lett. 15:127–133.

Hiley, C. R., and E. D. Bird. 1974. Decreased muscarinic receptor concentration in post-mortem brain in Huntington's chorea. Brain Res. 80:355–358.

Hokfelt, T., L. Skirboll, J. F. Rehfeld, M. Goldstein, K. Markey, and O. Dann. 1980. A subpopulation of mesencephalic dopamine neurons projecting to limbic areas contains a cholecystokinin-like peptide: Evidence from immunohistochemistry combined with retrograde tracing. Neurosci. 12:2093–2124.

Housman, D., and J. Gusella. 1982. Molecular genetic approaches to neural degenerative disorders. In Molecular Genetic Neuroscience, Schmidt, F. O., S. J. Bird, and F. E. Bloom (Eds.), pp. 415–422, New York, Raven Press.

Huntington, G. S. 1872. On chorea. Med. Surg. Reporter (Philadelphia) 26:317–321.

Iversen, L. L., E. D. Bird, E. G. Spokes, and S. H. Nicholson. 1978. Agonist specificity of GABA binding sites in human brain and GABA in Huntington's disease and schizophrenia. In GABA-Neurotransmitters, Alfred Benzon Symposium 12, pp. 179–190, Copenhagen, Munksgaard.

Kanazawa, I., E. D. Bird, J. S. Gale, L. L. Iversen, T. M. Jessel, O. Muramoto, E. G. Spokes, and D. Sutoo. 1979. Substance P: decrease in substantia nigra and globus pallidus in Huntington's disease. In Advances in Neurology, Chase, T. N., N. S. Wexler, and A. Barbeau (Eds.), Vol. 23, pp. 495–504, New York, Raven Press.

Klawans, H. L., Jr. 1970. A pharmacological analysis of Huntington's disease. Eur. Neurol. 4: 148–163.

Klawans, H. L., and R. Rubovits. 1972. Central cholinergic-anticholinergic antagonism in Huntington's chorea. Neurology 22:107–116.

Kuhl, D. E., M. E. Phelps, C. H. Markham, E. J. Metter, W. H. Riege, and J. Winter. 1982. Cerebral metabolism and atrophy in Huntington's disease determined by ^{18}FDG and computed tomographic scan. Ann. Neurol. 12:425–434.

Lakowicz, J. R., and J. R. Sheppard. 1981. Fluorescence spectroscopic studies of Huntington fibroblast membranes. Am. J. Hum. Genet. 33:155–165.

Lange, H., G. Thorner, A. Hopf, and K. F. Schroder. 1976. Morphometric studies of the neuropathological changes in choreatic diseases. J. Neurol. Sci. 28:401–425.

Liss, L., G. W. Paulson, and A. Sommer. 1973. Rigid form of Huntington's chorea: a clinicopathological study of three cases. In Advances in Neurology, Barbeau, A., T. N. Chase, and G. W. Paulson (Eds.), Vol. 1, pp. 405–424, New York, Raven Press.

Ljungdahl, A., T. Hokfelt, G. Nilsson, and M. Goldstein. 1978. Distribution of substance P-like immunoreactivity in the central nervous system of the rat. II. Light microscopic localization in relation to catecholamine-containing neurons. Neuroscience 3:945–976.

Lloyd, K. G., M. Shibuta, L. Davidson, and O. Hornykiewicz. 1977. Chronic neuroleptic therapy: tolerance and GABA systems. In Advances in Biochemical Psychopharmacology, Costa E., and G. L. Gessa (Eds.), Vol. 16, pp. 409–415, New York, Raven Press.

London, E. D., H. I. Yamamura, E. D. Bird, and J. T. Coyle. 1981. Decreased receptor-binding sites for kainic acid in brains of patients with Huntington's disease. Biol. Psychiatry 16: 155–162.

Lyon, I. 1863. Chronic hereditary chorea. Am. Med. Times 7:289–290.

McGeer, E. G., and P. L. McGeer. 1976. Duplication of biochemical changes of Huntington's chorea by intrastriatal injections of glutamic and kainic acids. Nature (London) 263:517–519.

McGeer, P. L., E. G. McGeer, and H. C. Fibiger. 1973. Choline acetylase and glutamic acid decarboxylase in Huntington's chorea. Neurology 23:912–917.

McLean, D. R. 1982. Failure of isoniazid therapy in Huntington disease. Neurology 32:1189–1191.

McLean, D. R., and T. Nihei. 1980. Biochemical markers for Huntington's chorea. Can. J. Neurol. Sci. 7:281–283.

Myers, R. H., D. Goldman, E. D. Bird, D. S. Sax, C. R. Merril, M. Schoenfeld, and P. A. Wolf. 1983. Maternal transmission in Huntington's disease. Lancet i:208–210.

Myers, R. H., J. J. Madden, J. L. Teague, and A. Falek. 1982. Factors related to onset age of Huntington's disease. Am. J. Hum. Genet. 34:481–488.

Myers, R. H., D. S. Sax, M. Schoenfeld, E. D. Bird, P. A. Wolf, J. P. Vonsattel, R. F. White, and J. B. Martin. 1983. Late onset of Huntington's disease. J. Neurol. Neurosurg. Psychiatry (in press).

Newcombe, R. G. 1981. A life table for onset of Huntington's chorea. Ann. Hum. Genet. 45: 375–385.

Newcombe, R. G., D. A. Walker, and P. S. Harper. 1981. Factors influencing age at onset and duration of survival in Huntington's chorea. Ann. Hum. Genet. 45:387–396.

O'Keefe, R., D. F. Sharman, and M. Vogt. 1970. Effects of drugs used in psychoses on cerebral dopamine metabolism. Br. J. Pharmacol. 38:287–304.

Okon, E., and Y. Koch. 1976. Localisation of gonadotropin-releasing and thyrotropin-releasing hormones in human brain by radioimmunoassay. Nature (London) 263:345–347.

Perry, T. L., S. Hansen, and M. Kloster. 1973. Huntington's chorea: deficiency of gamma-aminobutyric acid in brain. N. Engl. J. Med. 288:337–342.

Perry, T. L., J. M. Wright, S. Hansen, and P. M. MacLeod. 1979. Isoniazid therapy for Huntington's disease. In Advances in Neurology, Chase, T. N., N. S. Wexler, and A. Barbeau (Eds.), Vol. 23, pp. 785–796, New York, Raven Press.
Pettegrew, J. W., J. S. Nichols, and R. M. Stewart. 1979. Fluorescence spectroscopy on Huntington's fibroblasts. J. Neurochem. 33:905–911.
Reed, T. E., J. H. Chandler, E. M. Hughes, and R. T. Davidson. 1958. Huntington's chorea in Michigan. I. Demography and genetics. Am. J. Hum. Genet. 10:201–225.
Reisine, T. D., G. J. Wastek, R. C. Speth, E. D. Bird, and H. I. Yamamura. 1979. Alterations in the benzodiazepine receptor of Huntington's diseased human brain. Brain Res. 165:183–187.
Roberts, E. 1976. Gamma-aminobutyric acid and nervous system function – a perspective. In GABA in Nervous System Function, Roberts, E., T. N. Chase, and D. B. Tower (Eds.), pp. 515–539, New York, Raven Press.
Roberts, P. J., and S. D. Anderson. 1979. Stimulatory effect of L-glutamate and related amino acids on (^{3}H) dopamine release from rat striatum: An in vitro model for glutamate actions. J. Neurochem. 32:1539–1545.
Sakai, T., S. Mawatari, H. Iwashita, I. Goto, and Y. Kuroiwa. 1981. Choreoacanthocytosis. Clues to clinical diagnosis. Arch. Neurol. 38:335–338.
Sandberg, P. R., and G. A. Johnston. 1981. Glutamate and Huntington's disease. Med. J. Aust. 2:460–465.
Schoenfeld, M., R. H. Myers, B. Berkman, and E. Clark. 1984. Potential impact of a predictive test on the gene frequency of Huntington disease. Am. J. Med. Genet. 18:423–429.
Shaw, M., and A. Caro. 1982. The mutation rate to Huntington's chorea. J. Med. Genet. 19: 161–167.
Spindel, E. R., D. J. Pettibone, and R. J. Wurtman. 1981. Thyrotropin-releasing hormone (TRH) content of rat striatum: modification by drugs and lesions. Brain Res. 216:323–331.
Spindel, E. R., R. J. Wurtman, and E. D. Bird. 1980. Increased TRH content of basal ganglia in Huntington's disease. N. Engl. J. Med. 303:1235–1236.
Spokes, E. G. S. 1980. Neurochemical alterations in Huntington's chorea. A study of post-mortem brain tissue. Brain 103:179–210.
Spokes, E. G. S., and D. J. Koch. 1978. Post-mortem stability of dopamine, glutamate decarboxylase and choline acetyltransferase in the mouse brain under conditions simulating the handling of human autopsy material. J. Neurochem. 31:381–383.
Stevens, D. L., and M. Parsonage. 1969. Mutation in Huntington's chorea. J. Neurol. Neurosurg. Psychiatry 32:140–143.
Tarsy, D., N. Leopold, and D. S. Sax. 1974. Physostigmine in choreiform movement disorders. Neurology 24:28–33.
Tellez-Nagel, I., A. B. Johnson, and R. D. Terry. 1973. Ultrastructural and histochemical study of cerebral biopsies in Huntington's chorea. In Advances in Neurology, Barbeau, A., T. N. Chase, and G. W. Paulson (Eds.), Vol. 1, pp. 387–398, New York, Raven Press.
Tellez-Nagel, I., M. Vidal, M. P. Rojas, and R. Guiloff. 1970. Electron microscopic observations in a cerebral biopsy of Huntington's disease. In Proceedings of the Sixth International Congress of Neuropathology, Masson et Cie (Eds.), pp. 1055–1056, Paris.
Vegter-van der Vlis, M., W. S. Volkers, and L. N. Went. 1976. Ages of death of children with Huntington's chorea and their affected parents. Ann. Hum. Genet. 39:329–334.
Vessie, P. R. 1932. On the transmission of Huntington's chorea for 300 years – the Bures family group. J. Nerv. Ment. Dis. 76:553–573.
Vestergaard, P., P. C. Baastrup, and H. Petersson. 1977. Lithium treatment of Huntington's chorea: A placebo-controlled clinical trial. Acta Psychiatr. Scand. 56:183–188.
Volkers, W. S., L. N. Went, M. Vegter-van der Vlis, P. S. Harper, and A. Caro. 1980. Genetic linkage studies in Huntington's chorea. Ann. Hum. Genet. 44:75–79.
Walker, D. A., P. S. Harper, C. E. C. Wells, A. Tyler, K. Davies, and R. G. Newcombe. 1981. Huntington's chorea in South Wales. A genetic and epidemiological study. Clin. Genet. 19: 213–221.
Wallace, D. C. 1972. Huntington's chorea in Queensland. A not uncommon disease. Med. J. Aust. 1:299–307.
Walsh, F. X., E. D. Bird, and T. J. Stevens. 1982. Monoamine transmitters and their metabolites in the basal ganglia of Huntington's disease and control post-mortem brain. In Gilles de la Tourette Syndrome, Chase, T. N., and A. J. Friedhoff (Eds.), pp. 165–169, New York, Raven Press.

Wastek, G. J., L. Z. Stern, P. C. Johnson, and H. I. Yamamura. 1976. Huntington's disease: Regional alteration in muscarinic cholinergic receptor binding in human brain. Life Sci. 19: 1033–1040.

Welsh, M. J., C. H. Markham, and D. J. Jenden. 1976. Acetylcholine and choline in cerebrospinal fluid of patients with Parkinson's disease and Huntington's chorea. J. Neurol. Neurosurg. Psychiatry 39:367–374.

Westfall, T. C. 1974. Effect of muscarinic agonists on the release of ^{3}H-norepinephrine and ^{3}H-dopamine by potassium and electrical stimulation from rat brain slices. Life Sci. 14:1641–1652.

Westphal, A. 1905. Case report. Zbl. Nervenheilk 28:674–675.

Wexler, N. S. 1979. Genetic "Russian roulette": The experience of being "at risk" for Huntington's disease. In Genetic Counseling: Psychological Dimensions, Kessler, S. (Ed.), pp. 199–220, New York, Academic Press.

Wright, H. H., C. N. Still, and R. K. Abramson. 1981. Huntington's disease in black kindreds in South Carolina. Arch. Neurol. 38:412–414.

5 The Hereditary Ataxias

Håvard Skre

Department of Neurology, Akershus Central Hospital, University of Oslo, and Institute of Medical Genetics, University of Oslo, Oslo, Norway

Tor S. Haugstad

Department of Neurology, Akershus Central Hospital, University of Oslo, Oslo, Norway

Kåre Berg

Institute of Medical Genetics, University of Oslo, and Department of Medical Genetics, City of Oslo, Oslo, Norway

---

The authors' work was supported by grants from the Norwegian Research Council for Science and the Humanities and Melzer's Fund for Medical Research.

Spinocerebellar Ataxia in Japan
Olivopontocerebellar Atrophy
Onset, Symptoms, and Signs
Course, Duration, and Prognosis
Epidemiology
Pathology
Biochemistry
Diagnosis
Variants and Transitional Forms
Therapy
Genetics
Linkage Studies
Cerebello-olivary Atrophy
Onset, Symptoms, Signs, Course, and Pathology
Biochemistry
Diagnosis
Therapy
Genetics
Genetics of OPCA and LCCA in Japan
Familial Spastic Paraplegia
Onset, Symptoms, and Signs
Course and Duration
Prognosis
Epidemiology
Pathology
Biochemistry
Diagnosis
Therapy
Genetics
Variants of FSP
ATAXIAS INVARIABLY EXHIBITING AUTOSOMAL RECESSIVE INHERITANCE
Friedreich's Ataxia
Onset, Symptoms, and Signs
Course, Duration, and Prognosis
Epidemiology
Pathology
Biochemistry
Diagnosis
Therapy
Genetics
Marinesco-Sjögren Syndrome
Symptoms, Findings, Course, and Duration
Prognosis
Epidemiology

- *Pathology*
- *Biochemistry*
- *Diagnosis*
- *Therapy*
- *Genetics*

Deficiency of Serum Low-Density Lipoprotein and Ataxia

Bassen-Kornzweig Syndrome (Abetalipoproteinemia)

- *Onset, Symptoms, and Signs*
- *Pathology*
- *Biochemistry*
- *Diagnosis*
- *Treatment*
- *Genetics*

Hypobetalipoproteinemia in "Friedreich's Ataxia"

Ataxia-Telangiectasia

- *Onset, Symptoms, and Signs*
- *Course and Prognosis*
- *Epidemiology*
- *Pathology*
- *Biochemistry and Immunology*
- *Diagnosis*
- *Therapy*
- *Genetics*

Refsum's Disease

- *Onset, Symptoms, and Signs*
- *Course, Duration, and Prognosis*
- *Epidemiology*
- *Pathology*
- *Biochemistry*
- *Diagnosis*
- *Treatment*
- *Genetics*

SOME OTHER ATAXIAS WITH METABOLIC ABNORMALITIES

Kearns-Sayre Syndrome

- *Onset, Symptoms, Signs, and Prognosis*
- *Pathology*
- *Biochemistry*
- *Diagnosis*
- *Therapy*
- *Genetics*

Leigh's Syndrome

- *Onset, Symptoms, Signs, and Prognosis*
- *Pathology*
- *Biochemistry*
- *Diagnosis*

INTRODUCTION

Traditionally, neurologists use the term *hereditary ataxia* (HA) to describe a category of genetic disorders of the central nervous system (CNS) where the main symptoms are clumsiness and/or unsteady gait. HA is also called *spinocerebellar degeneration*. Degeneration of certain spinal tract and brain stem nuclei as well as cerebellar degeneration are typical features of autopsy. Classification is based upon a combination of clinical and pathological characteristics. Gradual atrophy of affected tracts and nuclei without any evidence of cellular reaction or abnormal deposits are common to the spinocerebellar degenerations. Gowers (1902) introduced the concept of "abiotrophy" to describe such a process, but the term *systemic degeneration* is most widely used today. The latter concept was introduced by van Bogaert in his exposé on "*les maladies systematisées*" (1949). He proposed the transneuronal degeneration as an explanation for the extensive degenerations in the spinocerebellar conditions, and this concept has been adopted by many workers in the field (Eadie, 1975a, 1975c).

Typical of systemic degeneration is the "dying back process." This process of wasting of neurons from the end organ toward the cell body was first observed by Mott in 1896 (quoted by Coërs and Woolf, 1959) in peripheral neuritis. Gowers (1902) elaborated this theory in his discussion of the abiotrophies. Greenfield (1954) appears to have been the originator of the expression "dying back of the neuron" and its application to the spinocerebellar degenerations, but Mott (1907, quoted by Cavanagh, 1964) was aware of how the neurons of peripheral nerves and spinal tracts decayed in Friedreich's ataxia. He noted that the tracts of Goll degenerated more than those of Burdach, that the pyramidal tracts were less markedly affected in the cervical region of the cord than in the thoracic, and that sensory loss always began distally and affected discriminatory aspects earlier and more extensively than those of pain and temperature, which are generally agreed to be subserved by small-diameter fibers.

In recent years cerebellar neuronal connections, physiology (Hallet, 1979; Llinas and Simpson, 1981), and transmitter chemistry (Ito, 1978; Llinas and Simpson, 1981; Ottersen and Storm-Mathisen, 1984) have been extensively revealed, which have been a prerequisite for the new insights gained in the pathophysiology of cerebellar disorders. Figures 5.1 and 5.2 provide an overview of this knowledge.

Classification

In later years, heredity has been emphasized as an important parameter in the classification of HA, besides clinical and morphological signs (Becker, 1966a; Pratt, 1967; Konigsmark and Weiner, 1970; Weiner and Konigsmark, 1971; Eadie, 1975a, 1975c), but uncertainty as to the definitions of different

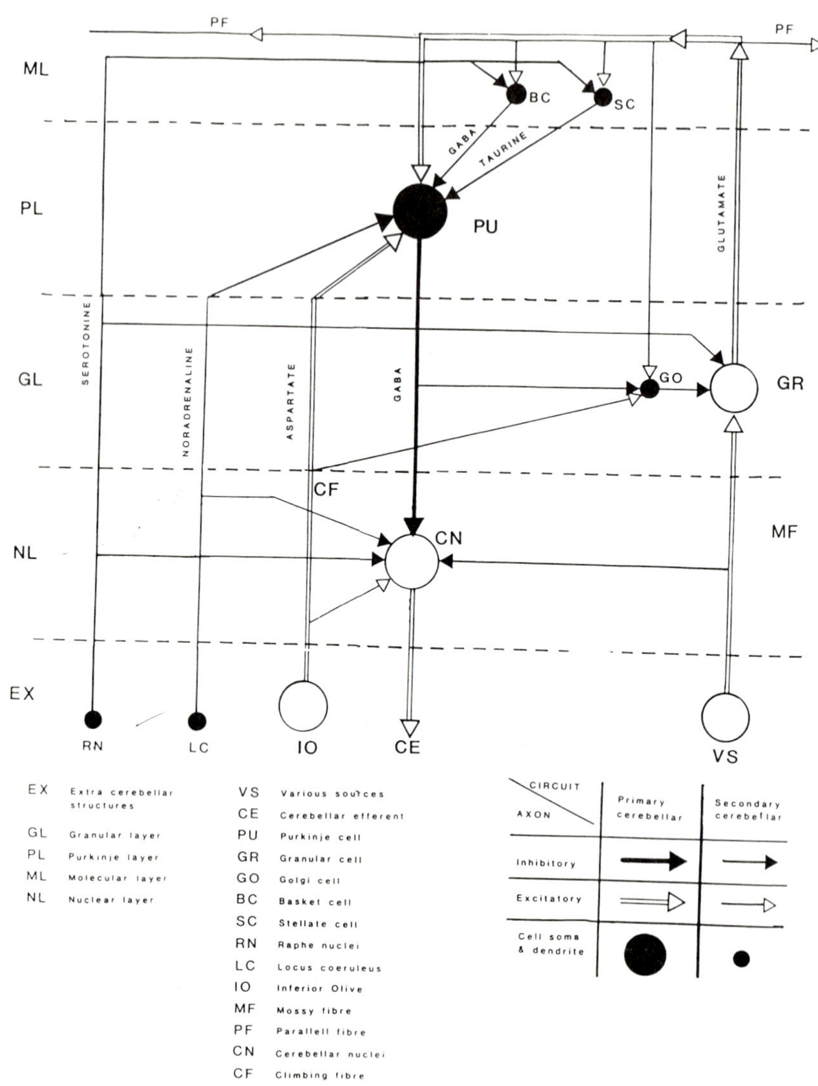

FIGURE 5.1. Basic neuronal connections in the cerebellum with respective transmitter candidates. The neurons of the primary cerebellar circuit are shown with **large circles** and **arrows**, whereas **small circles** denote inhibitory interneurons. (Adapted from Ito M. [1978] Recent advances in cerebellar physiology and pathology. Advances in Neurology. 1978: 21: 59–84.

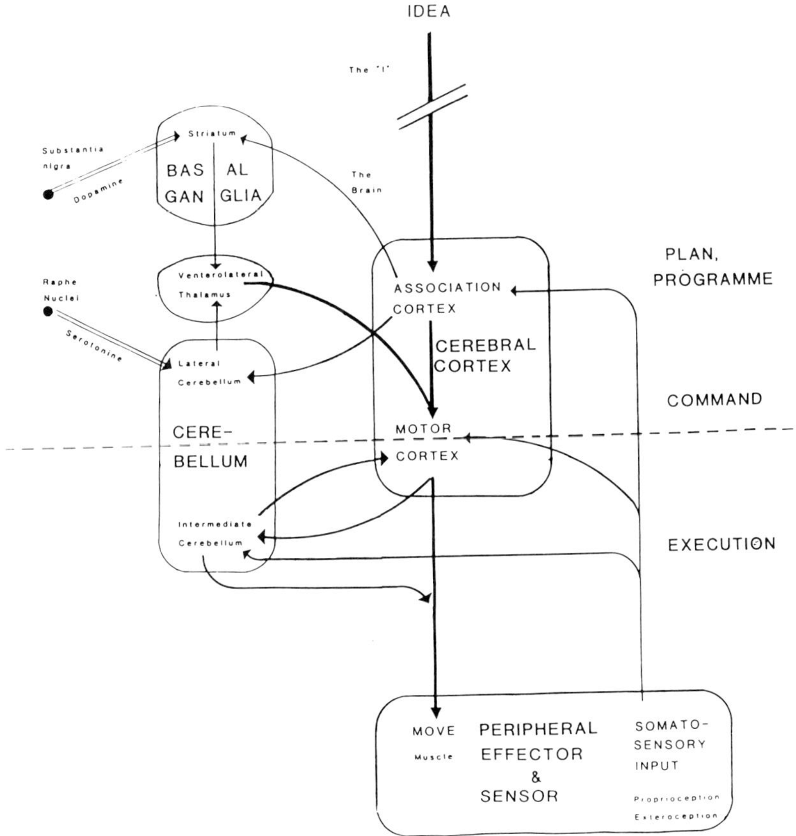

FIGURE 5.2. Diagram showing pathways in planning, executing, and controlling voluntary movement. (Adapted from Eccles J. [1977] Cerebellar function in the control of movement. In: Rose FC (ed). Physiological aspects of clinical neurology. Blackwell Scientific Publications, Oxford, pp 157–178.) See text.

disease entities will continue until more is known about the basic pathophysiology. This applies especially to the "classical," late onset ataxias with a predominant cerebellar pathology.

Other classification principles have been suggested recently by Baraitser (1982), with emphasis on associated signs.

Recently, Harding (1983) presented a comprehensive classification system for the hereditary ataxias and paraplegias based on different clinical and pathogenetic criteria, but with less emphasis on heredity than in the present classification.

Present Classification

The following classification is a modified version based on clinical, pathological, and genetic criteria. Well-defined nosological entities are listed separately. The classification is largely in agreement with what is generally accepted by workers in the field today (Eadie, 1975a, 1975b, 1975c, 1975d, 1975e; Kondo and Hirota, 1980; Kondo and Sobue, 1980; Kondo, Hirota, and Katagiri, 1981). Kondo and co-workers studied HA on a large scale in Japan from 1975 through 1978, using a nationwide reporting system. They conducted extensive clinical as well as family investigations.

I. Ataxias with a heterogeneous genetic background
   A. Spinocerebellar ataxias (SCA-Marie's)
   B. Olivopontocerebellar ataxias (OPCA–Dejerine-Thomas, Mentzel)
   C. Cerebello-olivary degeneration or late cortical cerebellar atrophy (Holmes, LCAA–Marie-Foix-Alajouanine)
   D. Familial spastic paraplegia (FSP)
II. Ataxias that invariably exhibit autosomal recessive inheritance
   A. Friedreich's ataxia
   B. Marinesco-Sjögren's syndrome
   C. Bassen-Kornzweig syndrome or abetalipoproteinemia
   D. Ataxia-telangiectasia
   E. Refsum's disease
III. Some other ataxias with metabolic abnormalities

In the Japanese study a total of 844 cases were collected, of which 765 belonged to category I. The remaining 79 cases then fell within category II, with most cases (52) accumulating in the Friedreich's ataxia group.

Rosenberg (1980) regards the Machado-Joseph disease a primary disorder of the basal ganglia, and hence it will be excluded from the following review (cf. also Coutinho and Andrade, 1978).

PHENOTYPIC VARIATION

One of the enigmas of HA is the phenotypic variation seen within families and among families. This variation, which includes quantitative and qualitative clinical aspects and so-called associated signs, has created complications for classification of these disorders. The variation of phenotypes is most apparent in large pedigrees. Typical examples are the great differences in clinical expression observed in patients from the Schut-Haymaker family (Schut and Haymaker, 1951) and in a large Danish pedigree with OPCA cases observed by Pedersen (1980). The anticipation phenomenon reported in some kindreds with dominant disease (Stern 1960) – and believed to be a sampling error—also, in some cases, reflects phenotypic variation. Schut and Böök (1953) found earlier and more severe disease development in the affected progeny of female than of male OPCA cases in the Schut-Haymaker family. Modifying genes, nongenetic host-related factors, or environmental influences may be of importance in this context (Bickerstaff, 1950).

Nonspecific Neurological Traits in Some Population Groups

Clinical examination of a randomly selected population sample of adults (n = 373) revealed an increase in pathological signs with age (Skre, 1972). The symptoms of aging were rather stereotyped, consisting of minor pareses, deformities, and proprioceptive disturbances in feet and hands; but slight ataxic signs were also observed. The extent of individual involvement varied within wide limits, but few were unaffected past the age of 65. Differences existed in the pattern of clinical signs between the two sexes, middle-aged females exhibiting more signs of pyramidal tract involvement, neuromuscular affection, and hyporeflexia than males; males, on the other hand, had far more signs of peripheral neuropathy than females.

In this study the different signs were rated and scored, and a total sum score for each individual was calculated (Figure 5.3). A standard neurological examination was done in every case, and findings (signs) were then rated through a combination of rank order scale and/or topographic extension scale, forming the basis for scores. This system resulted in scores that were in agreement with the degree of pathological changes and topographic distribution of signs—high scores indicating severe pathological changes, low scores indicating slight pathological changes, and zero score indicating the "normal" state. Qualitative discrimination within this scoring system was obtained by the definition of certain categories of signs, each pertaining to major physiological units of the nervous system. Scores within each category were pooled and in this way represented a subtest score. The subtest score pattern of an individual would reflect his or her clinical "profile." In all, the scoring system comprised nine such categories of subtests:

1. Disturbance of higher CNS activity, taking in consciousness, affective status, and cognition;
2. Oculoauditive signs;
3. Deformities: hammer toes, claw fingers, clubfeet, excavated palms, pes varus, scoliosis, and other deformities;
4. Neuromuscular signs, including bulbar motor nerves;
5. Loss of sensation, including proprioceptive modalities;
6. Hyporeflexia;
7. Ataxia, including spinal ataxia as well as cerebellar signs;

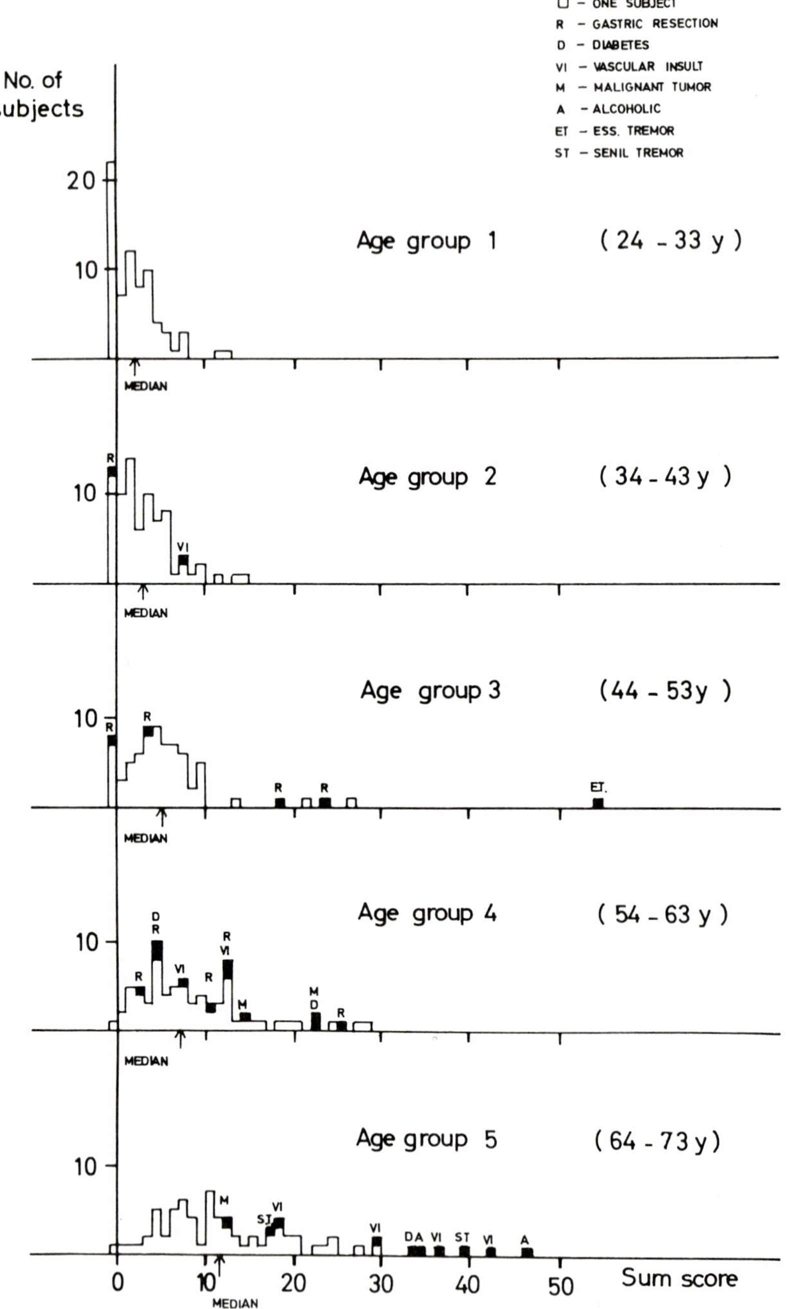

FIGURE 5.3. Distribution of sum scores in a random adult population. (© 1972 Munksgaard International Publishers Ltd., Copenhagen, Denmark.)

8. "Pyramidal" signs, including spasticity, exaggerated tendon reflexes, spastic paresis, and plantar inversion; and
9. Basal ganglia signs, including noncerebellar tremor.

Each subtest consisted of single tests, the number of which varied between the subtests (range, 4–14; mean 4.6). Each single test would give a maximum score of 4 or less (Skre, 1972). In order to eliminate age and sex-specific score differences, a weighting procedure was introduced (Skre, 1974a). The weighting was a multiplication of the test score with the inverse mean score figures for the particular single test in the corresponding age group and sex in the normal population sample. Weighting was not performed when the mean score figure for a particular test in the normal population was ≤1. Thus the weighting was both graded according to rank order and patterned according to regional subdivisions for subtests involving topographic distribution ratings. A figure of 3.5 was arbitrarily chosen as the limit of normality. Corrected sum scores equal to, or above, this level were regarded as abnormal. Of the random adult population 6.2 percent fell within this category. As a rule their condition caused few problems and was usually unnoticed by themselves. This state was defined as nonspecific neuropathy ("unspecific neuropathy" = Un), since patients with HA and other specific neurological diagnoses were excluded. An age-specific distribution of individuals in the normal population sample with high corrected sum scores showed a peak around the late fifties, slightly decreasing later (Skre, 1974a). The significance of this drop is unknown. See the Appendix for details of the scoring and weighting procedures.

Scoring of Neurological Signs in HA Families

Similar clinical examination with rating of signs was also performed in 502 unaffected close relatives of 255 patients with different hereditary ataxias/neuropathies (Skre, 1974a, 1975b) and revealed increased numbers of individuals with high scores as compared with the normal population (37 percent versus 6.2 percent). The aggregation of oculoauditive signs, deformities, and other traits in the relatives of patients with CNS system disorders has been noted previously by several authors (Hodskins and Yakovlev, 1930; Curtius, Störring, and Schönberg, 1935; Klein, 1937; Spillaine, 1940; André van Leeuven, 1948; Schut and Böök, 1953; McNutt, Klingman, and Harlan, 1960; Tyrer and Sutherland, 1961; Skre, 1963, but uncertainty exists about the interpretation of these findings. (See Table 5.1.)

Nonspecific neuropathy and other traits in the families of HA cases might be interpreted as:

1. A premanifestation state of an inherited neurological system disorder,
2. A heterozygous state of a recessively inherited system disorder, or
3. Conditions different from HA, of a genetic or nongenetic nature (Skre, 1974a).

TABLE 5.1. Mental and Oculoauditive Signs in Examined HA-Family Members and in Controls (age matched)

Sign	Cases with Signs						Test for Difference: HA + U versus C	
	HA		U		C			
	Number	Percent	Number	Percent	Number	Percent	χ^2	p
Feeblemindedness	36	18.4	17	4.7	3	1.0	22.30	<0.001
Dementia	57	29.1	17	4.7	0	0.0	38.67	<0.001
Epilepsy	14	7.1	6	1.7	1	0.3	8.34	<0.005
Psychiatric diagnosis	46	23.5	55	15.2	10	3.4	36.29	<0.001
TRD	9	4.6	5	1.4	0	0.0	7.44	<0.01
OA	11	5.6	1	0.3	2	0.7	2.48	<0.2
Cataract	8	4.1	4	1.1	3	1.0	0.84	<0.4
Ocular "myopathy"	36	18.4	22	6.1	6	2.1	19.10	<0.001
Squints	15	7.7	22	6.1	4	1.4	11.53	<0.001
Neural hearing loss	9	4.6	11	3.0	7	2.4	0.87	<0.4
Total number of cases	196	100	361	100	291	100		

TRD = tapetoretinal degeneration
OA = optic atrophy
HA = hereditary ataxia
U = nonaffected family members
C = control
p = probability
Source: Skre (1975c).

Family investigations of kindreds with Charcot-Marie-Tooth disease (CMT), Friedreich's ataxia, hereditary spastic paraplegia (HSP), and spinocerebellar ataxias (SCA, types 1, 2A, and 2B) uncovered evidence in support of hypotheses 2 and 3 (Skre, 1974b, 1974c, 1974d, 1975b). Significant clinical differences were found between these two manifestation types of Un, the presumed heterozygous Un state revealing a more specific clinical pattern than the polygenic Un state (Table 5.2) (Skre 1974b, 1974c, 1974d, 1980). The use of subtest scores as parameters in factorial analysis (Wiener, 1962) showed significant differences between Un occurring in families with recessive CMT and in families with dominant CMT segregation patterns, respectively (Skre, 1974b).

Hypothesis 1 could be excluded in most instances through the genetic analyses of pedigrees and kinships, showing almost complete penetrance of HA in this sample—Friedreich's ataxia and spinocerebellar ataxia, type 1, being exceptions (see Table 5.20). Statistics on the prevalence of Un in families with system disorders and in a normal population sample gave further evidence for hypothesis 2 that nonspecific neuropathy of this category might be a heterozygous manifestation of recessive system disorder, whereas Un of the 3 category, aggregating preferably in families where the main disorder segregated as an autosomal dominant trait, probably represented a polygenic state

TABLE 5.2. Clinical Differences between Un Segregating in Dominant HA Families and Recessive HA Families

Signs[a] in Un cases from	Prevalence of Specific Signs in Unaffected Relatives of SCA Patients Showing High Sum Scores: number of subjects with signs			Test of H_0 for Distribution of Signs in Un_D versus Un_{R1+R2} p^b
	SCA_D Family	SCA_{R1} Family	SCA_{R2} Family	
Ataxic phenomena, any type, including gait disturbances	0	6	6	0.023
Cerebellar signs, any type, such as bradykinesia, hypotonia, adiadochokinesia, dysmetria, asynergia, and/or rebound phenomenon	1	4	11	0.044
Total number investigated[c]	10	14	19	

SCA_D = spinocerebellar ataxia, type 1 (dominant)
SCA_{R1} = spinocerebellar ataxia, type 2A (recessive)
SCA_{R2} = spinocerebellar ataxia, type 2B, or cerebellar type (recessive)
[a]Signs present were subclinical.
[b]Fisher-Irwin test.
[c]Some Un subjects had both categories of phenomena present.
Source: Skre (1980).

(Skre, 1975b) (Table 5.3). In an additional study it was shown that mental disorders and ocular signs aggregated in Un cases of this type, whereas tapetoretinal degenerations, dementia, epilepsy, and neural hearing loss were preferably associated with Un in recessive families (Skre, 1975c) (Tables 5.4 and Table 5.5). These signs were also frequently observed in HA cases.

Associated Traits Introduced through Assortative Mating—Consequences for the HA Phenotypes

Families where HAs segregate as autosomal dominant traits through several generations may be at high risk for modifying genetic influences, which may have relevance for phenotypical variation. In these families assortative mating for disabilities may occur (Skre, 1975b, 1975c). Assortative mating is a selection process, which would probably operate through the social isolation and rejection of certain families by the community. This was actually observed in isolates in Norway by Ørbec and Quelprud (1954) for Huntington's disease. The stigmatized families in return tried to hide their sick and deny any illness. Members of such families were possibly left with the company of people with similar or other handicaps. In all likelihood such social discrimination would preferably work only in static, isolated populations.

TABLE 5.3. Segregation Patterns of Un Families with System Disorders (HA)

1. Distribution of persons with Un and healthy people (N) among 1st- and 2d-degree relatives of index cases with system disorders[a]

	Families where the main disorder segregates as:			
	Recessive Traits		Autosomal Dominant Traits	
	Un	N	Un	N
1st degree	96	130	32	74
2d degree	40	44	6	39
	$\chi^2=0.66$, $0.4<P<0.5$[b]		$\chi^2=4.77$, $0.025<P<0.05$	

2. Distribution of people with Un and healthy people (N) among 1st- and 2d-degree relatives of Un probands in the progeny of patients with autosomal dominant system disorders

	1st-Degree Relatives		2d-Degree Relatives	
	Un	N	Un	N
Observed	10	36	5	38
Expected[c]	11	35	5	38

[a]From Skre (1975a).
[b]Suggestive of heterozygous transmission of the trait.
[c]On a hypothesis of polygenic inheritance (Edwards, 1960).

TABLE 5.4. Distribution of Mental Signs and Epilepsy in Different Categories of Persons in HA Kindreds and in Controls (age matched)

	Manifest HA		Un in Recessive Families		Un in Autosomal Dominant Families		Normal Persons in HA Families		Controls	
Sign	Number	Percent	Number	Percent	Number	Percent	Number	Percent	Number	Percent
Feeblemindedness	36	18.4	5	4.3	2	5.1	10	4.5	3	1.0
Dementia	57	29.1	5	5.3	1	2.6	11	5.0	0	0.0
Epilepsy	14	7.1	2	2.1	0	0.0	4	1.8	1	0.3
Psychiatric diagnosis	46	23.5	14	14.9	10	25.6	27	12.2	10	3.4
Number of cases with at least one sign	98	50.0	19	20.2	11	28.2	46	20.7	14	4.8
Total number of cases*	196	100	94	100	39	100	222	100	291	100

*Six Un cases whose classification as such depended on mental signs and/or epilepsy were excluded.
Source: From Skre (1975a).

TABLE 5.5. Distribution of Oculocochlear Signs in Different Categories of Persons in HA Kindreds and in Controls (age matched)

Sign	Manifest HA		Un in Recessive Families		Un in Autosomal Dominant Families		Normal Persons in HA Families		Controls	
	Number	Percent	Number	Percent	Number	Percent	Number	Percent	Number	Percent
Squints	15	7.7	4	4.5	3	7.7	13	5.9	4	1.4
Cataract	8	4.1	0	0.0	1	2.6	3	1.4	3	1.0
Ocular myopathy	36	18.4	5	5.6	3	7.7	10	4.5	6	2.1
TRD	9	4.6	3	3.4	1	2.6	0	0.0	0	0.0
Optic atrophy	11	5.6	0	0.0	0	0.0	1	0.5	2	0.7
Neural hearing loss	9	4.6	4	4.5	0	0.0	7	3.2	7	2.4
Number of cases with at least one sign	69	35.2	8	9.0	7	17.9	29	13.1	21	7.3
Total number of cases*	196	100	89	100	39	100	222	100	291	100

TRD = tapetoretinal degeneration
*Eleven Un cases whose classification as such depended on oculoauditive signs were excluded.
Source: From Skre (1975a).

As mentioned before, neurologists have noted the aggregation of associated traits in patients with HA or in their close relatives. These traits include mental disorders, feeblemindedness, squints, digit anomalies, and skeletal deformities (see Skre, 1963, 1975c). As already mentioned (Skre, 1975b), they also include a largely genetically determined nonspecific neuropathy (Un) prevalent in the general population, mainly consisting of spinal and/or peripheral nerve involvement. Through assortative matings these traits may be introduced stepwise generation by generation into families where autosomal dominant HAs segregate, with a possibility of additive genetic influences and modifying effects through the generations. Systematic examination of spouses married to HA cases was not performed in this study since this hypothesis of assortative mating had not been formulated at the time, but a random six spouses had nonetheless been included. Half of these spouses had a Un state; another squinted, whereas the two remaining were normal (Skre, 1975b).

Polygenically determined conditions are the results of the action of several genes at different loci segregating independently (Falconer, 1965). Manifestation occurs when a sufficient number of genes and other factors (depending on the degree of heritability) aggregate in an individual, and the manifestation is a matter of quantity as opposed to what is the case of monogenically determined conditions. As a result of independently segregating polygenes, the Un state would manifest at random in the HA families; however, since dominant HA may be regarded as the main selecting force, HA cases and their first-degree relatives would be at the highest risk of accumulating these polygenes. The consequences for the HA phenotype would be unpredictable, but the chances of clinical modifications and aberrating disease developments like the anticipation phenomenon are present. These interpretations are also recognized by other workers in the field (Schoenberg, 1978).

Hypothetically, the number of polygenes determining the Un state may reach a critical level when their abundance causes manifest disease segregating in a pattern mimicking a dominant trait. Un and the main disease could then both contribute to assortative mating, according to the above hypothesis. This could explain the non-Mendelian HA segregation pattern observed in Japan (Kondo and Sobue, 1980; Kondo, Hirota, and Katagiri, 1981) and elsewhere. These considerations are important for classification of hereditary ataxias and for the study of their inheritance. This applies particularly to adult onset spinocerebellar ataxias.

ATAXIAS WITH A HETEROGENEOUS GENETIC BACKGROUND

Spinocerebellar Types

Marie's *hérédoataxie cérébelleuse* (1893) was a clinical concept based on reports by Fraser (1880), Nonne (1891), Brown (1892), and Klippel and Durante (1892) describing cases of dominantly inherited ataxia with spasticity, eye symptoms, and late onset as distinct from the already known Friedreich's ataxia, which lacks such features. Although cerebellar signs were present, little cerebellar involvement was observed at autopsy (see review by Greenfield, 1954; Eadie, 1975b) apart from the cases observed by Klippel and Durante (1892). According to Holmes (1907a), *spastic ataxia* would be a better term for these conditions since dorsolateral spinal degeneration was the most prominent finding.

Common to most cases in this category of hereditary ataxias are adult onset of symptoms, slow progression, normal life span, and very few pathoanatomical reports. Very little is known about their pathophysiology and biochemistry. Heredity is usually, but not always, autosomal dominant.

Brown's Spastic Ataxia

According to Eadie (1975b p. 365), the hereditary spastic ataxia (Brown, 1892) "appears to occupy a relatively central position in the clinicopathological spectrum of the hereditary spinocerebellar atrophies." However, later reports on SCA cases fail to comply with the criteria set by Brown for spastic ataxia through his original description from 1892. However, this excellent report marks a major step in the development of the concept of hereditary ataxias. In this family 22 cases were observed through four generations. Later on, an additional case was observed in the fifth generation, and three cases came to autopsy. The main clinical feature was an initial gait disturbance beginning in the third decade. Later on, clumsiness of the arms and dysarthria evolved in most cases and were, in some cases, the first signs. The disorder ran an even and slowly progressive course, but the degree of disability varied. In the majority of cases spasticity was found, as well as increased tendon reflexes. Both the clumsiness of the hands and the dysarthria were considered to be of a cerebellar type. Among other findings, visual failure was prominent, and optic atrophy and ptosis were noted in some patients. The pathoanatomical findings in three cases were a degeneration of the posterior columns increasing upward and continuing into the medullary lemnisci (Brown, 1892). Degeneration of the dorsal spinocerebellar tracts and cell loss in Clarke's columns were observed in the cervical cord. The pyramidal tracts and anterior columns were intact, but in one of the three cases the medullary pyramids "were slightly less dense than normal." Cell loss was seen in the dentate nucleus in another. Otherwise, the CNS was without abnormalities.

Variants of Spastic Ataxia

The condition described by Klippel and Durante (1892) may be regarded as a varient of spastic ataxia, but in addition these cases exhibited signs such as nystagmus and sensory loss in the legs, and later autopsy revealed abnormalities of the cerebellar cortex (loss of granular cells). Disease onset was in the fifth decade. Affected persons were observed through three generations in this family, and an autosomal dominant pattern of inheritance was established, as in Brown's family. Of other reports on cases with hereditary spastic ataxia, those by Boller and Segarra (1969) and Taniguchi and Konigsmark (1971) deserve attention since they include pathoanatomical descriptions in addition to clinical data. The cases differed somewhat with regard to the extent of brain stem involvement and sensory affection but were similar in other respects. Boller and Segarra (1975) proposed the term *spino-pontine degeneration* to describe their cases as distinct from other forms of SCA. They include the case investigated by Taniguchi and Konigsmark in their entity. A review of the sparse literature dealing with pathological findings in these conditions has been published by Kondo, Hirota, and Katagiri (1981).

Spinocerebellar Ataxias in Norway

Onset, Symptoms, and Signs. Investigations of SCAs in western Norway (Skre, 1974d, 1980) revealed similar clinical features as mentioned above in three different categories of affected but also some major differences, which will become evident.

Figure 5.4 shows diagrammatically the age at disease onset in the three categories of SCA, with peaks in the fifth decade. The mean age at onset was 40 years in categories 1 and 2A, 36 years in category 2B.

Table 5.6 shows the initial symptoms in these patients; and, as expected, unsteady gait is the predominating complaint in all groups. In category 1, weakness in feet comes next, whereas clumsiness is the principal second sign in the two other categories.

Table 5.7 summarizes the clinical features in these patients. As expected, signs of disturbed gait and coordination as well as cerebellar signs predominate. There are distinct differences between categories 1 and 2A, which represent cases with spinocerebellar symptomatology, and category 2B, which represents cases with a predominantly cerebellar type of disorder. However, statistical analyses of the distribution of signs showed significant differences between all three categories of patients. Incoordination in hands prevailed in categories 2A and B, whereas spasticity and pyramidal tract affection were significantly more prevalent in categories 1 and 2A. This was also the case for neuromuscular affection. Ocular signs were significantly more prevalent in patients in category 2 than in category 1 patients, the predominant sign being ophthalmoplegias (Skre, 1974d) (Table 5.7).

Course and Duration. Disease development was reported as steadily progressing in almost all patients, but in category 2A some patients reported a remittant disease course, and there were other features reminiscent of multiple sclerosis such as sphincter disturbances and paresthesias. In a few patients in category 1, no apparent progression of the disorder had been observed for several years.

Prognosis. In this study no excess mortality was found in autosomal dominant SCA (category 1) compared with a normal population. A single case died at 73 from pneumonia. In the recessive SCA types (categories 2A and B) mor-

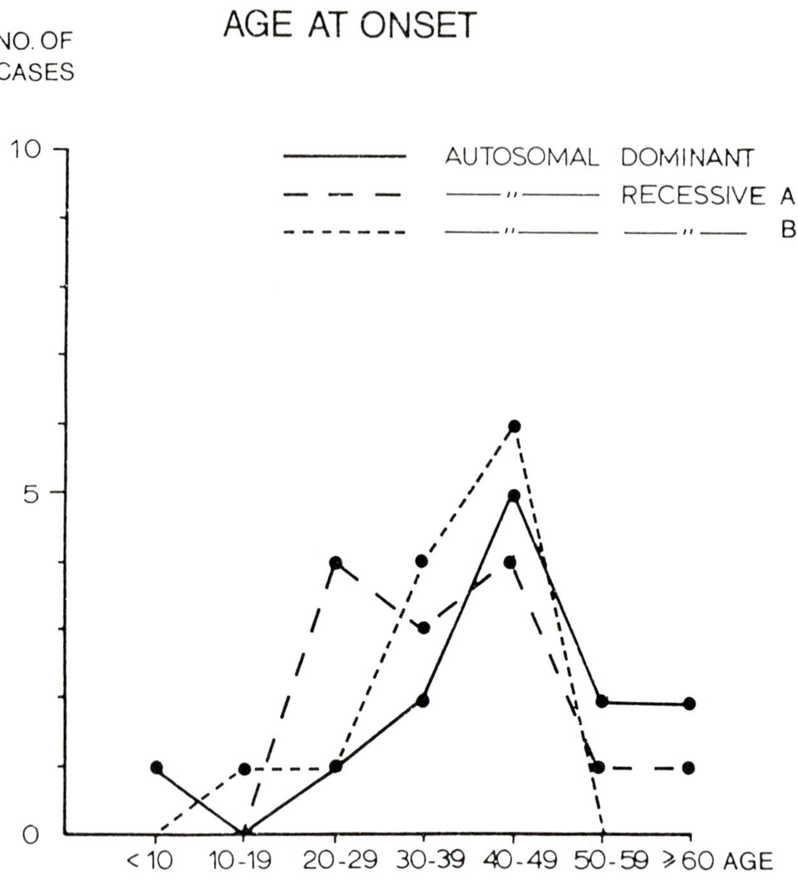

FIGURE 5.4. Age at onset of spinocerebellar ataxia. (Skre, 1983. By permission of Churchill Livingstone, Edinburgh.)

tality was increased, and 30 percent and 50 percent, respectively, of the patients had died at an age lower than 70, at the end of the observation period. Mean age at death was 41.5 and 51 years, but the figure of 41.5 (category 2A) may be misleading since all fatal cases but one died between the ages of 45 and 51 in this category, the last case at 27 after 7 years of illness (Case III of Skre and Løken, 1970) (Figure 5.5). Mean age of the surviving patients was then 58.9 and 59.7 years, respectively. In these categories (2A and B), the prognosis was also less favorable regarding physical handicap and self-reliancy compared with category 1 patients.

Epidemiology. Estimated prevalences and gene frequencies, as well as ascertainment probabilities for these disorders, are given in Table 5.8. No doubt, the unusually high prevalence of the cerebellar type of ataxia (2B) in this region, despite the relatively low estimated gene frequency (4.8×10^{-4}), re-

TABLE 5.6. Symptoms Reported by SCA Patients as the First Sign(s) of the Disease

	Autosomal Dominant SCA		Autosomal Recessive SCA					
	Type 1[a]		Type 2A[a]		Type 2B[b]		Total Sample	
Symptom	Number	Percent	Number	Percent	Number	Percent	Number	Percent
Unsteady gait	5	46	7	64	11	91.7	23	67.6
Clumsiness	1	9	3	27	2	16.7	6	17.6
Weakness in feet	3	27	1	9	0	0.0	4	11.8
Epilepsy	0	0	2	18	1	8.3	3	8.8
Spastic gait	1	9	0	0	1	8.3	2	5.9
Paresthesia in legs	0	0	1	9	0	0.0	1	2.9
Numbness in legs	0	0	0	0	1	8.3	1	2.9
Dysarthria	0	0	0	0	1	8.3	1	2.9
Intention tremor	1	9	0	0	0	0.0	1	2.9
Total number	11	100	11	100	12	100	34	100

SCA = spinocerebellar ataxia
[a]Spinocerebellar forms.
[b]Predominantly cerebellar.

TABLE 5.7. Clinical Signs in Spinocerebellar Ataxias

	Type 1		Type 2A		Type 2B	
Clinical Sign	Number	Percent	Number	Percent	Number	Percent
Unsteady gait	7	64	11	100	12	100
Clumsiness	6	55	11	100	11	92
Cerebellar signs	10	91	10	91	9	75
Nystagmus/dysarthria	6	55	9	82	11	92
Ocular "myopathy"	0	0	6	55	4	33
Spasticity	8	73	5	45	2	17
Babinski's sign	8	73	8	73	4	33
Neuromuscular signs	6	55	8	73	3	25
Parkinsonian signs	4	36	7	64	2	17
Hyperkinesia	0	0	2	18	5	42
Dementia	4	36	9	82	7	58
Epilepsy	0	0	2	18	2	17
Total number investigated	11		11		12	

Type 1 is inherited as an autosomal dominant trait, whereas types 2A and 2B are inherited as autosomal recessive traits.

Source: From Skre (1980).

flects the very high consanguinity rate in the small genetic isolates where these patients lived. All patients stemmed, in fact, from only three large kindreds. The pedigree of one of them is shown in Figure 5.6.

Pathology. Autopsy was performed in a case of type 2A spinocerebellar ataxia. He was the only affected in this particular family, but since his parents were first cousins, single-gene heredity was assumed (Skre and Løken, 1970, case III) (see Figure 5.5). Atrophy of the columns of Goll and Burdach and also of the pyramidal and dorsal spinocerebellar tracts were found (Figure 5.7). In the brain stem some cell loss was observed in the pontine and olivary nuclei, whereas the Punkinje cells and granular layer in the cerebellar cortex, as well as the dentate and brachium conjunctivum, had degenerated to some extent. These changes as well as some paling of the white matter and glial proliferation were probably anoxic changes from epileptic fits. The patient's condition was characterized by continuous myoclonic seizures in the final stage. Autopsy also disclosed a neuromuscular atrophy and cell loss in the anterior horns of the spinal medulla. The cerebellar affection in this case is difficult to classify but is probably nonspecific. The primary changes observed were dorsolateral spinal degeneration and spinal muscular atrophy.

Biochemistry. No significant biochemical changes have until now been found in these conditions.

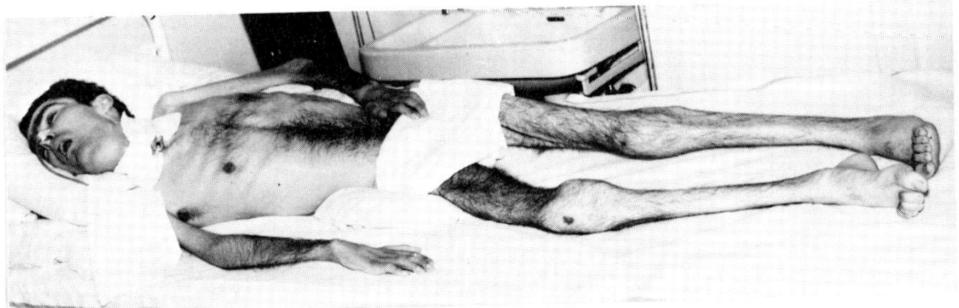

FIGURE 5.5. Case III of Skre and Løken (1970) in the final stage, showing extensive muscle wasting and deformities in the hands and feet. Spinocerebellar ataxia type 2A. (© 1970 Munksgaard International Publishers Ltd., Copenhagen, Denmark.)

TABLE 5.8. Prevalence of Spinocerebellar Ataxia in Western Norway

Diagnosis	Number of Cases Alive in the Region on January 1, 1968	Estimated Prevalence (per 100,000) on January 1, 1968 (population 725,000)	Ascertainment Probability	Gene Frequency
Spinocerebellar ataxia				
Autosomal dominant (type 1)	11	6.4	0.214	3.9×10^{-5}
Autosomal recessive (type 2A)	9	1.8	0.7	2.5×10^{-3}
Predominantly cerebellar type, recessive (type 2B)	7	1.2	0.778	4.8×10^{-4}

Source: Skre (1980).

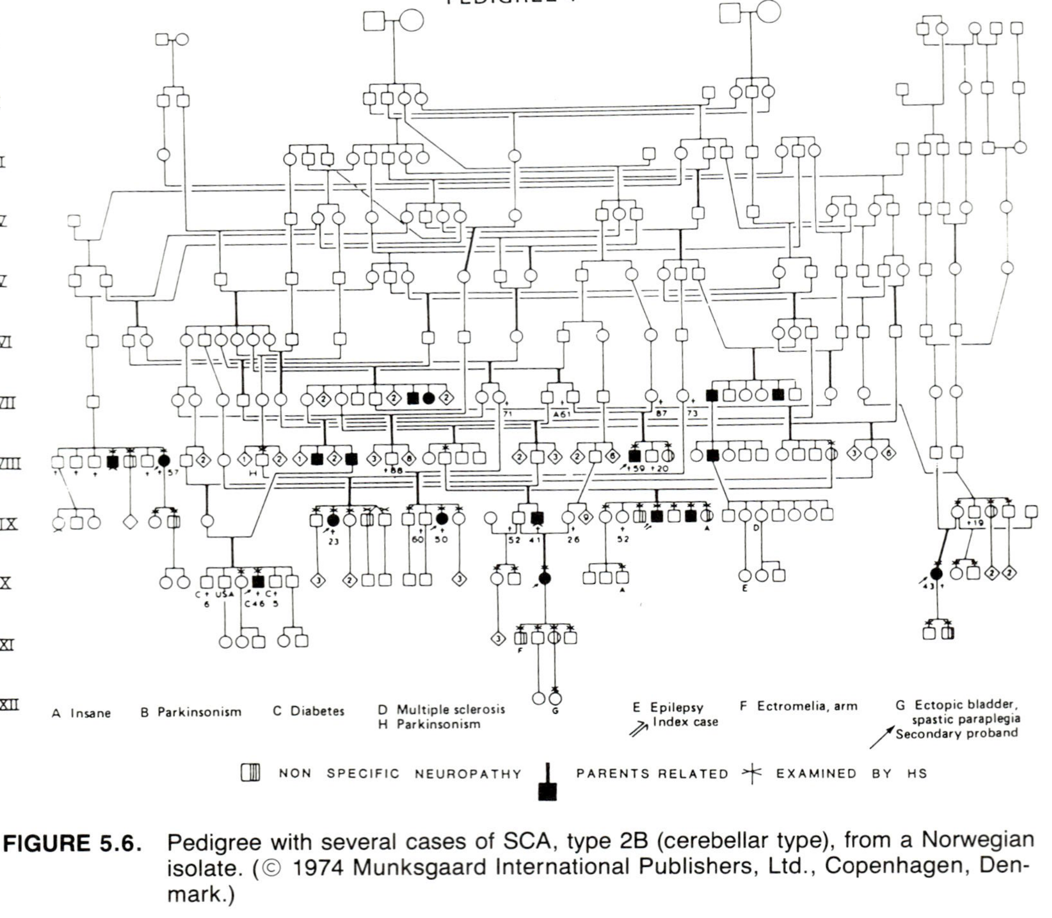

FIGURE 5.6. Pedigree with several cases of SCA, type 2B (cerebellar type), from a Norwegian isolate. (© 1974 Munksgaard International Publishers, Ltd., Copenhagen, Denmark.)

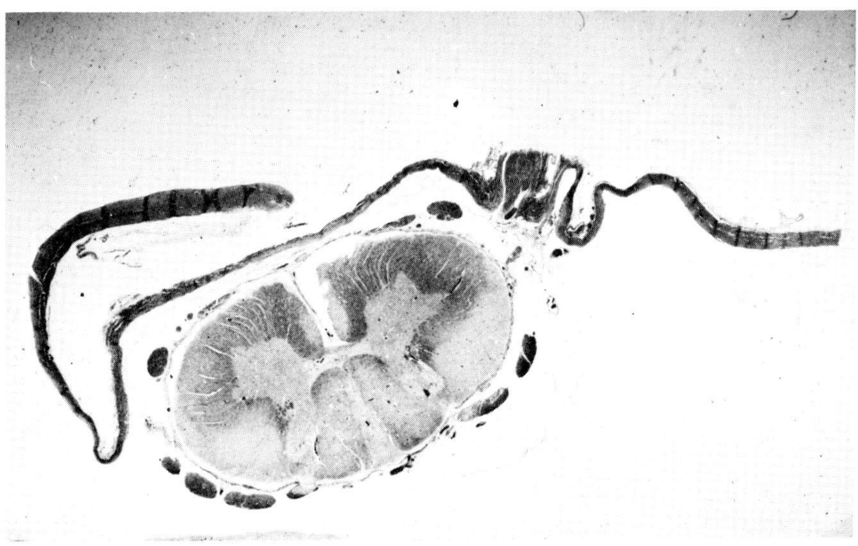

FIGURE 5.7. Cross section of cervical cord in Case III showing atrophy of dorsolateral columns and of the anterior pyramidal tracts. Weil myelin strain. (Courtesy of Dr. K. Skullerud, Institute of Pathology, Rikshospitalet, University of Oslo, Oslo.)

Diagnosis. The diagnosis is one of exclusion. Multiple sclerosis and neurosyphilis are excluded through specific tests. Computerized axial tomography of the brain may reveal cerebellar atrophy and hence distinguish between spinocerebellar and primary cerebellar degenerations. Neurophysiological investigations may be helpful, notably tests of conduction velocity in peripheral nerves. McLeod and Evans (1981) found slowing of sensory conduction in 9 of 19 patients with SCA, whereas all patients had some slight reduction in the motor nerve conduction velocity in the lateral popliteal nerve. Visual-evoked responses were tested in 17 patients with spastic ataxia by Livingstone et al. (1981), but abnormal responses were found in only 3 patients (whereas two thirds of patients with a Friedreich ataxia exhibited such responses).

The differentiation between spinocerebellar ataxias and ataxias with a predominantly cerebellar or olivopontocerebellar degeneration is hardly possible on a clinical basis. Probably some of the cases described as spinocerebellar ataxia may in fact belong to the cerebellar system degenerations,which are discussed later. Definitive knowledge would require pathoanatomical investigation. The matter will be further examined in the chapters that follow.

Therapy. No treatment is available except training schemes, antibiotics against respiratory tract infections, and so on. In the final stages some patients develop epilepsy, which requires special treatment.

Genetics. Autosomal dominant inheritance is the prevailing mode of inheritance in SCA. Exceptions to this rule are the cases described by Klippel and Durante (1892) and some of the Norwegian cases. No sign of X-linked inheritance was observed in the Norwegian series, and the sex ratio was even. This was not the case in the study of Kondo and co-workers (Kondo and Sobue, 1980; Kondo, Hirota, and Katagiri, 1981), who found male preponderance in SCA (see below).

Spinocerebellar Ataxia in Japan

In their comprehensive study on spinocerebellar disorders in Japan, Kondo and co-workers (Kondo and Sobue, 1980; Kondo, Hirota, and Katagiri, 1981) were able to distinguish between three hereditary types of SCA on the basis of statistical data and segregation analyses.

In a few families autosomal recessive inheritance seemed probable since there was a high consanguinity rate (18.2 percent) among parents, and the Mendelian ratio was 0.27. Disease onset was in all cases below 20 years of age. However, the cases did not differ clinically from other categories.

The larger proportion of cases either were from families with autosomal dominant segregation patterns or were sporadic, possibly with a polygenic background (127 and 110 cases, respectively). Different ocular signs and nystagmus were rather common in the dominant SCA as well as muscular hypotonia, but clinical differences did not amount to significant proportions among the three categories. Thus clinical information gave no support to the genetic classification, which was based only on statistical data. However, age at onset was higher in the sporadic cases than in the autosomal dominant ones (50.2 versus 32.5 years). For dominant SCA 60 percent penetrance was observed in females, and 100 percent in males. No evidence for X-linked inheritance of SCA was found. Kondo and his associates held that only a few of the sporadic SCA cases could be explained on a hypothesis of new mutations or offspring of nonmanifesting women. In their view the majority of these sporadic cases had a multifactorial etiology similar to what they suggested for OPCA and LCCA. Phenotypically, these disorders intermingle, since clinicians have never been able to define clear-cut boundaries between them.

Olivopontocerebellar Atrophy

Onset, Symptoms, and Signs. Eadie (1975c, 1975d) distinguishes between the sporadic olivopontocerebellar atrophy (OPCA) of Dejerine-Thomas and the hereditary Mentzel type of OPCA. Pathologically, very little differences are observed (Greenfield, 1954), and the conditions also present with similar clinical pictures, although age at onset is later (mean, 50; range, 17–66) in the sporadic form than in the hereditary form (mean, 35; range, 14–73). The two types will be described jointly, but differences will be mentioned when appropriate.

Unsteady gait may be the first sign, followed by clumsiness, dysarthric speech, Parkinsonian signs (tremor, rigidity), or hypotonia. A static tremor in head and arms may develop, and in some cases extensor plantar responses are seen. Dementia may occur as well as sphincter disturbances. In the Mentzel type of OPCA, dystonic movements, spasticity, kyphoscoliosis, pes cavus, and various ocular signs have been reported. The latter include optic atrophy, macular degeneration, paresis on lateral gaze, diplopia, and slow eye movements. In a number of families (Weiner et al., 1967; Harding, 1982) retinal degeneration has been recorded. Konigsmark and Weiner (1970) claim that such cases belong to a separate category within the OPCA range (OPCA III). Harding supports such a distinction though without consenting to the rest of their classification system, which includes five subtypes of OPCA (Table 5.9) (Konigsmark and Weiner, 1970). A surprisingly high prevalence of autonomic disturbances was observed in OPCA by Kondo, Hirota, and Katagiri (1981).

Course, Duration, and Prognosis. The disease is slowly progressive, with a mean duration of 13 years, ranging from 4 to 23 years. This estimate is based on reports of autopsied cases, whereas others affected may have a more prolonged course. Death usually comes from intercurrent diseases such as pneumonia. Some of the patients develop bulbar motor paresis in the later stages of the disease; in some cases spinal muscular atrophy has also been observed (Eadie, 1975d).

Epidemiology. OPCA is a rare disorder, but the diagnosis could easily be missed since it depends on autopsy in most instances. Some of the cases reported as spinocerebellar ataxia (clinical diagnosis) may in fact be cases of OPCA. Skre (1963) found 5 cases of spinocerebellar ataxia fitting into such a frame in a population totaling 100,000, and Kondo, Hirota, and Katagiri (1981) found 180 cases in Japan in the years 1975 through 1978.

Pathology. According to Eadie (1975c), "It is generally accepted that the pathology of OPCA begins in the terminal axons of the olivocerebellar and pontocerebellar fibres, and later involves the cerebellar cortex by a process of trans-synaptic atrophy" (p. 421). Invariably, there is a great loss of neurons in the pontine and olivary nuclei, whereas cerebellar degeneration, mostly affecting the hemispheres, varies. Purkinje and granular cells suffer most, but the degeneration also includes the dentate nuclei in some cases. Changes have also been found in the basal ganglia, thalamus, substantia nigra, posterior columns, Clarke's columns, spinocerebellar tracts, pyramidal tracts, and anterior horn motor cells, with decreasing frequency in that order.

TABLE 5.9. Classification of Late Onset Cerebellar Degenerations

Present Classification	Original Papers: Author	Original Papers: Mode of Inheritance	Greenfield, 1954 Pratt, 1967	Konigsmark and Weiner, 1970 Weiner and Konigsmark, 1971		Eadie, 1975
Olivoponto-cerebellar atrophy (OPCA)	Mentzel, 1891	D	Mentzel type D	I	Mentzel type D	Mentzel type D (R)
	Dejerine and Thomas, 1891	S	Dejerine-Thomas type S (R/D)	II	Recessive type	
				III	OPCA + retinal degeneration, D	
				IV	Schut-Haymaker type D	Dejerine-Thomas S
				V	OPCA + dementia + ophthalmoplegias, D	OPCA with additional features D (R)
Cerebello-olivary atrophy (LCCA)	Holmes, 1907a	R?	Holmes type D/R	I	Dominant LCCA	Holmes type D/R
	Marie, Foix, and Alajouanine, 1922	S	Marie, Foix, and Alajouanine type S (R/D)	II	Recessive LCCA (excluding Holmes cerebello-olivary atrophy)	Intermediate types, OPCA-LCCA D (R)
				(III & IV	Congenital cerebellar abnormalities)	Marie-Foix-Alajouanine type S

D = autosominal dominant
R = autosomal recessive
S = sporadic (symptomatic)

Biochemistry. Perry et al. (1977) and Perry et al. (1978) were able to show that patients with OPCA (type I of Konigsmark and Weiner, 1970, or Mentzel type of Eadie, 1975d and Greenfield, 1954) had low levels of aspartate in their cerebellar cortices, and in four of five patients the gamma-aminobutyric acid (GABA) content of the dentate nucleus was also diminished. In the fifth patient no degeneration of Purkinje cells was observed. The findings are consistent with the assumptions that aspartate is the neurotransmitter of climbing fibers, which are the first to degenerate in OPCA, and that GABA goes with Purkinje cell depletion, being their transmitter substance.

Animal experiments have shown that the antinicotinamide substance 3-acetylpyridine, which competes with the physiological niacin for incorporation into nicotinamide-adenine dinucleotide (NAD), an important enzyme in the pyruvate dehydrogenase complex, destroys the inferior olives with their climbing fibers as well as some brain stem motor nuclei but leaves other cerebellar systems intact (Perry et al., 1978). In such animals cerebellar aspartate content was lowered (Nadi et al., 1977). Malfunction of the pyruvate dehydrogenase complex has been found in several disorders of the CNS, including the hereditary ataxias (Blass, 1979). It is an important enzyme system, which converts pyruvate to acetyl-CoA, the fuel of the Krebs cycle. The resulting energy deficit might then be compensated for by a greater demand for certain amino acids such as glutamate and aspartate, which in turn have consequences for their availability in the CNS.

Whereas aspartate is believed to be the transmitter of climbing fibers, glutamate is assumed to transmit excitatory impulses from the granular cells (Herndon, 1978). Kainic acid mimics this action while blocking the receptors, causing a permanent depolarization, flooding of the cells (Purkinje, basket, and Golgi) with sodium, and finally cell death. The pathology is reminiscent of cerebellar degeneration in man. Recently, Plaitakis, Berl, and Yahr (1982) have put forward a hypothesis for the pathology of OPCA with a recessive inheritance pattern (type II of Konigsmark and Weiner, 1970). Cultured fibroblasts from patients revealed a defect in the enzyme glutamate dehydrogenase (GDH), which catalyzes the interconversion of glutamate and alpha-ketoglutarate by a process requiring NAD phosphate. Investigations of seven patients from six families disclosed reduced catabolism of glutamate with accumulation of the amino acid in serum as compared with controls. The authors find reason to believe that a similar accumulation will occur in glutamatergic neuron systems like that in the cerebellar cortex. The resulting long-term effect could be an overstimulation similar to that seen in animal experiments with kainic acid and, finally, cell death.

Diagnosis. Typical clinical features including unsteadiness, dysarthria, and clumsiness, with a slowly progressive course and onset in adult life, will provide the diagnosis in most instances when exclusions have been made for multiple sclerosis, syphilis, and cerebellar tumor. The later addition of extrapyramidal or Parkinsonian signs should confirm this assumption, but in recent years more direct evidence of pontocerebellar degeneration has become available through computerized tomography (Figure 5.8) (Pedersen and Glydensted, 1978). Pedersen and Trojaborg (1981) found abnormal evoked responses—visual evoked potentials (VEP) and somatosensory evoked potentials (SEP)—in about half of the patients with OPCA. McLeod and Evans (1981) found mild to moderate reduction of sensory conduction velocity in peripheral nerves in 19 patients with OPCA or cerebello-olivary ataxia; and in sural biopsies from 5 such patients there was a size reduction of myelinated fibers in 3.

Ogawa et al. (1982) hold that cerebrospinal fluid (CSF) GABA content is significantly decreased in OPCA as distinct from what is found in spinocerebellar ataxia (SCA-Marie).

Variants and Transitional Forms. OPCA cases with impaired vision, severe dementia, or both, may be regarded as variants of the disease (Eadie, 1975d). Weiner et al. (1967) reported dominantly inherited OPCA with blindness caused by retinal degeneration and/or optic atrophy. The patients also had internuclear opthalmophlegia and developed a dementia late in the course of the disease. Optic atrophy and dementia were also seen by Carpenter and Schumacher (1966) in an affected father, and three affected children died in infancy of a similar disorder. Autopsy revealed changes typical of OPCA in the father but only cerebellar cortical atrophy in the children. Carter and Sukavaja (1956) described a family with an affected father, five affected sons, and an affected daughter. Disease onset was in mid-life in the father but in childhood or adolescence in the children. Prominent signs were ataxic gait, but later on Parkinsonian signs and dementia developed as well as extensor plantar responses. Two of the children came to autopsy, one showing a diffuse cortical degeneration in the cerebellum, the other showing olivopontocerebellar affection. The pathology in this family as well as in the Carpenter-Schumacher family gives conflicting evidence in relationship to the traditional pathoanatomical classification.

Therapy. Treatment with the serotonine precursor 1-5-HTP (1-5-hydroxytryptophan) was tried by Trouillas et al. (1982) in a case of OPCA with a moderately beneficiary effect. This treatment was also tried in other cerebellar syndromes (see below). Aside from this effort, no attempts have been made so far to treat OPCA with neurotropic drugs. Treatment of these patients has been confined to different training schemes, physical activation, and antibiotics. If Parkinsonian signs develop, traditional treatment has not proven very helpful.

Genetics. Heredity is almost invariably autosomal dominant in the Mentzel type OPCA (Figure 5.9). One of the most famous examples of this kind was presented by Schut and Haymaker where ataxia occurred in 41 members of a family totaling 343 in six generations (Schut, 1950; Schut and Haymaker,

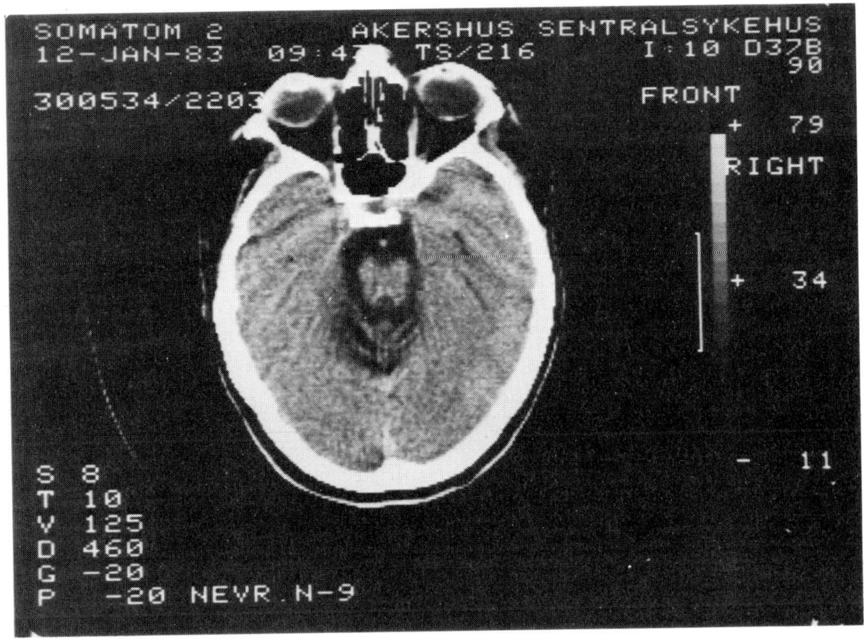

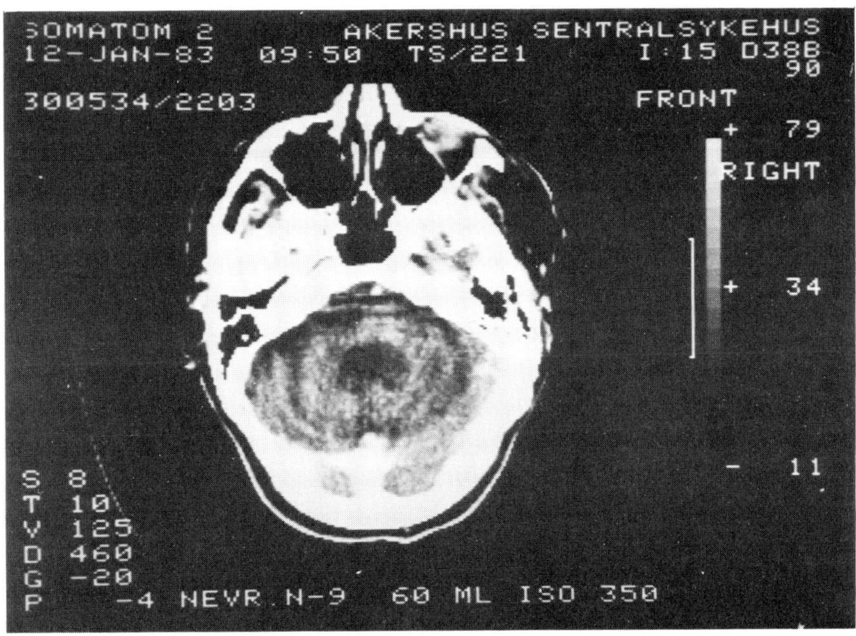

FIGURE 5.8. Computerized axial tomography of patient with OPCA showing shrinkage of brain stem, large fourth ventricle, and cerebellar atrophy (sporadic case). (**A**) Pontine level. (**B**) Medullary level. (Courtesy of the X-ray Department, Akershus Central Hospital, University of Oslo.)

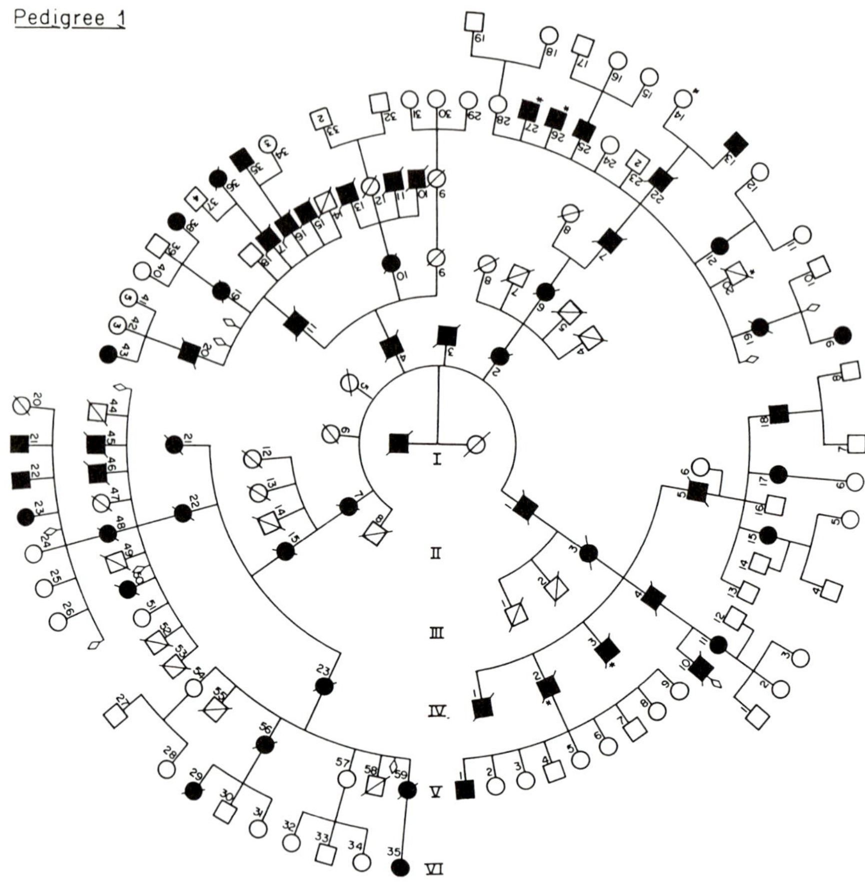

FIGURE 5.9. Pedigree of family CA 1 of Pedersen (1980) with 57 affected members in six generations. **Black symbols** denote OPCA cases. **Arrows** point to probands. Family members with **asterix** emigrated. © 1980 (Courtesy of Dr. L. Pedersen, Rigshospitalet, University of Copenhagen, and Munksgaard International Publishers Ltd., Copenhagen, Denmark.)

1951). Some cases of the sporadic Dejerine-Thomas OPCA may be new mutations or autosomal recessive manifestations. Konigsmark and Weiner (1970) regard the families described by Fickler (1911) and Winkler (1923) with an autosomal recessive inheritance pattern a separate disease entity (OPCA, type II). However, Kondo and Sobue (1980) reject a single-gene hypothesis for both types, presenting evidence for multifactorial liability with a threshold effect.

Linkage Studies. Linkage studies have been done in several large kindreds with dominant OPCA. Yakura et al. (1974) found evidence for linkage between the histocompatibility antigen (HLA) locus on chromosome 6 and ataxia. Later studies have been conflicting because such linkage could not be confirmed by van Rossum, Veenema, and Went (1981) in a pedigree with the Schut-Haymaker type of OPCA (type IV of Konigsmark and Weiner, 1970), whereas Nino et al. (1980) came to the opposite conclusion through the extensive investigation of a large pedigree totaling about 600 persons in six generations. They were unable to classify their cases according to the Konigsmark-Weiner system since the pathology complied with both type I and type IV criteria. Some 47 persons developed cerebellar ataxia, of which 24 were examined by the authors. Blood samples for HLA typing were collected from 73 members of the two last generations. A lod score of 1.97 was obtained at recombination fraction 0.29. Adding information from the families of Yakura et al. (1974) and Jackson et al. (1977), a combined maximum lod score of 4.681 was obtained for recombination fraction 0.22. In addition, Morton et al. (1980) excluded close linkage and found recombination frequencies of 0.15 and 0.30 in males and females, respectively, from the accumulated data from nine pedigrees with OPCA, type I. Other markers could probably help in a more precise localization of the locus. Pedersen et al. (1980) found evidence for linkage with the HLA locus in three of their four families investigated in Denmark with OPCA, but there was no evidence for linkage in the fourth family; the authors concluded that this might point to genetic heterogeneity in OPCA. At present, linkage analyses are of very limited practical use – for example, for the detection of nonmanifesting heterozygotes – because of the looseness of the linkage to HLA. Such analyses may, however, become useful in assigning kindreds to various disease categories.

Cerebello-olivary Atrophy

Holmes (1907b) described a cerebellar ataxia in four sibs, starting in the fourth decade. The parents and two other sibs were healthy. The last of the affected sibs came to autopsy after a disease course lasting 35 years, which is a very long survival time.

Onset, Symptoms, Signs, Course, and Pathology. Clinically, gait disturbances and, later, dysarthria and nystagmus are typical features of the cerebello-olivary atrophies. However, they differ from OPCA in the rarity of extrapyramidal and spinal signs (spasticity, lower motor neuron affection, sensory sign). Dementia is rather common in later stages as are sphincter disturbances. Ocular signs are rare in the present condition. Epilepsy preceded ataxia in two brothers described by Thorpe (1935). An overview of the symptomatology in this as well as in other cerebellar ataxias is given by Eadie (1975a, 1975b, 1975c, 1975d, 1975e) (Table 5.10). The average onset is somewhat later for cerebello-olivary atrophy than for OPCA, according to Eadie (1975a), 46 years as against 35 (range, infancy-70). Disease duration (autopsied cases) is 17 years (range, 5–35), death usually resulting from respiratory tract infec-

TABLE 5.10. Clinical Features in Late Onset Cerebellar Degenerations

Criterion	Olivopontocerebellar Atrophy	Cerebello-olivary Atrophy
Sex	M = F	M = F (except for one family)
Age of onset (years)		
Mean	35	46.1
Range	14–73	Infancy–70
Duration (years)		
Mean	13	17.3
Range	4–23	5–35
Unsteady gait	All cases	All cases
Arm clumsiness	Frequent	Frequent
Dysarthria	All cases	All cases
Dysphagia	A few cases	A few cases
Extrapyramidal disorder	Moderately frequent and fairly severe	Infrequent and mild
Pyramidal signs	Some cases	Rare
Lower motor neuron involvement	Occasional and late	Not recorded
Dementia	Some cases	Some cases
Incontinence	Some cases	Some cases
Nystagmus	Some cases	Some cases

M = male
F = female
Source: Eadie (1975a).

tions. Several authors regard the late cortical cerebellar atrophy (LCCA) described by Marie, Foix, and Alajouanine in 1922 (sporadic case) as principally the same condition as Holmes's cerebello-olivary degeneration. Both conditions are primary degenerations of the paleocerebellar cortex (lobus anterius with vermis), whereas changes in other systems (inferior olives, dentate) may be transsynaptic degenerations and secondary to the primary change (see Eadie, 1975a). However, the etiology in LCCA is probably heterogeneous. Pratt in his review from 1967 reported on both dominant and autosomal recessive LCCA cases from the literature, whereas Mancall (1975) regards all sporadic cases of this type as an acquired condition different from system atrophies. However, a combined toxic-hereditary etiology cannot be excluded, as in the case of Wernicke's encephalopathy. Similarities exist between these conditions because both alcoholism and malnourishment may contribute to the clinical pictures, whereas the cerebellar pathologies are indistinguishable (Victor, Adams, and Mancall, 1959; Victor, 1976).

Biochemistry. The pathogenetic mechanisms in LCCA are unclear, but it seems to be unequivocally settled that Wernicke's encephalopathy is the result of insufficient access of cellular metabolism to thiamine (vitamin B_1). Apart from this, little is known about the precise biochemical mechanism underlying peripheral and central neuronal dysfunction. One may, however, suspect that enzyme systems relying on thiamine as a cofactor are affected. Owing to the similarities among LCCA, Wernicke's encephalopathy, and Leigh's syndrome, it is tempting to suggest a further scrutiny of the pyruvate dehydrogenase complex also in these syndromes (Victor, 1976; Lancet, Editorial, 1979)—all of which may belong to the mitochondrial dysfunction group of disorders.

The experiments with 5-hydroxytryptamine (5-HT, or serotonin) and the decarboxylase inhibitor beneserazide in LCCA patients reported by Trouillas et al. (1982) provide a promising new principle in the treatment of degenerative cerebellar ataxias.

It has been known for some time that nonmedullary axonal connections exist between the medullar and pontine raphe nuclei and the inferior olives, cerebellar cortex, and dentate nuclei and that these neurons are monoaminergic (Hökfelt and Fuxe, 1969; Bloom, Hoffer, and Siggins, 1971; Hoffman and Sladek, 1973; Chan-Palay, 1975; Shinnar, Maciewicz, and Shofer, 1975; Bobillier et al., 1976; Chan-Palay, 1977). Chan-Palay (1979) found ataxia in thiamine-deficient rats and that these animals developed specific degenerations in various 5-HT neuronal systems, particularly in the cerebellopetal raphe systems. Wiklund, Björklund, and Sjölund (1977) showed that nonmedullary monoaminergic fibers from raphe nuclei converged to parts of the inferior olives receiving spinal afferents and projecting to the paleocerebellum, that is, those parts of the cerebellum that subserve walking and erect position. Chan-Palay (1977) maintained that the effects exerted by serotoninergic fibers in experimental animals were in the form of slowly working neuromodulation and not ordinary synaptic transmission, similar to what is suggested for the dopaminergic influence in the basal ganglia. Trouillas et al. (1982) propose a parallelism in man (Figure 5.10). The beneficiary effect of serotonin treatment on cerebellar symptoms has been inferred from these and other animal experiments. For instance, harmaline-induced cerebellar tremor (Headley, Lodge, and Duggan, 1976) can be temporarily abolished by local application of serotonin to the inferior olives, where this tremor is supposed to originate (Guidotti, Biggio, and Costa, 1975).

Diagnosis. A clinical distinction between Holmes's cerebello-olivary ataxia and LCCA is impossible. The differentiation between cerebello-olivary ataxias and OPCA is also difficult because discrimination between the two conditions is mainly based on pathoanatomical findings. The presence of Parkinsonian signs and/or spinal pathology may point in favor of the latter diagnosis. Probably computerized tomography is of less value in this context, although it may be helpful in assessing a diagnosis of cerebellar degeneration.

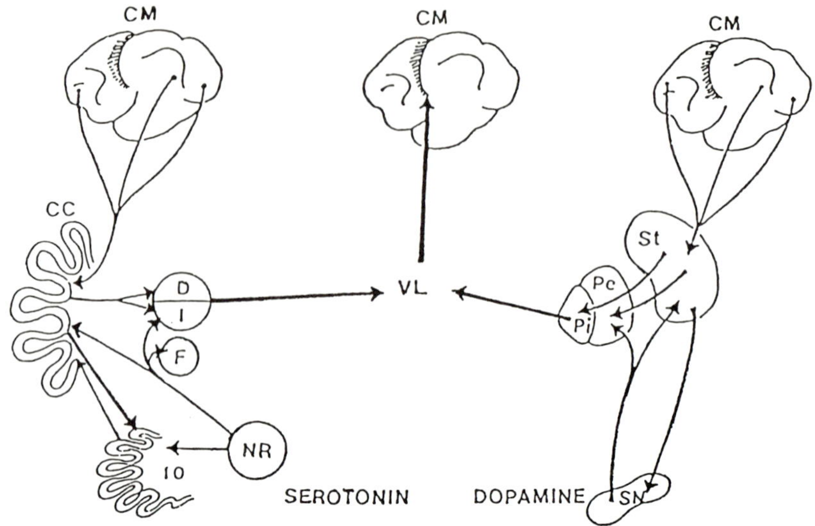

FIGURE 5.10. Comparison of the serotoninergic cerebellar system and the striatal dopaminergic system: **CM** = motor cortex; **CC** = cerebellar cortex; **IO** = inferior olive; **NR** = raphe nuclei; **DI** = dentate and interpositus nuclei; **F** = fastigial nuclei; **St** = striatum; **Pe** = Putamen; **Pi** = pallidum; **SN** = substantia nigra; **VL** = ventrolateral thalamic nucleus. (Trouillas et al., 1982. Courtesy of Masson S. A., Paris.)

Therapy. The efforts of Trouillas et al. (1982) to develop a rational therapy in cortical cerebellar degenerations (mentioned above) are interesting. No other treatment can reverse symptoms.

Genetics. The original cases described by Holmes (1907b) were familial, suggesting an autosomal recessive mode of inheritance. However, surveys in later years report an autosomal dominant mode of transmission in most families (Pratt, 1967; Weiner and Konigsmark, 1971; Eadie, 1975a). Some sporadic cases of LCCA may in fact be autosomal recessive manifestations or new mutations.

Genetics of OPCA and LCCA in Japan

In Japan, Kondo and Sobue (1980) found a male-to-female ratio of 112/52 and 115/65 for LCCA and OPCA, respectively. They also found a crude sib recurrence rate (males only) for all ages. Given that one brother was affected, the overall rates were 7.5 percent and 9.4 percent, respectively, for the two conditions; given that two brothers were affected, the rates were 25 percent and 20 percent, respectively; and given that three sibs were affected, the rates were 37.5 percent and 22.2 percent, respectively, for LCCA and OP-CA. Familial aggregation of cases was most prominent with disease onset in the fifth decade (Table 5.11), decreasing with later onset. An increased risk to offspring of probands of the sex less often affected was also observed in this series, since 7.5 percent of sibs of male probands were affected as opposed to 18.5 percent of female probands in LCCA. (In the Schut-Haymaker family, disease onset was earlier and more severe in the offspring of female probands [Schut and Böök, 1953].) These patterns are compatible with multifactorial etiologies (Table 5.12).

The ascertainment and classification of cases within the spinocerebellar ataxias have always posed problems. Usually, the selection of cases has been biased, since "rare" symptoms, "new variants," large pedigrees, and medical tradition have influenced both authors and editors. The original description of a heredodegenerative nervous disorder may not necessarily be the typical expression of a certain genotype. In the Japanese study ascertainment of cases was carried out in a systematic and unbiased way, with the use of uniform clinical criteria; analyses of family patterns came next. Pathological verification was not possible in an epidemiologic study of this kind and magnitude; neither was total ascertainment. However, the Japanese study, nonetheless, is of great importance for the understanding of the genetics and epidemiology of spinocerebellar and cerebellar ataxias since emphasis was on maximum ascertainment of cases in a large population. In this way selection bias was avoided.

The haphazard way in which cerebellar ataxias segregate may comply with a polygenic etiology, as does the variation in phenotypes, disease onset,

TABLE 5.11. Crude Sib Recurrence Rates in LCCA and OPCA, Male Only, in Japan

	LCCA	OPCA
Rates given one brother affected (in percent)		
All ages	7.6	9.4
20–39	7.4	9.3
40–49	10.9	13.9
50–59	7.1	7.1
60+	0	0
Rates given two brothers affected (in percent)		
All ages	25.0	20.0
Rates given three brothers affected (in percent)		
All ages	37.5	22.2

LCCA = Late cortical cerebellar atrophy
OPCA = Olivopontocerebellar atrophy
Source: Kondo and Sobue (1980).

TABLE 5.12. Expected Patterns under a Multifactorial Liability Hypothesis with a Threshold Effect: Patterns Observed in LCCA and OPCA in Japan

Expected Pattern	LCCA	OPCA
Sib recurrence rate is about the roof of the prevalence rate	?	?
Risk decreases with decreasing blood relations	Yes?	Yes?
Concordance rate is higher in monozygotic than dizygotic twins	?	?
Risk increases when ages at onset are younger	Yes	Yes
Risk increases when more sibs are already affected	Yes	Yes
Risk increases in sibs of probands with less affected sex	Yes	No
Risk increases when probands have severer symptoms	?	?

LCCA = Late cortical cerebellar atrophy
OPCA = Olivopontocerebellar atrophy
Source: Kondo and Sobue (1980).

and severity. There is a tendency toward familial aggregation and vertical transmission in polygenic conditions, which may mimic autosomal dominant heredity with reduced penetrance. However, several large affected pedigrees described within the spinocerebellar range are most likely caused by single-gene mutations, which is also substantiated by the existence of linkage with the HLA locus in OPCA.

Familial Spastic Paraplegia

Familial spastic paraplegia (FSP) was recognized as a heredofamilial entity through the writings of Strümpell and Lorrain around the turn of the century. Its main feature is spasticity, as distinct from the spinocerebellar disorders where ataxia is the major finding (Strümpell, 1880; Lorrain, 1898).

Onset, Symptoms, and Signs. Table 5.13 shows the symptoms claimed as the first sign of the disease in 34 cases of FSP from western Norway. Note that about a fourth of the patients mentioned unsteady gait as the first notice of disease. In many instances spastic gait was observed earlier by family members than by the patients themselves. This may explain why this symptom is only mentioned as the start of the disorder by every second patient.

Spastic signs and exaggerated reflexes dominate the clinical picture in these patients, as will be evident from Table 5.14 and Figure 5.11. In these respects no differences exist between the two hereditary types of FSP in western Norway. This is also in accord with what is observed by others (Sutherland, 1975). However, the involvement of the upper limbs is more frequently observed in autosomal recessive than in autosomal dominant FSP; and this is the case with additional phenomena such as neuromuscular affection, ataxic signs, and dementia (Table 5.15).

TABLE 5.13. Familiar Spastic Paraplegia (FSP) in Norway: Symptoms Claimed by FSP Patients as the First Signs of the Disease

Symptom	Autosomal Dominant		Autosomal Recessive		Total Sample
	Number	Percent	Number	Percent	
Spastic gait	11	47.8	4	36.4	15
Unsteady gait	6	26.1	3	27.3	9
Weariness in feet/legs	4	17.4	1	9.1	5
Bradykinesia (spastic)	1	4.3	1	9.1	2
Ankle clonus	1	4.3	0	0	1
Tremor of hands	0	0	1	9.1	1
No information	2	8.7	1	9.1	3
Total number investigated	23		11		34

Source: Skre (1974c).

TABLE 5.14. Findings in Familiar Spastic Paraplegia Patients

Sign	Autosomal Dominant		Autosomal Recessive		Total Sample
	Number	Percent	Number	Percent	
Spasticity in legs	22	95.7	9	81.8	31
Spastic gait	17	73.9	11	100	28
Exaggerated tendon reflexes in legs	23	100	10	90.0	33
Extensor plantar response	23	100	10	90.9	33
Spasticity in arms	3	13	8	72.7	23
"Fat bottle calves"	5	21.1	7	63.6	12
Paresis/atrophy of muscles in legs	16	69.6	9	81.8	25
Paresis/atrophy in hand muscles	3	13	3	27.3	6
Proximal/axial or bulbar paresis localizations	6	26.1	6	54.5	12
Pes cavus	8	34.8	5	43.5	13
Scoliosis	5	21.7	2	18.2	7
Total number investigated	23		11		34

Source: Skre (1974c).

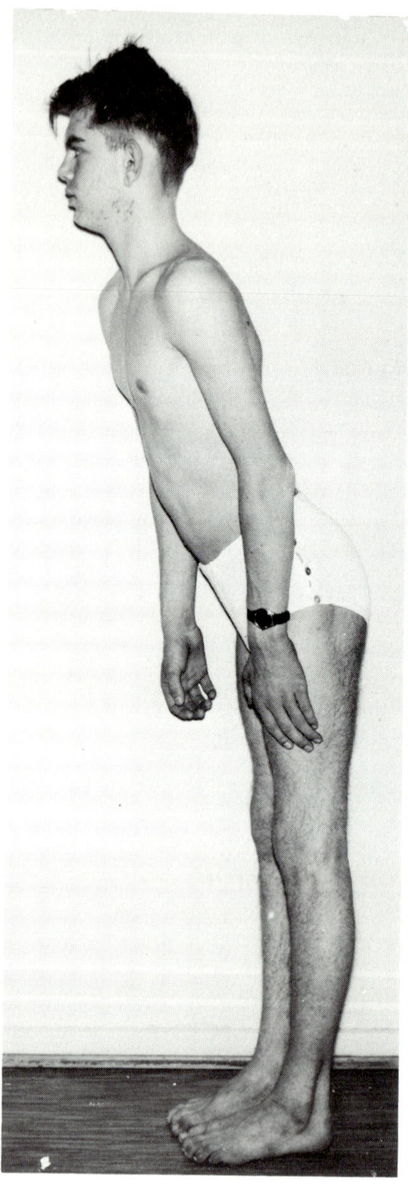

FIGURE 5.11. Boy aged 18 with familiar spastic paraplegia. Flexion contractures in the hips prevent erect posture. Moderate pes cavovarus and hammer toes are present. (Skre, 1983. By permission of Churchill Livingstone, Edinburgh.)

TABLE 5.15. Familiar Spastic Paraplegia: Associated Signs

Clinical Sign	Type				Total	H_0p*
	D	Percent	R	Percent		
Romberg test positive	8	35	7	64	15	0.112
Incoordination (arms)	5	21	5	45	10	0.155
Intention tremor	4	17	7	64	11	0.011
Cerebellar signs	6	26	8	73	14	0.013
Dementia	4	17	5	45	9	0.096
Retinitis pigmentosa	0	0	4	36	4	0.007
Total number investigated	23		11		34	

D = dominant
R = recessive
*Fisher-Irwin test.
Source: Skre (1980).

Retinitis pigmentosa was seen in four cases with an otherwise typical familial spastic paraplegia (three sibs plus a sporadic case, her parents being first cousins). A more varied phenotype was seen in secondary cases than in probands.

Disease onset may vary. Becker (1966b) mentions infantile, juvenile, and adult onset types; Harding (1981b) gives figures for disease onset where there is a clear-cut aggregation of cases in an early infantile group and in a late adult group, whereas onset in early and middle adult life has a more even distribution. She calculated a mean onset age of 20.5 ± 18 years. Similar findings were observed by Skre (1974c) (Figure 5.12). However, in this series a juvenile onset type also was observed in the dominant FSP cases. The infantile onset cases within autosomal dominant FSP families occurred along with cases with an adult onset type, which was the rule. All first probands belonged to the latter category of patients, whereas cases with a very early debut all belonged to the last generation (Figure 5.13, case V, 1). Sampling error can hardly explain the phenomenon, as observed in these families.

Course and Duration. Even progression of symptoms was the rule in almost 60 percent of the Norwegian cases, whereas about 30 percent reported an insidious or arrested development. Accelerating progression was present in three cases, whereas one case reported a remittant course. In patients with early onset disease, walking abilities were delayed, and a diagnosis of cerebal palsy was often made. In many instances patients retained walking abilities for decades despite disease progression, but some became dependent upon wheelchairs after shorter periods. Flexion contractures in hips and knees may develop as well as a scissor gait. No cases of total dependency were observed in the Norwegian study. Similar findings were reported by others (Sutherland, 1975; Harding, 1981b).

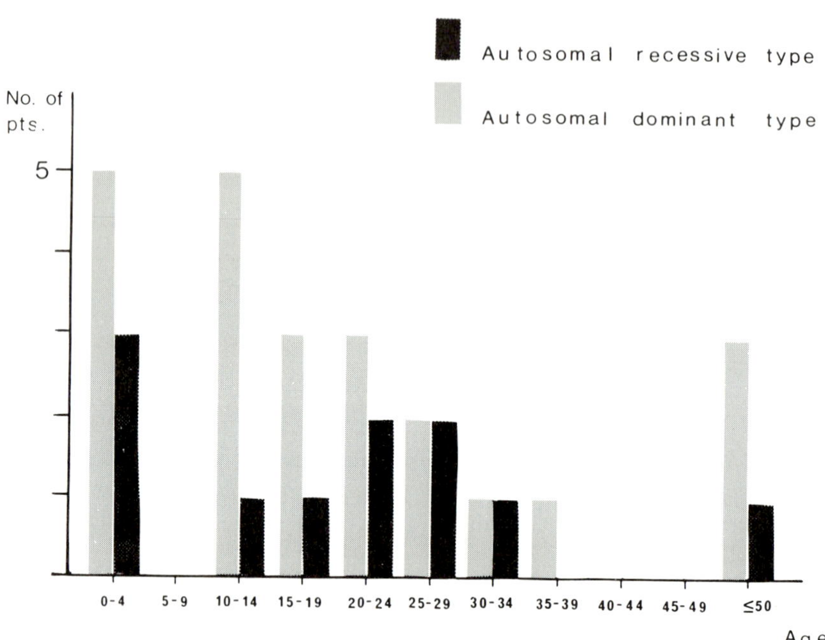

FIGURE 5.12. Age at onset of FSP.

Prognosis. Mortality figures are not raised in FSP (Skre, 1974c; Harding, 1981b). However, Bell and Carmicheal (1939) reported a mean age at death of 55 years and a mean duration of the disease of 19 years in their survey of 199 dominant FSP cases. However, classification problems may influence such findings, especially in a study that includes cases from the literature.

Epidemiology. FSP is a rare disorder. Very few population surveys have been undertaken, but from 1960 through 1966 FSP was investigated in Guam and seven cases were uncovered, giving a prevalence rate of 18.4/100,000 (Chen, Brody, and Kurland, 1968). Skre (1980) estimated, in 1968, the prevalence in western Norway to be 10/100,000 for dominant cases and 1.9/100,000 for autosomal recessive cases (gene frequency $= 9.7 \times 10^{-5}$ and 7.3×10^{-4}, respectively).

Pathology. The common neuropathological finding in this disorder is degeneration of the pyramidal tracts from their origin and downward through the spinal cord. Frequently, there is also some degeneration in the dorsal columns and slight loss of neurons in Clarke's columns. Loss of anterior horn motor cells and of Betz's cells in the precentral gyrus is also seen in some cases, but more extensive abnormalities in the CNS are unusual (Behan and Maia, 1974).

Biochemistry. Apart from its genetics, little is known about the etiology and pathophysiology of FSP. Hall et al. (1979) contend that spasticity may arise from segmental spinal postsynaptic disinhibition. They found decreased levels of glycine in feline spastic cord but increased levels of serine, which is thought to be a precursor of glycine. Glycine is a postsynaptic inhibitory transmitter substance involved in the mediation of motor impulses. Bank, Pizer, and Pfendner (1978) reported hyperglycinemia and hyperglycinuria in four cases who presented with spastic paraplegia from childhood (all were males, of which three were brothers). One of these cases had in addition slight cerebellar and sensory disturbances as well as optic atrophy.

Diagnosis. A diagnosis of FSP should be suspected in the absence of space-occupying intraspinal processes, cerebral palsy, or metabolic/nutritional disorder. Multiple sclerosis may present with a slowly progressive paraplegia but is easily ruled out through specific tests.

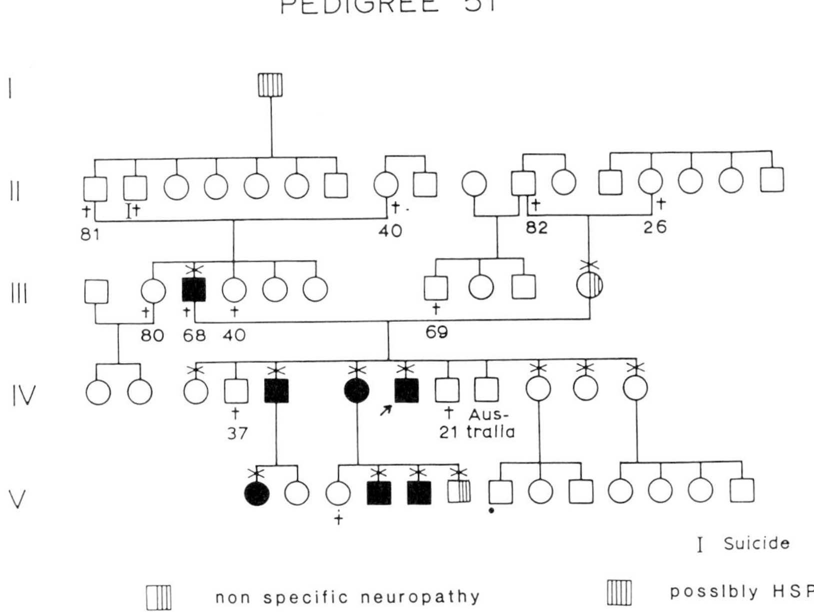

FIGURE 5.13. Pedigree 51 showing cases of autosomal dominant FSP and anticipation phenomenon in case V,1. **Arrow** indicates the proband. (© 1974 Munksgaard International Publishers Ltd., Copenhagen, Denmark.)

The discrimination against spinocerebellar ataxia may pose problems, although ataxia never dominates in FSP as it does in SCA. A considerable phenotypical variation is seen in FSP. Sutherland (1975) has proposed a clinical subclassification, with a "pure" Strümpell type corresponding to the original descriptions, and types with associated features. These features include pes cavus, tendon contractures, optic atrophy, impairment of vibration sense, some ataxia, and impaired intelligence. Dementia, extrapyramidal features, retinal degeneration, muscle atrophy, and cardiac disorder are also seen but only in secondary cases from these families. Evidently, rather wide phenotypic variation is often observed. This raises doubts about the validity of a strict classification. The selection of probands, being a chance event, may erroneously determine what is to be regarded as typical of the condition.

The problem inherent in the classification of familial spastic paraplegia was exemplified by Landau and Gitt (1951), who reported a family totaling 283 persons through seven generations of which 21 were affected. The most frequent finding was spastic paraplegia, but signs such as muscle atrophy and cerebellar or basal ganglia were also observed, spasticity not being a constant finding in all secondary cases.

Neurophysiological tests may be applied in FSP. Electromyograms (EMGs) may show signs of lower motor neuron affection in some cases (Skre, 1974c; Malin, 1976; Harding, 1981b), whereas nerve conduction studies reveal normal or borderline low velocities in motor fibers. Skre and Malin maintain that recessive cases show more abnormalities than autosomal dominant ones. Dimitrijevic et al. (1982) found normal conduction velocities in sensory nerves in 16 patients with a "pure" Strümpell-type FSP but abnormal evoked cortical responses to peroneal sensory stimulation in all patients and in some younger individuals at risk. This was interpreted as caused by a dying-back process of the first sensory neurons in the posterior columns. Visual evoked responses have also been tested in FSP but are of limited value (Pedersen and Trojaborg, 1981).

Therapy. Glycine therapy was attempted by Stern and Bokanjic (1974) in seven spastic patients, with a moderate beneficiary effect. GABA agonists are also given with a similarly slight effect. In fact, no effective treatment exists, but training schemes are useful in maintaining ambulation in these patients.

Genetics. Autosomal dominance is the prevailing mode of inheritance in FSP, but autosomal recessive cases also occur as do X-linked recessive cases (Sutherland, 1975). Harding (1981b) collected 22 families with a "pure" Strümpell-type FSP, in which heredity was autosomal recessive in 3 and autosomal dominant in 19. No X-linked case was observed in any of these families. Such was the case also in the Norwegian study (Skre, 1974c), comprising eight families. However, in dominantly inherited FSP some male preponderance may be observed (Harding, 1981b).

Variants of FSP. Refsum and Scillicorn (1954) described familial cases of spastic paraplegia with extreme muscle wasting reminiscent of amyotrophic lateral sclerosis.

Cross and McKusick (1967a, 1967b) described the so-called Mast and Troyer syndromes occurring in the extensively inbred Old Order Amish community. Both of these disorders were inherited as autosomal recessive traits resulting in spasticity with onset in childhood. The Troyer syndrome displayed additional signs such as dysarthria, distal muscle wasting and mild cerebellar symptoms, whereas the Mast syndrome, apart from spasticity, consisted of intellectual deterioration and dysarthria and, in the later stages, of extrapyramidal and cerebellar signs.

In Sweden, Kjellin (1959) described families displaying hereditary spastic paraplegia appearing in early life was a progressive course, distal amyotrophy, mental retardation, and central retinal degeneration. Evidently, the disorder was inherited as an autosomal recessive trait. More than 200 affected families have been reported from Sweden (Kjellin, 1975).

Sjögren-Larsson syndrome (1957) is a congenital disorder with oligophrenia, ichthyosis, and spasticity. It is probably an autosomal recessive disorder.

Gilman and Horenstein (1964) described hereditary dystonic paraplegia with amyotrophy and mental deficiency in two sibs. However, in several other family members in two generations only asymptomatic neurological signs were present, consisting of extensor plantar responses and brisk tendon reflexes in the lower extremities.

ATAXIAS INVARIABLY EXHIBITING AUTOSOMAL RECESSIVE INHERITANCE

Friedreich's Ataxia

In a series of five papers from 1863 through 1877, Nicholaus Friedreich, a professor of medicine in Heidelberg, described a fairly distinct clinical syndrome in which neurologic signs such as gait ataxia, clumsiness, and dysarthria were prominent in nine members of three sibships. Although there have been controversies about the strict limitations of the clinical syndrome, there is general agreement that Friedreich's ataxia (FA) represents a fairly well defined nosological entity. However, the genetic homogeneity is being disputed in light of emerging genetic and biochemical data. FA is a good example of a classical problem that through the advent of recently developed biochemical and genetic tools now seems to be brought toward a solution, thanks to the meticulously planned and conducted Quebec Cooperative Study on Friedreich Ataxia (Barbeau, 1979).

Onset, Symptoms, and Signs. The first complaint in the majority of patients is gait ataxia (90 percent—Skre, 1975a; 72 percent—Harding, 1981a), followed by generalized clumsiness and intention tremor in the upper extremities.

The patients develop dysarthria and hypotonic muscle weakness. Some may present themselves with a tendency to trip, with a scoliosis, or even in some patients with cardiac symptoms. In the later stages many develop cardiac failure, and 15 percent develop a clinical diabetes. There are certain features of the Friedreich's diabetes which is not common in either juvenile or senile diabetes.

Examination reveals loss of vibration and position senses in the lower extremities. The deep tendon reflexes are invariably reduced or absent but may be better preserved in the upper extremities (Barbeau, 1978b). Over the first few years most patients develop pes cavus; some also develop kyphoscoliosis (Figure 5.14). Most patients eventually show inverted planar responses.

Associated neurological signs may be a nystagmus of central type, sometimes optic atrophy. In some patients some degree of dementia may develop, and cochleovestibular signs are noted, as well as lightning pains, usually in the lower extremities. Gradual loss of superficial sensation occurs and, in quite a number, distal muscular wasting of the extremities. Some patients suffer from epileptic fits.

The median age of onset is early in puberty. Disease onset at 20 years and later, or in early infancy (before 6 years), raises doubt about the diagnosis—even though some authors argue for a wider age distribution (Harding, 1981a).

Course, Duration, and Prognosis. The average course of disease, from detection to decease, is approximately 20 years. Death is usually caused by heart failure or respiratory tract infection. The introduction of antibiotics has increased the patients' life span considerably.

Harding (1981a) gives a good impression of the successive development of symptoms in her material of 90 families. The majority of patients seem to become chair-bound in the second and third decade of life, some 15 years after onset of disease.

Epidemiology. The prevalence of FA is about 1/100,000 in Scandinavia (Sjögren, 1943; Skre, 1975a). Gene frequency was estimated to be 8×10^{-5} in western Norway, the mutation rate being 1.6×10^{-5} (Skre, 1975a). Inbreeding accounts for such a low gene frequency, although prevalence is relatively high (Figure 5.15). Most cases investigated in the Quebec Cooperative Study belonged to 35 sibships whose ancestors could be traced back to 10 French immigrant couples who had come to Canada in the seventeenth century (Barbeau et al., 1976). Harding (1981a) finds the heterozygote frequency to be 1/110 in a British material from the London region.

Pathology. In FA, changes are found in the dorsal root ganglia and in the posterious columns, notably in the tracts of Goll. Changes are also found in the pyramide tract, in Clarke's column, and in the dorsal spinocerebellar tracts. Oppenheimer (1979) examined the brains of 15 cases of FA. The study re-

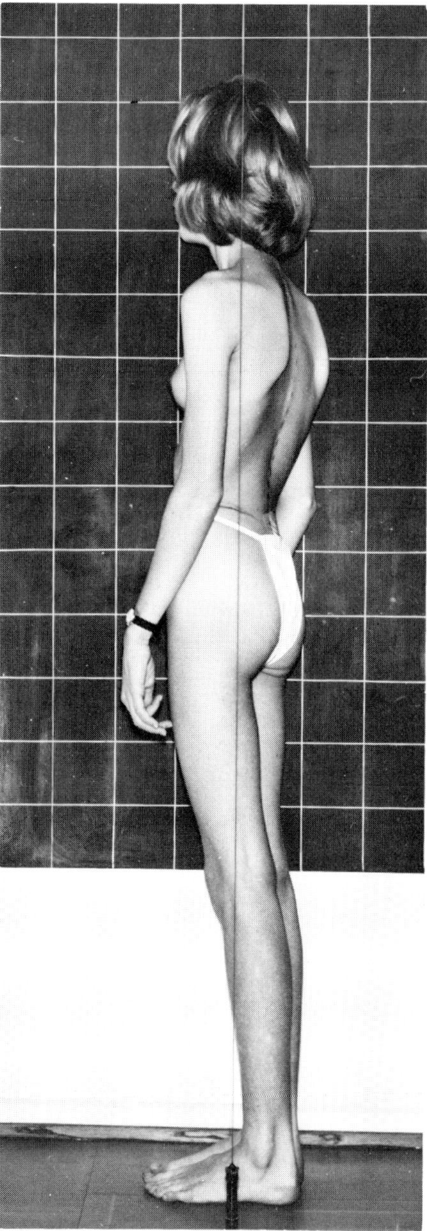

FIGURE 5.14. Girl aged 15 with Friedreich's ataxia. The picture shows an initial torsion scoliosis of the thoracic column with prominence of the right posterior aspect of the chest and winging of the right scapula. (Skre, 1983. By permission of Churchill Livingstone, Edinburgh.)

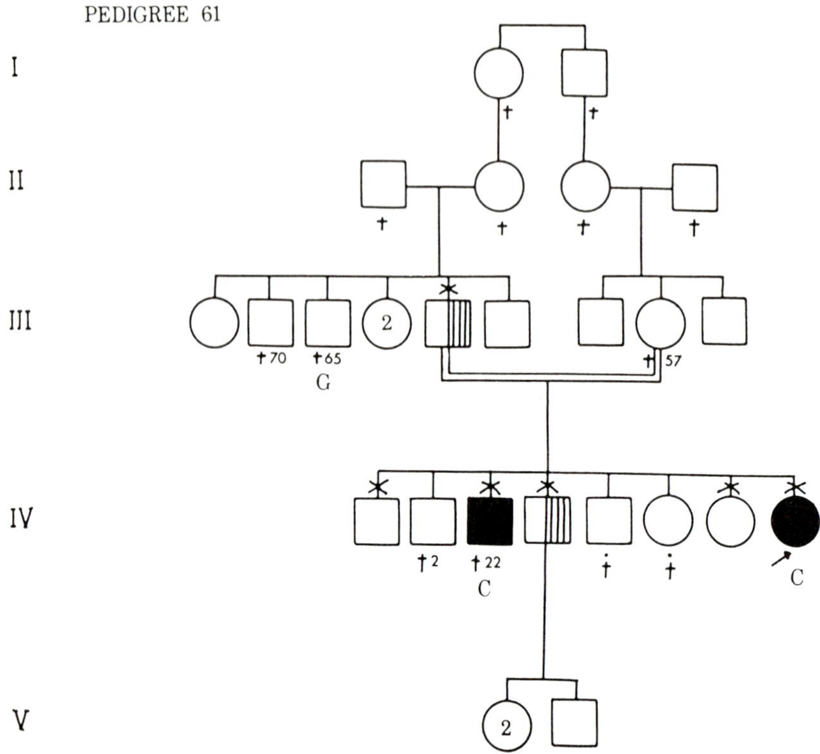

FIGURE 5.15. Pedigree 61 shows consanguinity of parents of two cases with Friedreich's ataxia. The proband (**arrow**) is shown in Figure 5.4. The letter **C** beneath the symbols denotes cardiomyopathy, whereas **G** denotes glioma of the brain. (Skre 1975a. © 1975 Munksgaard International Publishers Ltd., Copenhagen, Denmark.)

vealed varying degrees of degenerative changes in different structures, including the optic nerve, dentate nucleus, pallidum, subthalamic nucleus and the externe cuneate. Some loss of Purkinje cells was also observed in the anterior cerebellar lobe, but loss was thought to be the result of late changes probably due to hypoxia in the terminal stages, where cardiac failure seems to be prominent. Enlargement of the heart as well as degenerative changes and interstitial fibrosis of the myocardium were found.

Biochemistry. The last decade has witnessed the rather impressive accumulation of data concerning biochemical abnormalities in patients with FA. This section briefly examines a few of the more important discoveries.

- *Amino acid metabolism.* The first abnormality noted in amino acid metabolism was an increased urinary excretion of taurine and beta-alanine and increased renal clearance of these and of aspartic acid (Lemieux et al., 1976).

Beta-alanine and taurine share the same transport system in the renal tubuli (Goldman and Scriver, 1967). Filla et al. (1978) showed normal transport of taurine over platelet members, as did Melancon et al. (1979) for fibroblasts, suggesting either different mechanisms of transport in different tissues or that the increased renal excretion is due to extrarenal factors. Lemieux et al. (1976) found that levels of taurine and aspartic acid in CSF were not decreased as expected (see Barbeau, 1978b).

Huxtable et al. (1979) found glutamate, aspartate, phenylanine, and GABA decreased in the cerebellar hemispheres and in the vermis in brains of FA patients. In contrast, the content of taurine was increased in the same areas. It is of interest to note that Barbeau et al. (1981) were able to create experimental ataxia by injection of taurine into the subarachnoideal space in rats.

- *Carbohydrate metabolism*. The association of diabetes with FA has long been recognized. Harding (1981a) finds 10 percent of FA patients with manifest diabetes. Other authors report the incidence to be higher. However, if glucose loading is performed, the percentage with pathologic response increases substantially, still leaving some with normal response. Tolis et al. (1980) state that there is no quantitative defect in insulin release in FA, although release appears to be slower; Draper, Shapcott, Larose, Stankova, Levesque, and Lemieux (1979) find indications of normal insulin receptors and increased insulin binding in patients with abnormal glucose handling but without clinical diabetes. In response to glucose loading, they also find a hyperinsulinemic response in FA patients.

  Many centers find an in vivo defect of pyruvate handling (Kark, Blass, and Engel, 1974; Barbeau, 1978b) in response to glucose loading. This defect is not consistently found in in vitro tests of various tissues (fibroblast cultures, platelets, leukocytes, muscle cells), but where it appears it seems that the defect is localized in the E_3 (lipoamide dehydrogenase; see Figure 5.26) component of the pyruvate dehydrogenase complex (v.i.). Barbeau (1982) reviews the conflicting data on this aspect.

  Quite recently, Stumpf and collaborators (1982) have found greatly reduced mitochondrial malic enzyme (ME_m; see Figure 5.16) in fibroblasts from a group of FA patients (n = 6), compared to controls (n = 8). The mean activity in the FA group was 8.5 percent of that in the control group.
- *Lipid metabolism*.Butterworth et al. (1976) found that some of the typical FA patients had low levels of cholesterol in serum. Davignon et al. (1979) found a reduced high density lipoprotein (HDL) fraction in serum; the relative content of cholesterol and triglycerides increased in HDL, whereas the relative protein content was reduced. The same authors find striking reduction of the linoleic acid (18 : 2) content of the HDL esterified cholesterol but near-to-normal content of linoleic acid in the lipides and triglycerides of HDL. Walker, Chamberlain, and Robinson (1980) found no abnormalities of HDL linoleic acid content in FA patients in a British material, but they did not examine the relative contents of fatty acids in the different fractions of HDL.

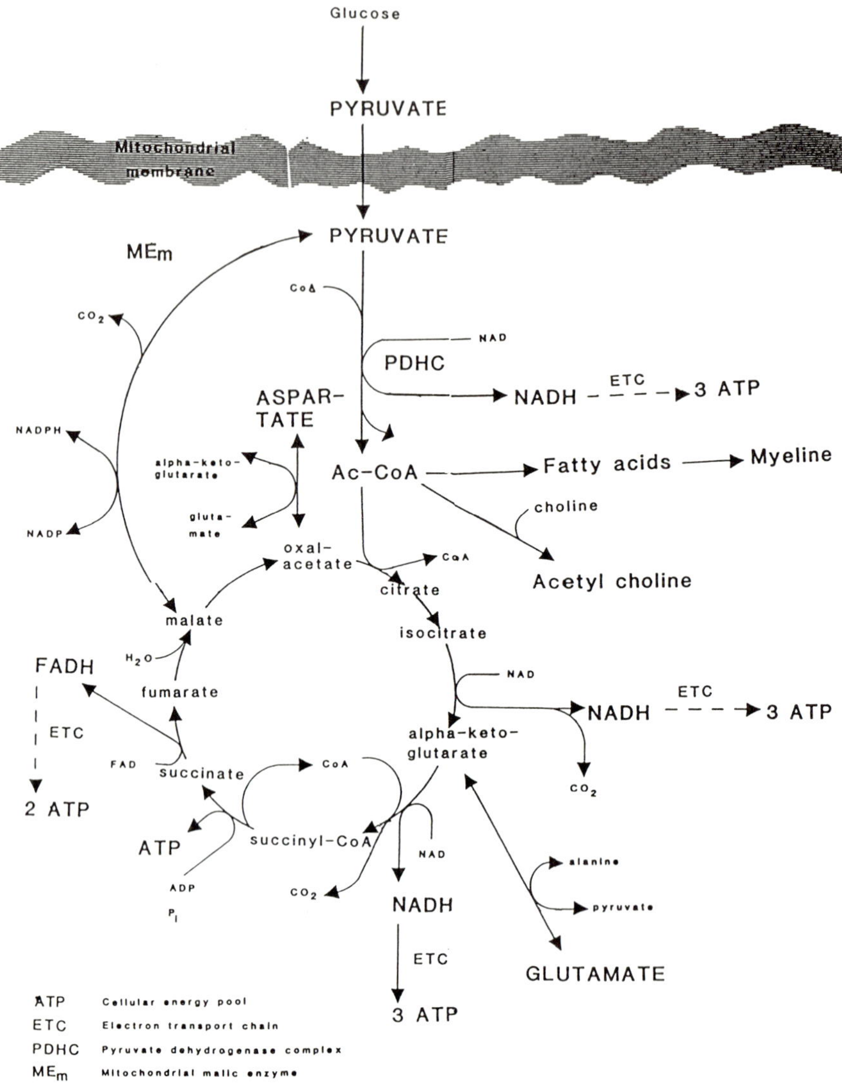

FIGURE 5.16. Glucose/pyruvate metabolism as related to the pathophysiology of Friedreich's ataxia. Note the relationships to **fatty acid** synthesis and **acetylcholine** production, as well as to the amino acids **aspartate** and **glutamate**. Further details about the electron transport chain are shown in Figure 5.25. Draining from the Krebs cycle of transmitter amino acids prevents regeneration of oxalacetate and a slowing down of energy metabolism. **ME_m** Mitochondrial malic enzyme provides restoration of mitochondrial malate and oxalacetate concentrations.

- *Miscellaneous.* Sanchez-Casis, Cote, and Barbeau (1976) demonstrated granular deposits of calcium salts and iron in the muscle cells of the heart. In view of the interplay between taurine and calcium (v.i.), this may be of importance.

  Many patients with FA in the Canadian material show total and unconjugated bilirubin levels in serum above normal limits. This could not be verified in a British material of 20 patients with FA (Barbeau 1978b; Purkiss et al., 1981). Thus hyperbilirubinemia may be a chance association, or linkage, with ataxia, or there may be two phenotypically similar forms of FA.

Diagnosis. A rather strict definition of what should be called "typical" FA is given by Geoffroy et al. (1976) and reporduced in Table 5.16. These criteria seem to be accepted by many authors.

The electroencephalogram (EEG) may reveal abnormal discharges, usually spike or spike-waves. Measurements of the conduction velocity in peripheral nerves reveal moderately reduced velocity in motor fibers but considerable reduction in the sensory conduction velocity. Somatosensory evoked potentials (SEPs) show characteristic changes, as loss of amplitude of peripheral and cervical potentials, and dispersed and delayed cortical potentials (Jones, Baraitser, and Halliday, 1980). Jones and colleagues also find that SEP abnormalities in FA are sufficiently characteristic as to distinguish the condition from peripheral neuropathy and isolated central degenerative disease as well as from multiple sclerosis.

Respiratory deficit of restrictive nature is noted (kyphoscoliosis). There is no evidence of centrally mediated respiratory failure in the early stages of disease (Cote et al., 1979). The cardiomyopathy is usually of a symmetric, concentric hypertrophic type in the early stages, whereas asymmetric septal hypertrophy represents a later stage of disease. Careful investigation with electrocardiogram (EKG) and echocardiography will demonstrate abnormalities in almost all cases. A considerable number show glucose intolerance. This intolerance does not seem to be primarily due to lack of insulin production (Tolis et al., 1980), neither to defect of the insulin receptors (Draper et al., 1980). Possible mechanisms will be discussed later.

Therapy. Scoliosis is treated by use of specially designed molded wheelchairs or by early surgical intervention. Bracing is now recommended at any scoliosis over 20°, and surgical intervention over 40°.

A theoretically possible consequence of reduction of pyruvate dehydrogenase activity is a slowing down of the production of acetylcholine in cholinergic fibers (v.i.). This possibility has led to various therapeutic approaches such as administration of the centrally active cholinesterase inhibitor physostigmine. Kark, Blass, and Spence (1977) found improved ataxia scores after a single dose; it also seems that sustained long-term administration is beneficial (Perlman et al., 1980). Precursors to acetylcholine, such as choline and

TABLE 5.16. Characteristic Features of "Typical" Friedreich's Ataxia

1. Always inherited in an autosomal recessive fashion
2. Age of onset before end of puberty
3. Ataxia first of lower limbs, then of all four limbs, progressing relentlessly without remission and always accompanied by muscle weakness
4. Presence of dysarthria very early in the disease
5. Absence of vibratory and/or position sense in the lower limbs
6. Absence of deep tendon reflexes in the lower limbs
7. Progressive development, within two years after onset, of pes cavus and kyphoscoliosis
8. Presence of a progressive cardiomyopathy, generally of the hypertrophic type, which is best detected by a combination of vectocardiography and echocardiography
9. Absence of sensory nerve conduction (evoked potentials) in the lower limbs with a considerable slowing in the upper limbs

Source: Barbeau (1982).

lecithin, have been tried with various success, as reported by different centers (Livingstone and Mastaglia, 1979; Chamberlain et al., 1980; Livingstone, Mastaglia, Pennington, and Skilbeck, 1981; Pentland et al., 1981; Barbeau, 1982).

Intraventricular injection of glutamate and aspartate led to reversal of cerebellar signs in the 3-acetylpyridine-produced experimental ataxia (De Michele et al., 1980). This may be of importance because the pathogenesis of FA seems to have aspects in common with this particular experimental ataxia. As described earlier, Trouillas et al. (1982) have administered a combination of 5-HTP and benserazide (an inhibitor of the intestinal amino acid decarboxylase) to a group of patients with spinocerebellar disorders. In a total of 11 out of 17 cases with FA, they registered a clear decline in symptom score.

Genetics. Typical FA is inherited as an autosomal recessive trait. In a Japanese population, Kondo and Hirota (1980) found the genetic ratio—calculated with the improved Weinberg method—to be 0.244; the material did not differ significantly from 0.25 ($p<0.05$), again indicating an autosomal recessive inheritance. There is frequently a high degree of consanguinity between parents. The gene frequency was, again, estimated to 8×10^{-5} in western Norway (Skre, 1975a). The inconsistent presence of cardiomyopathies, diabetes, biochemical abnormalities, and clinical manifestations such as distal muscular wasting and age of onset point toward a considerable genetic heterogeneity. Harding (1981a) shows clustering of diabetes in certain sibships. The eight sibships studied by Bouchard et al. (1979), all descendants of a single couple, show a remarkable clinical homogeneity and signs regarded as atypical for FA, among them myopia. Whether this global heterogeneity combined with intrafamilial homogeneity is due to different mutant genes, specific modifying genes, or nongenetic modifying factors awaits further genetic and biochemical data for clarification. No evidence of genetic linkage of FA with any other trait or polymorphism has so far been reported, and the chromosomal location of FA is thus not yet possible.

The only mode of expression of the genetic information given in cellular DNA is through transcription to mRNA (messenger ribonucleic acid) and translation to protein by the ribosomes. So it is obvious that any genetic abnormality in humans must be reflected in some protein, either structural proteins (membranes, plasma proteins, and the like) or enzyme proteins, at some point in the life cycle.

For the time being, a reasonable candidate to the claim of being "the defect" in FA is mitochondrial malic enzyme. Other candidates could be membrane proteins in nerve cells or intestinal mucous cells, plasma apoproteins for the lipoproteins, or possibly enzymes regulating the synthesis of long-chain fatty acids. Thus it seems obvious that there is still a long way to go before "real, rational" therapy for FA can be offered (see "Emerging Pathogenetic Classification").

Marinesco-Sjögren Syndrome

Marinesco-Sjögren syndrome is a rather well-defined nosological entity. It was first described by Marinesco and associates in 1931 already in four members of a family. Although they observed the familial occurrence of the disorder, its genetic basis was only concluded through Sjögren's study from 1950. A thorough historical and clinicopathological account was presented by Mahloudji in 1975.

Symptoms, Findings, Course, and Duration. All cases present with nonspecific cataracts from early infancy on, whereas the neurological signs may appear somewhat later in childhood. Usually, the psychomotor development is retarded, but improvement of abilities is seen until five years of age (Todorov, 1965). Later, ataxia becomes evident and the condition gradually deteriorates until complete disability in most cases in adolescence or early adulthood. Muscular weakness with wasting is observed in more than half of the cases and contributes to the disability (Todorov, 1965).

On examination ataxic signs including nystagmus/dysarthria are found in all cases. Muscular atrophy, weakness, and hypotonia may be found in about two of three cases. Spastic phenomena are rare, but exaggerated tendon reflexes and a positive Babinski sign may be seen in about half of the cases. On the other hand, sluggish or abolished reflexes are also frequently observed. Romberg's sign is present; and as the disorder progresses, walking difficulties grow worse, leading eventually to complete helplessness. In one of the cases described by Skre et al. (1976), indifference to the pain was observed.

Growth is retarded, resulting in low stature. Infantilism is also noted by some workers (Alter and Talbert, 1962; Todorov, 1965) (Figure 5.17A).

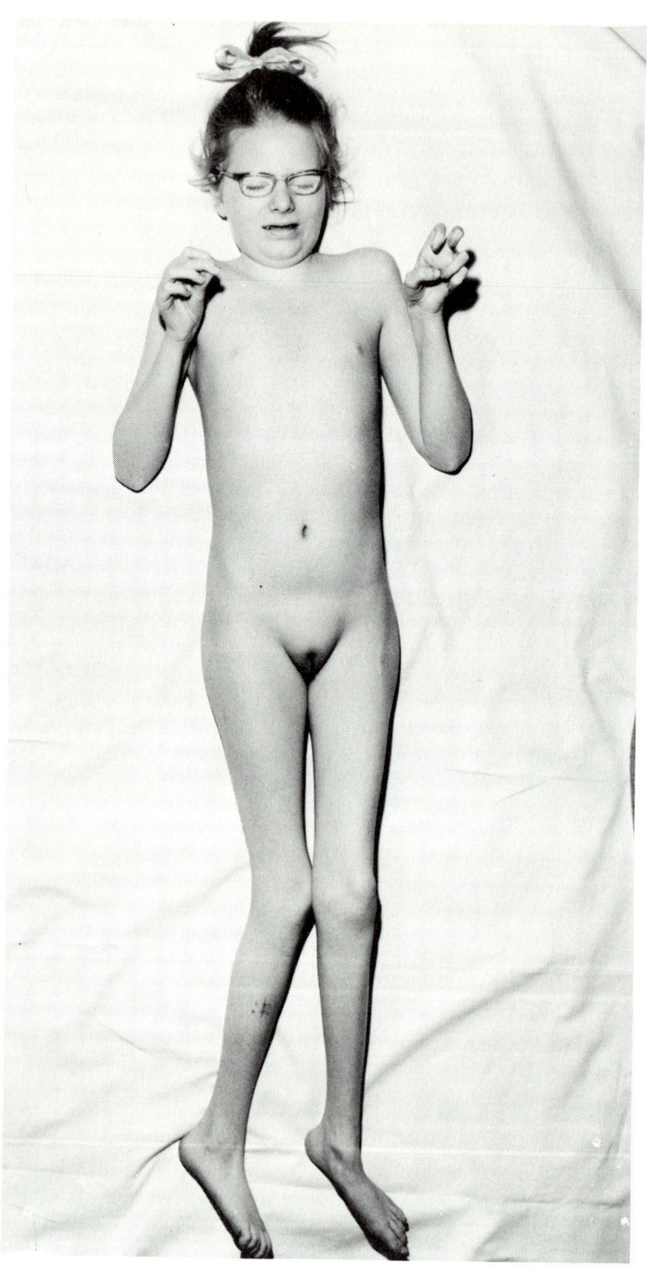

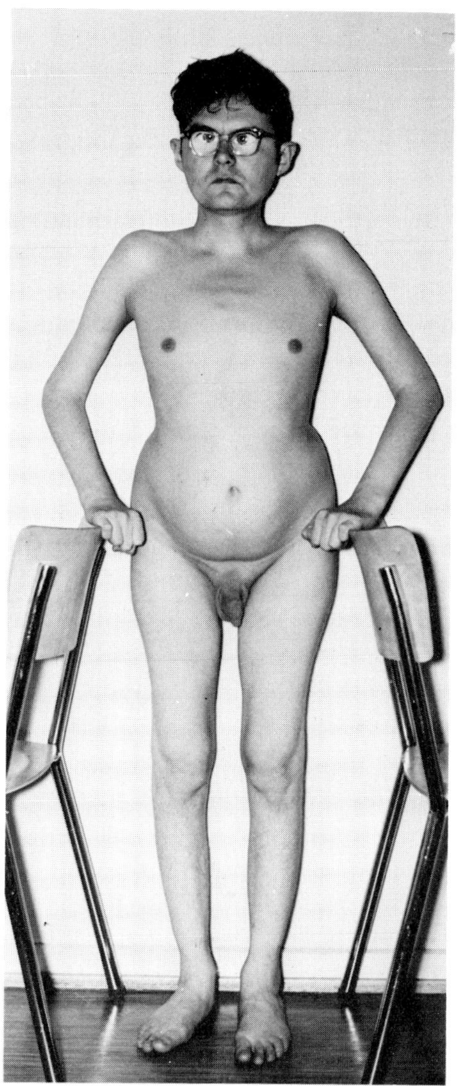

FIGURE 5.17. (**A**) A girl (Case XII,6 in Pedigree 87) 15 years old with Marinesco-Sjögren syndrome. Note infantilism, genua valga, and pes plano-valgus. Unable to stand or walk. (**B**) Man 28 years old, same diagnosis. Needs support to stand and walk. Hypogonadism, short stature, strabism, muscular atrophy in limb girdles and extremities. (Skre et al., 1976. © 1976 Munksgaard International Publishers Ltd., Copenhagen, Denmark.)

Psychological testing reveals IQ levels well below normal ranges, but profound oligophrenia is rare (Todorov, 1965; Andersen, 1965). Pneumoencephalography reveals gross cerebellar atrophy (Skre et al., 1976).

EMG studies are conflicting. While some report changes compatible with a myogenic disorder (Mahloudji, 1975), neurogenic lesion is observed by others (Serratrice, Gastaut, and Dubois-Gambarelli, 1973; Skre et al., 1976).

Ocular signs in addition to cataracts and nystagmus include strabismus (Figure 5.17B), upward gaze paresis, or ptosis (Todorov, 1965; Mahloudji, 1975).

Most patients display deformities of some kind such as pes planus, genua and cubita valga, and kyphoscoliosis. Dentation may be irregular and features deviating. Some patients present with tachycardia, hypotension, and epilepsy (Mahloudji, 1975). Skre and associates (Skre et al., 1976; Skre and Berg, 1977) found hypergonadotropic hypogonadism in nine cases from two families; hypogonadism was also observed by Ionescu (1970).

Prognosis. Prognosis is dubious. Patients usually end up totally crippled in a wheelchair or bed within 10 to 15 years after onset of disease.

Epidemiology. In all, about 100 cases have been described in Caucasians. Cases have also been described in the American black (Alter and Talbert, 1962). Some 17 cases have been reported from Norway, most from the western and northern parts of the country, where there is a rather high consanguinity rate (Andersen, 1965; Skre et al., 1976; Skre and Berg, 1977) (Figure 5.18).

Pathology. Pathoanatomical reports on a few cases show a marked cortical cerebellar and olivary atrophy of a similar type as in Holmes's cerebello-olivary degeneration (Mahloudji, 1975) (Figures 5.19 and 5.20). In addition, degenerative changes were seen in peripheral nerves, with retrograde changes in anterior horn motor neurons (Skre et al., 1976, Case XI,3, Pedigree 87), but muscle specimens showed changes more in accord with a muscular dystrophy (Figures 5.21 and 5.22). This conflicts with neurophysiological findings. However, pseudomyopathic changes are seen in other chronic neuromuscular atrophies such as Charcot-Marie-Tooth disease.

Biochemistry. The autosomal recessive inheritance and the well-defined phenotype with disease onset in childhood point to some inborn error of metabolism as the primary defect in this disorder. Recently, Borud, Torp, and Purkiss (1979) have demonstrated abnormal urinary polysaccharides in 9 patients with the disorder, indicating abnormal complex carbohydrate metabolism in the affected. The findings may indicate some abnormality of lysosomal enzymes similar to that found in the mucopolysaccharidoses, but further research is necessary before any final conclusion can be drawn.

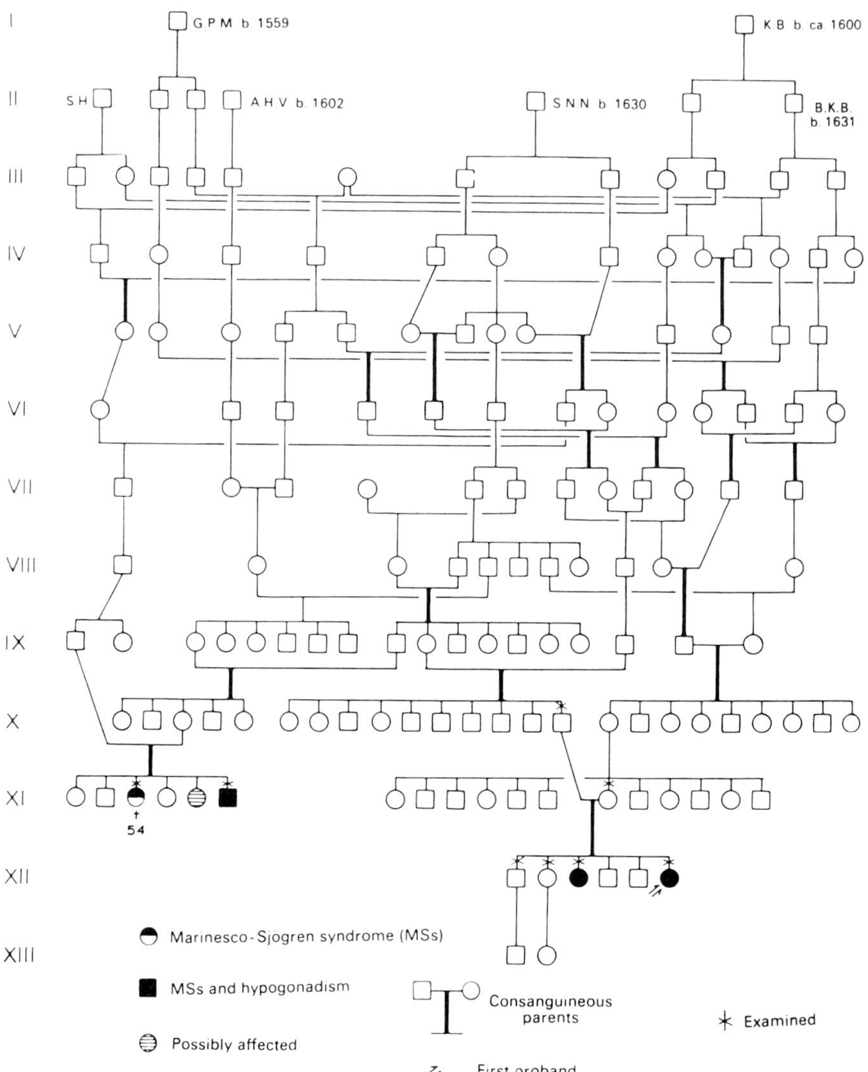

FIGURE 5.18. Pedigree showing extensive inbreeding through generations (genetic isolate in western Norway) and several cases of Marinesco-Sjogren syndrome. In three of the four examined cases a hypergonadotrophic hypogonadism was found. Case XI,3 came to autopsy. Compare Figures 5.19–5.22. (Skre et al., 1976. © 1976 Munksgaard International Publishers Ltd., Copenhagen, Denmark.)

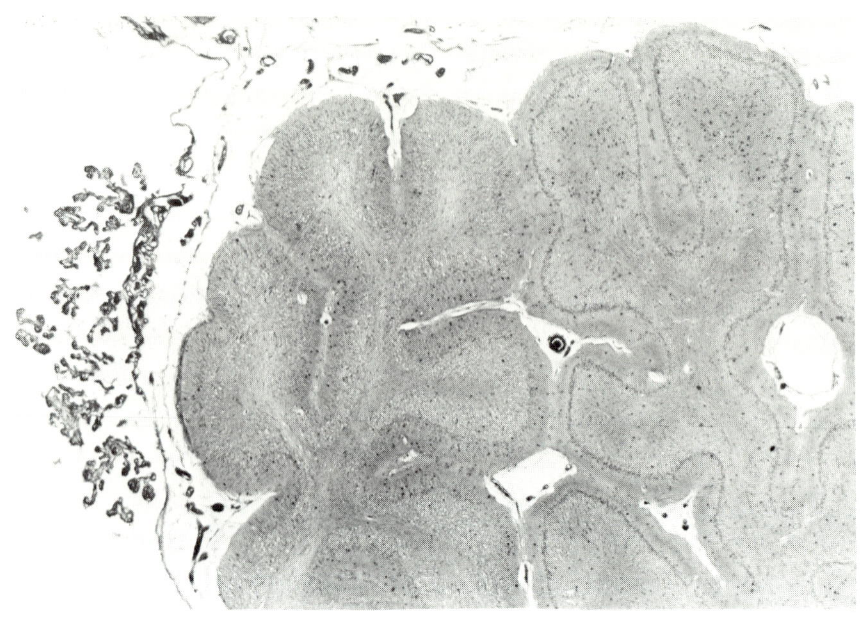

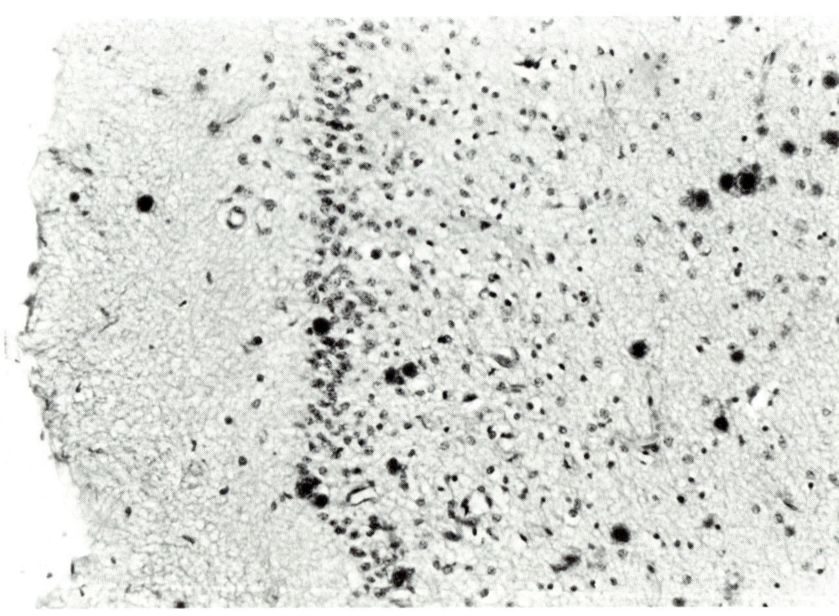

FIGURE 5.19. (**A**) Case XI,3 (see Figure 5.18). Sagittal section through vermis of the cerebellum into the fourth ventricle (note the choroid plexus). Severe atrophy of the cerebellar cortex. (Hematoxylin-erythrosin-saffran stain, × 17). (**B**) The same section at higher magnification (× 175). Complete absence of Purkinje cells and severe reduction of neurons in the molecular and granular layer. Note gliosis and numerous corpora amylacea. (Courtesy of Professor A. Myking, The Gade Institute of Pathology, University of Bergen.)

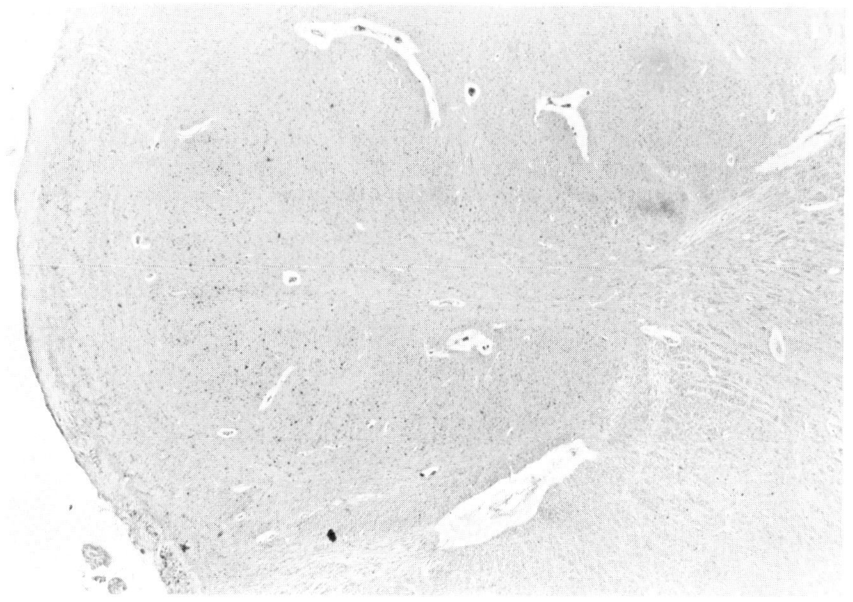

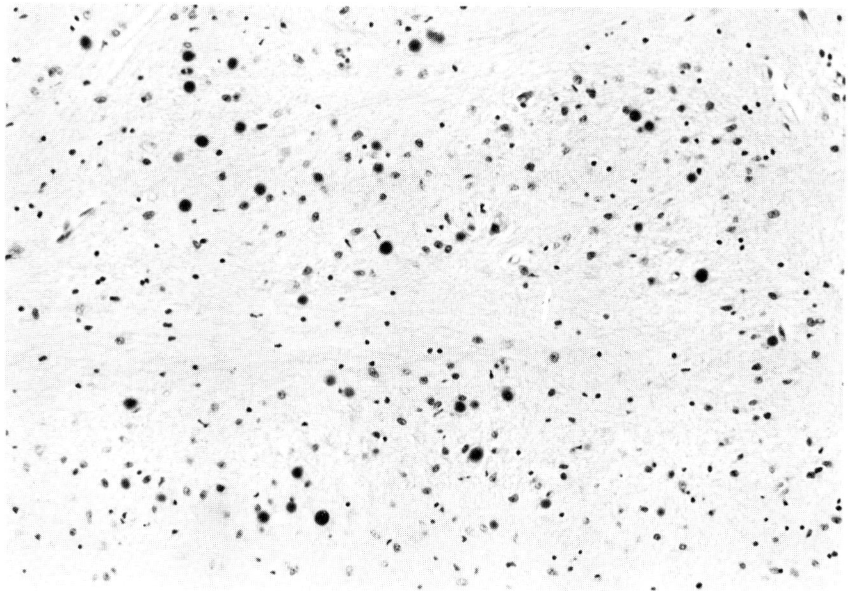

FIGURE 5.20. (**A**) Case XI,3. Cross section of medulla oblongata through the inferior olivary nucleus. (Hematoxylin-erythrosin-saffran stain, ×22.) Compare with Figure 20**B** showing the same section at higher magnification (×175). Complete absence of neurons, gllosls, and numerous corpora amylacea. (Courtesy of Professor A. Myking, The Gade Institute of Pathology, University of Bergen.)

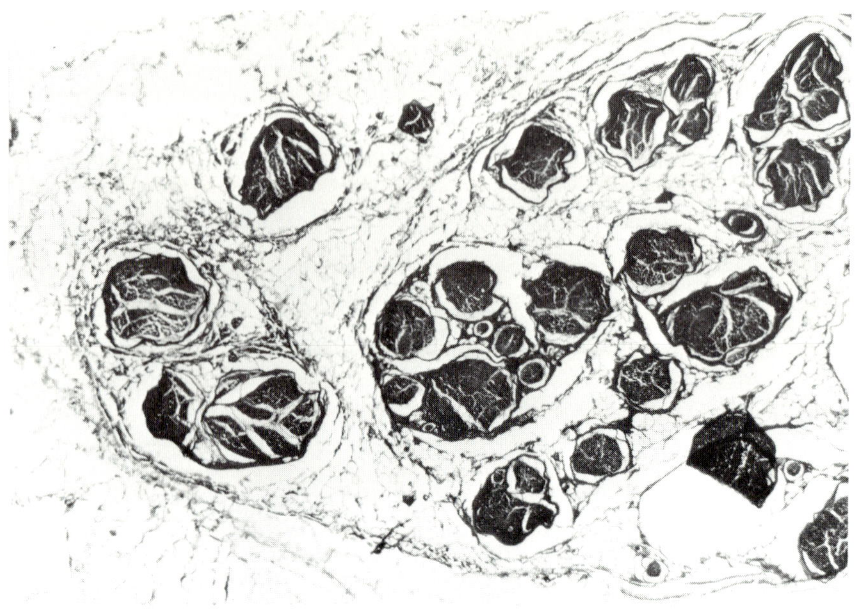

FIGURE 5.21. Case XI,3. Cross section of the sciatic nerve. Severe atrophy and fatty replacement of nerve bundles. (Hematoxylin-erythrosin-saffran stain, × 17.) (Courtesy of Professor A. Myking, The Gade Institute of Pathology, University of Bergen.)

Diagnosis. The combination of congenital cataracts and infantile progressive ataxia should make the diagnosis apparent. A CAT scan should be helpful in detecting the cerebellar atrophy.

Therapy. Only palliative treatment is available. Physiotherapy and proper management of distributed bowel and bladder functions is essential. No causal treatment is available at present.

Genetics. Marinesco-Sjögren syndrome is genetically determined, and aggregation of cases in sibships provides evidence for recessive inheritance. A high consanguinity rate in parents of cases has been reported, which is in agreement with recessive inheritance with a low frequency of the gene responsible for the disease.

Chromosomal analyses have revealed normal karyotypes in these patients (Skre and Berg, 1977). Further clarification of the biochemical defect in patients could perhaps make it possible to detect heterozygotes and enable prenatal diagnosis to be made in pregnancies at risk.

Deficiency of Serum Low-Density Lipoprotein and Ataxia

Bassen-Kornzweig Syndrome (Abetalipoproteinemia)

Bassen and Kornzweig (1950) reported a progressive neurological disorder in two young patients exhibiting ataxia, retinitis pigmentosa, and acanthocytosis. Hyphocholesterolemia was observed by Jampel and Falls (1958), and Salt et al. (1960) found a deficiency of low-density lipoprotein (LDL), or beta-lipoprotein, in serum and a histological abnormality in the small bowel mucosa.

Onset, Symptoms, and Signs. The disorder is rare and manifests itself in early childhood as an intestinal disorder much like celiac disease. This is later followed by clumsiness and ataxia. The full syndrome is usually attained in the third year of life. The neurological disorder runs a slowly progressive course with the addition of cerebellar signs, intention tremor, loss of vibration sense and tendon reflexes, muscular hypotonia and weakness, and nystagmus of an oscillating type (as in disorders of the retina). The intestinal disorder is characterized by abdominal distention, steatorrhea, and a defect

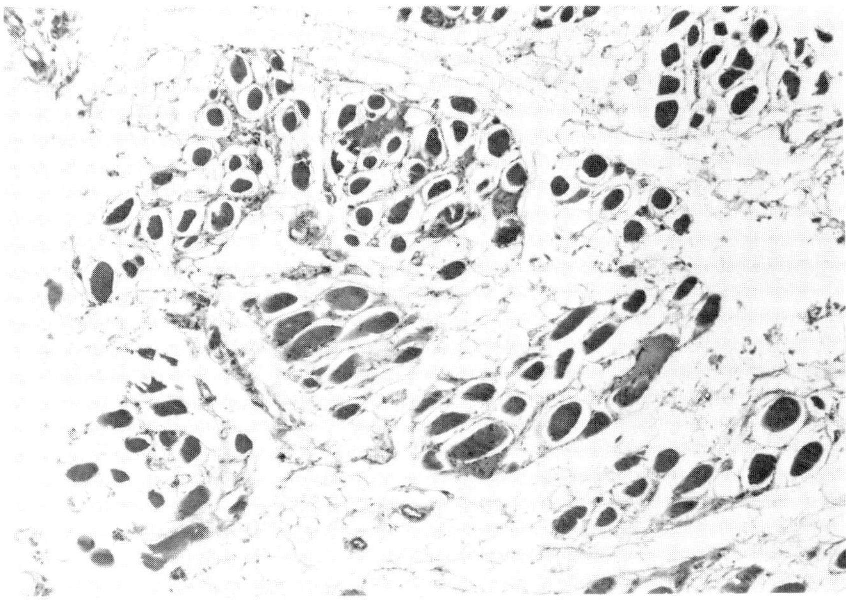

FIGURE 5.22. Case XI,3. Section from the diaphragm. Advanced stage of muscular atrophy. The histological pattern is more compatible with a primary myopathy than with a neurogenic atrophy. (Hematoxylin-erythrosin-saffran stain, × 70.) (Courtesy of Professor A. Myking, The Gade Institute of Pathology, University of Bergen.)

in the absorption of vitamins A, E, and K. Many patients are bedridden by age 20 (Salam, 1975).

Pathology. Pathoanatomically there is a loss of Purkinje and granular cells in the cerebellum, loss of anterior horn cells, and degeneration of the posterior and lateral columns, including the spinocerebellar tracts (Salam, 1975). The pathogenic mechanism is believed to be an inability to synthesize certain proteolipids, which may lead to a lipid membrane defect in several organs (Herbert, Gotto, and Fredrickson, 1978).

Biochemistry. Patients who lack LDL also lack the ability to form chylomicrons. More than 90 percent of the protein of LDL is apoB, which is also a major protein of chylomicrons and very low-density lipoproteins (VLDL). It is believed that most likely some failure of the synthesis of apoB—or of the intracellular assembly of apoB-containing lipoproteins—is the primary defect in abetalipoproteinemia, but this has not been established conclusively (Herbert, Gotto, and Fredrickson, 1978). LDL abolition may have a widespread effect on cell membranes. Fredrickson, Goldstein, and Brown (1978) maintain that LDL is taken up by specific receptors at the cell membrane and that cholesteryl esters may regulate the cellular cholesterol synthesis, and possibly the membrane sterols. The total lacl of LDL could cause damage to several cell systems in Bassen-Kornzweig disease and be responsible for the complicated clinical picture seen in this disorder. Lack of vitamin E has been focused on in later years as a cause of the neuroretinal signs in abetalipoproteinemia. It depends on chylomicrons for its absorption and LDL for its transport and is undetectable in serum of patients (Muller and Lloyd, 1983). Vitamin E–deficient chickens develop ataxia, and myopathies and neurological deficits have been found in other species (Muller and Lloyd, 1983).

Diagnosis. The diagnosis should be suspected from the clinical picture, and it is secured by detection of absence of serum LDL, most convenient by immunoelectrophoresis (see Berg, 1979). Lipid droplets are regularly seen in small bowel cells from the tips of the villi. After a fatty meal there is failure of fat absorption into the mucosa cells and lack of chylomicrons in the blood. Steatorrhea is present from birth and so is acanthocytosis, whereas the pigmentary retinal degeneration develops toward the end of the first decade along with the CNS changes.

Treatment. In recent years dietary regimens with the restriction of triglycerides containing long-chain fatty acids have been tried to relieve the gastrointestinal symptoms of infants with Bassen-Kornzweig syndrome. It seems apparent that these patients are unable to cope with this type of fat since they lack the ability to form chylomicrons. Triglycerides eventually accumulate in the mucosa cells and cause the abnormal appearance seen at biopsy. However, medium-chain triglycerides are normally absorbed into the portal system without the formation of chylomicrons and have been tried with a good result as a substitute (Herbert, Gotto, and Fredrickson, 1978). In addition, a supplement of fat soluble A and K vitamins should be given.

Treatment with vitamin E (Muller, Harries, and Lloyd, 1970) has been credited with improvement observed in retinal and neuromuscular abnormalities present in this disorder. Muller, Lloyd, and Bird (1977) reported similar results after prolonged treatment with vitamins A, E, and K and fat restrictions. Other reports seem to confirm these results (Azizi et al., 1978; Illingworth, Connor, and Miller, 1980).

Genetics. Inheritance is autosomal recessive, and parents tend to be related. No chromosomal abnormality has been detected.

Hypobetalipoproteinemia in "Friedreich's Ataxia"

In 1974 Aggerback, McMahon, and Scanu described low serum cholesterol and hypobetalipoproteinemia in four members of a sibship of seven, possibly also in the mother. The proband and two other affected sibs had neurological symptoms reminiscent of Friedreich's ataxia, with ataxic gait, sluggish tendon reflexes, extensor plantar responses, and pes cavus. Some dysarthria was present in the proband as well as sensory loss in the extremities, as in polyneuropathies. Conduction velocity in peripheral nerves was borderline low.

No abnormality was found in the small bowel mucosa, and there were no signs of retinitis pigmentosa or acanthocytosis.

Neurological signs in hypobetalipoproteinemia are often confined to muscular abnormalities, but a few patients may have retinopathy. Signs of the disorder may be present in the heterozygous state. In the homozygous state, which is rare, the disorder may resemble an abetalipoproteinemia, although it is considered to be different from this condition. Genetically, the disorder is considered to be an autosomal dominant disorder with limited penetrance (Herbert, Gotto, and Fredrickson, 1978).

Ataxia-Telangiectasia

Ataxia-telangiectasia (AT) was described in 1941 in a single case reported by Louis Bar (Louis-Bar syndrome). It was further characterized in the 1950s, especially through the writings of Boder and Sedgwick, who had reviewed 101 cases by 1963.

Onset, Symptoms, and Signs. As yet, AT is defined purely on clinical grounds even though immunologic and biochemical studies are increasing our understanding of probable pathogenetic mechanisms.

The patient is brought to medical attention by slow development in locomotion and fine motor skills. Others may present themselves with a high frequency of sinopulmonary infections.

Incoordination of the limbs, tremor, and dysdiadochokinesia is present in the typical patient, along with hyperkinesias (choreoathetosis and orofaciomandibular dyskinesias), and dysarthria. The muscles are hypotonic, and tendon reflexes are abolished. Pursuit movements of eyes are jerky. Intellectual decline starts in prepuberty. Comparatively late manifestations are also the characteristic telangietatic lesions in the conjunctiva, ears, and exposed parts of the neck. Even other manifestations of phakomatous type may appear (Salam, 1975). Some show hypoplastic genitalia.

Course and Prognosis. Ataxia starts in early infancy and leads to complete disability. The hyperkinesias are usually evident from five years, as are the cutaneous lesions. Intellectual decline begins at about 10 years, but this may in part be due to sensory deprivation in a chronically ill child. The patients are usually bedridden by the age of 12 to 20 years. Life expectancy is dramatically reduced owing to infections or malignancies of various kinds. Patients often die from tumors of which reticuloendothelial neoplasias are dominating, but gliomas are no rarity.

Epidemiology. No unequivocal data are as yet present as to incidence, prevalence, and gene frequencies. Probably the condition is more common than the relatively few available reports may indicate. The diagnosis is easily missed since the cutaneous manifestations are often inconspicuous. The child may therefore pass as a case of cerebral palsy or Friedreich's ataxia.

Pathology. The predominant abnormality within the CNS is seen in the anterior cerebellar lobe, which is atrophic. Histologically, there is a widespread atrophy of Purkinje, basket, and granule cells. Degeneration is also present in the posterior columns and the spinocerebellar tracts (Solitaire and Lopez, 1967; Aguilar et al., 1968).

Axonal degeneration is seen in peripheral nerves. Intracytoplasmic inclusion bodies have been observed in Purkinje cells, dorsal root ganglia, and the pituitary gland (see Salam, 1975). The thymus is either absent or hypoplastic (Boder and Sedgwick, 1963), and changes have been found in ovaries and pituitary acidophils (Bowden, Danis, and Sommers, 1963; Solitaire and Lopez, 1967).

A hypothesis of viral etiology for the CNS abnormalities has been proposed (Salam, 1975), whereas the immunodeficiency observed may be the primary defect. However, this explanation fails to account for all abnormalties—for instance, the cutaneous manifestations or insulin resistance.

Biochemistry and Immunology. Thieffry et al. (1961) demonstrated a deficiency of IgA in patients with AT. In 1971 McFarlin et al. (1971) was able to show this to be due to decreased synthesis. Regarding the immune system, it has further been noted that AT patients demonstrate abnormalities of both the cellular and humoral immune systems. They have absent or embryonic thymus glands, in vivo defects in delayed hypersensitivity, and cell-mediated immunity (Peterson, Cooper, and Good, 1966). In vitro, T cell proliferation in response to mitogens and antigens is decreased, as is proliferation in response to the autologous mixed lymphocyte culture. In the humoral system, additional abnormalities have been found, such as IgE deficiency (Amman et al., 1969) and lowered synthesis of low-molecular weight IgM, alone or in combination. In vivo and in vitro studies of antibody responsiveness show defects, and it is postulated that this could be due to defective T helper cell activity (Misiti and Waldmann, 1981). Kaufman and Miller (1977) found cytotoxic autoantibodies to brain and thymus in the serum of four of six AT patients.

Patients show increased serum levels of alpha-fetoprotein and carcinoembryonal antigens; but no fetal erythrocyte isoenzyme or abnormal erythrocyte protein values have so far been demonstrated (Richlind, Boder, and Teplitz, 1982). These results suggest abnormal endodermal development in these patients. Altered collagen formation is also observed (McReynolds et al., 1976). Bar et al. (1978) discovered extreme insulin resistance in AT due to insulin receptor insufficiency. Abnormal sensitivity to gamma radiation and certain cytotoxins has been shown, as well as reduced DNA repair capacity (Paterson et al., 1976; Ford, Houldsworthy, and Lavin, 1981; Coquerelle and Weibezahn, 1981).

Diagnosis. Diagnosis is rather easy if cutaneous manifestations are present in addition to ataxia. Lack of IgA or reduced concentration of this immunoglobulin is an important and constant finding. There is also a lymphopenia and an increased sedimentation rate regularly.

Therapy. Therapy is symptomatic and directed against the physical handicap and the secondary complications such as infections and neoplasms.

Genetics. The pattern of inheritance is probably autosomal recessive since affected relatives are confined to sibs. Pratt (1967) reported few cases of parental consanguinity in this disorder, whereas Kiran et al. (1973) observed consanguinity in 14 of 23 families. The latter observation gives credibility to a hypothesis of Mendelian inheritance.

Refsum's Disease

In 1946 Refsum described a clinical disorder that he called "heredopathia atactica polyneuritiformis" based upon observations from 1937 to 1943 of four patients from two families. Two of these patients died, and pathoanatomical studies revealed fatty deposits in neurons and degeneration of fiber tracts.

These patients were considered by the neuropathologist Cammermeyer (1946) to have suffered from a form of lipidosis. In 1963 Klenk and Kalhke discovered an abnormal fatty acid—3,7,11,15-tetramethylhexadecanoic acid (phytanic acid)—in serum and organ lipids from a patient who had died from this disorder; similar findings have been found in several patients with the typical syndrome (Refsum, 1975).

Onset, Symptoms, and Signs. In many instances patients are unable to tell when symptoms appeared and what the first sign was. Retrospectively, many find that night blindness could have been the start or that unsteady gait had manifested at a very early age, usually at night.

Skin dryness, itching, or desquamation were early symptoms in Refsum's original Cases 3 and 4 (Refsum, 1975). Later, other symptoms prevailed such as difficulties in walking and hearing.

Many patients notice tingling, numbness, or sensory loss in fingers, hands, and feet; also weakness in these parts.

Typical findings in these patients are restricted fields of vision, night blindness, and a retinitis pigmentosa of the salt and pepper type. Retinal vessels seem atrophic and discs pale. Visual acuity is reduced. There is central hearing loss. Patients have a sensorimotor polyneuropathy with flaccid muscles and hypo- or areflexia. In some hyperalgesia may be present in feet. Gait tends to be broad based and unsteady, whereas ataxia is less conspicuous in hands.

Pupils were small in Refsum's first patient; and in the patient described by Nordhagen and Grøndahl (1964), and later by Skre et al. (1969), iridodonesis was observed. Cataracts may develop later in the course of the disease. Anosmia is a rather frequent finding.

Xeroderma or ichthyosis is often observed. The CSF protein level is invariably elevated, but there is no pleocytosis.

ECG abnormalities are prominent. There are increased P-Q intervals as well as changes of the S-T segment and T wave in several leads.

X ray shows disturbed epiphyseal growth leading, in some cases, to irregular articular surfaces or to retarded growth (Skre et al., 1969).

Disease onset is insidious, but, as mentioned, often appears very early in life. Some patients are unable to provide any definite statements on this point. In some cases the disorder may have an acute or subacute debut corresponding to the intermittent disease progression usually experienced. In the patient described by Skre et al. (1969) symptoms appeared at the age of four with severe affection of locomotion and specific senses, but the parents held that some unsteadiness had persisted since early infancy. On the other hand, symptoms may appear in adult life; in a patient reported by Gautier et al. (1973) the debut of symptoms came only at 52 years.

Course, Duration, and Prognosis. Typical to the condition is acute exacerbations of symptoms with more or less complete remissions later. Skre et al. (1969) reported severe loss of hearing, night blindness, and blurred vision at the age of four during such an exacerbation, and the patient never again attained a steady gait. His later disease course was serious, with sudden death at the age of 17. Sudden death has been observed in several cases (see Refsum's review, 1975) after a relatively short disease course. Fryer et al. (1971) carefully observed such a patient in her final stage. They found her suddenly apneic and pulseless; and eventually she died. Although an autopsy was carried out, her death could not be satisfactorily explained, but evidently it came during an exacerbation of the disease and while dietary treatment was attempted. In the patient mentioned previously (Skre et al., 1969) death came during an exacerbation when serum phytanic acid concentration skyrocketed to 232 mg/100 ml (Eldjarn, Stokke, and Try, 1977) (Figure 5.23). However, early onset of the disease does not ensure a severe disease course in every instance. In some patients the disorder may come to a standstill, perhaps for several decades; only partial clinical pictures are seen in others, not interfering with life span.

Prognosis seems to have improved considerably after the introduction of diet regimens.

Epidemiology. The disorder is rare. It is the most common in Scandinavia, the British Isles, and France, but it is also seen in Germany and in other countries in western Europe. A few cases have been described from the American continent, probably in Caucasians.

Pathology. The main pathoanatomical features are changes of the peripheral nerves and nerve roots characterized by so-called hypertrophic interstitial neuritis. However, degenerative changes in myelinated fibers are also seen in the CNS, notably around the inferior olivary nuclei but also as patchy degeneration accompanied by fat accumulation in macrophages in several other parts of the brain stem and around the dentate nucleus.

Lipid storage is evident in the leptomeninges, ependyma chorioideum, liver, and kidneys; fat accumulation is also prominent in the pallidum. Lipofuscin deposits in neurons are striking.

The hypertrophic neuritis of the peripheral nerves leads to the so-called onion bulb formation resulting in destruction of the neurons. Retrograde change in anterior horn motor cells is observed as well as muscular atrophy (for a review, see Cammermeyer, 1975).

Biochemistry. The principal biochemical defect in Refsum's disease is the lack of the enzyme phytanic acid alpha-oxidase. This results in a block in the degradation of phytanic acid (Figure 5.24), which then accumulates in fatty deposits in different parenchymatous organs of the body. Phytanic acid originates from foods like vegetables and bovine fats and is not synthesized in the body (Eldjarn, Stokke, and Try, 1977). Recently, Billimoria et al. (1982) showed that patients are able to metabolize approximately 10 mg phytanic acid a day by way of so-called omega oxidation—an alternative, but inefficient, metabolic pathway to alpha oxidation. Immunological mechanisms may

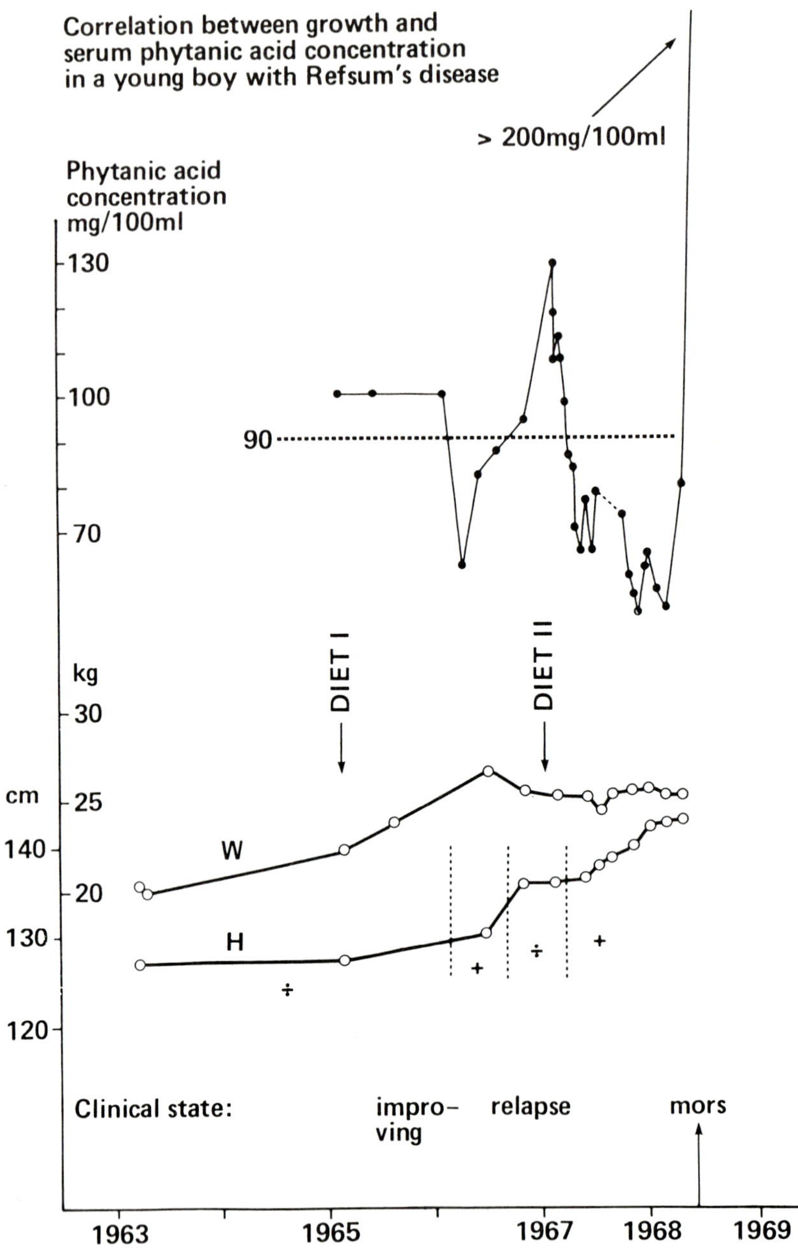

FIGURE 5.23. In this diagram the result of dietary regime in Refsum's disease is shown (Skre et al., 1969; Eldjarn, Stokke, and Try, 1977). When the serum phytanic acid level fell below approximately 90 mg/100 ml, growth resumed. Cardiac arrest occurred when the serum phytanic acid level rose above 200 mg/100 ml. (Modified from Skre et al., 1969. By permission of Churchill Livingstone, Edinburgh.)

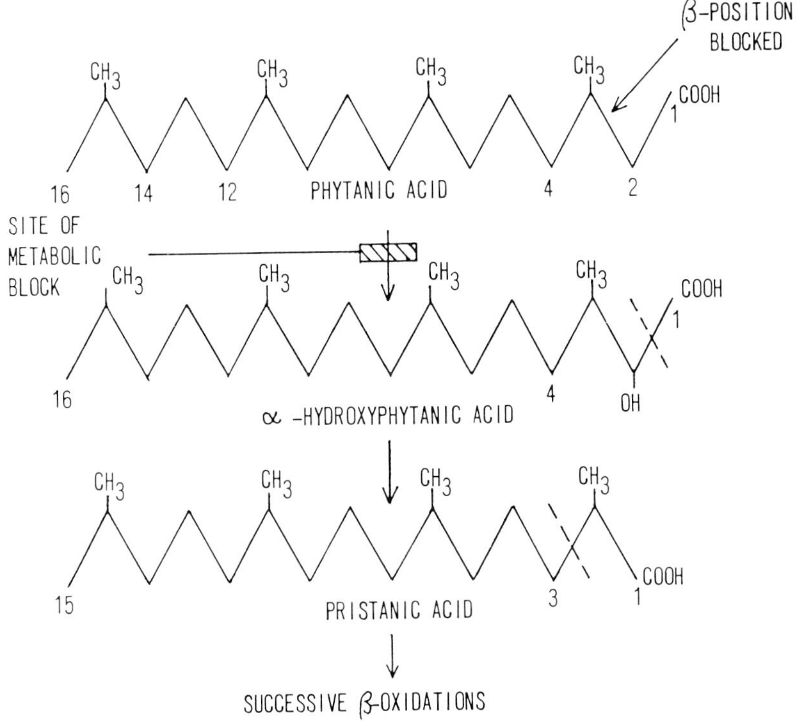

FIGURE 5.24. The normal degradation of phytanic acid. Since the beta position is blocked by a CH_3 group, normal betaoxidation cannot occur. In Refsum's disease alpha oxidation is impossible since the enzyme is lacking. In normal individuals phytanic acid is initially degraded through alpha oxidation, and so subsequent beta oxidation is possible.

play an important part in the pathophysiology of the neuropathy in Refsum's disease. Behan (1983) compares the exacerbations of the disease with Guillain Barré-syndrome, an autoimmune disorder. He contends that patients may become sensitized to peripheral nerve antigen from nonspecific nerve damage, since increased reactions of both humoral and cellular types are observed.

Diagnosis. A diagnosis of Refsum's disease may be easily missed if the clinical picture is incomplete. Eye symptoms may pass as some type of retinitis pigmentosa, and the polyneuropathy present is also an unspecific sign. The combination of these signs and of an additional albuminocytologic dissociation in the spinal fluid is very typical and gives the diagnosis in most cases—

but only in the presence of elevated levels of phytanic acid in the serum.

Cases of so-called Refsum's syndrome have been described, but in every instance atypical clinicopathological features were present and the phytanic acid level was normal (Refsum, 1975).

Scotto et al. (1982) found elevated serum phytanic acid in three children who were mentally retarded, presented with abnormal facial features, and showed retarded growth (in two of them). Typical to these children were onset of symptoms shortly after delivery with vomiting and hepatomegaly. The disorder was studied carefully with biopsy techniques and morphological and biochemical investigations. In mesenchymal cells (Kupffer's, machrophage) and hepatocytes they found inclusions containing lipid, glycogen, and trilaminar-structured lamellae, probably of a kind similar to that found in plant chloroplasts known to contain bound phytol. In serum from one patient was found hypobetalipoproteinemia and hypocholesterolemia; from another, hypoalphalipoproteinemia. All patients had osteoporosis and visual failure as well as retinitis pigmentosa and neural hearing loss. However, signs of peripheral neuropathy were sparse or nonexistent, conduction velocities were normal, and muscle and nerve biopsies were inconclusive. CSF protein was normal in two cases examined. The authors regard these cases as etiologically similar to, but clinically different from, Refsum's disease.

Usually, nerve conduction velocities are markedly reduced in Refsum's disease, but this is not a constant finding (Salisachs et al., 1982). Ultrastructural studies of peripheral nerves show nonspecific changes (Cammermeyer, 1975).

Treatment. Restriction of the intake of phytanic acid is possible, and patients placed on such a diet have improved considerably (Eldjarn, Stokke, and Try, 1977). However, a young boy described by Skre et al. (1969) died during such treatment. He refused the recommended restricted regimen, starving himself, which probably precipitated a mobilization of fat deposits and a rapid increase in phytanic acid concentration in the serum, eventually causing cardiac arrest (Figure 5.23). Lundberg, Lilja, and Try (1972) treated such a patient successfully with plasma exchange.

Genetics. The disorder segregates as an autosomal recessive trait. Consanguinity between parents is frequently observed as in other rare recessive disorders. Cultured fibroblasts show a moderate enzymatic deficiency in heterozygotes but a total metabolic block in homozygotes (Herndon, Steinberg, and Uhlendorf, 1969), which has important implications for genetic counseling.

SOME OTHER ATAXIAS WITH METABOLIC ABNORMALITIES

In this section we shall briefly deal with some conditions in which hereditary factors play a certain role, seemingly interconnected by a common pathogenetic concept in that they all may represent a state of mitochondrial dysfunction. In addition to the entities described below, other disorders could have been discussed, but they are sacrificed here for the purpose of clarity.

Still it deserves mention that there are other inherited disorders in which cerebellar symptoms occur. In some of these ataxia patients, we have a fair degree of knowledge about pathogenetic mechanisms, as in the hexosaminidase deficiency diseases (Tay-Sachs disease) presenting with cerebellar symptoms (Johnson, 1981); the spinocerebellar degeneration associated with cystic fibrosis and chronic childhood cholestasis, responding to vitamin E treatment (Elias, Muller, and Scott, 1981); the periodic ataxias (Hill and Dysart, 1975); and the rare cases of adrenoleukomyeloneuropathy presenting with ataxia, dementia, and cortical blindness (Marsden, Obesco, and Lang, 1982).

Abnormalities of the gamma-glutamyl cycle, which is important in the cellular transport mechanisms of certain amino acids, may give rise to clinical syndromes similar to those of the spinocerebellar disorders (Meister, 1978). One of these conditions was described 15 years ago by Jellum et al. (1970); further investigations have identified the metabolic block responsible for the condition as a 98 percent reduction in activity of the enzyme glutathione synthetase (Marstein et al., 1976).

Kearns-Sayre Syndrome

The field of *opthalmoplegia plus* (the term was coined by Drachman in 1968) is a complex one, and one is bound to conclude—like Drachman (1975) and Rosenberg et al. (1968)—that the only valid classification must depend on etiologic differences alone. The syndrome presented by Kearns and Sayre in their original paper from 1958 may serve as an example picked from a heap of syndromes classified by opthalmologists (and some neurologists) as (external) "opthalmoplegia plus" (Drachman, 1975; Rowland, 1975; Walton and Gardner-Medwin, 1981) and by some neurologists (Baraitser, 1982) as "ataxia plus," the *plus* representing what is perceived as additional symptoms (retinal degeneration, opthalmoplegia, ataxia and other CNS symptoms, heart conduction abnormalities, and peripheral neuropathy).

Onset, Symptoms, Signs, and Prognosis. The syndrome originally described by Kearns and Sayre (1958) presented with the following constellation of symptoms in addition to the external opthalmoplegia: pigmentary retinal degeneration (pigment clumps at some distance from the optic disc, differing from retinitis pigmentosa), impaired vision, progressive growth failure, delayed menstruation and secondary amenorrhea in the female, progressive heart block, skeletal muscular weakness, progressive cerebellar ataxia, and neural deafness. Less often, dementia and pyramidal signs are included in the syndrome. The disease often commences in childhood, leading a progressive course. Among fatal complications, asystolia on the basis of heart conduction block may be the most dangerous. No certain data are available as to life expectancy, but it is certainly reduced.

Pathology. Kearns and Sayre (1958) found siderosis of the globus pallidus and marked spongiform degeneration of the central white matter. Some patients show degenerative changes in the oculomotor nuclei, and some patients do not (Rowland, 1975); it is not yet clear whether the ocular myopathy should be considered myogenic or neurogenic in origin. Olson et al. (1972) describe characteristic findings in skeletal muscle biopsy, including clusters of normal and abnormal mitochondria and lipid droplets ("ragged red fibres"). This finding points toward the muscular symptoms being myogenic in origin. Identical mitochondrial findings have been made in biopsy of the cerebellum (Schenck et al., 1973). Degenerations of the posterior columns and ventral horn neurons have also been described (Stephens, Hoover, and Denst, 1958). The findings may be explained by a theory of mitochondrial dysfunction involving both muscular and neural tissue.

Biochemistry. DiMauro et al. (1973) have shown a decreased ability of the mitochondria to divert electrons from alpha-glycerophosphate into the electron transport chain via flavoprotein in some, but not all, patients with this disorder.

Diagnosis. As long as there is no general agreement as to classification, the diagnosis may become somewhat arbitrary but should hinge on prominent clinical features and muscular pathology showing "ragged red fibers." The protein content of CSF is generally elevated. Both Refsum's and Bassen-Kornzweig syndromes are possible differential diagnoses and should be excluded.

Therapy. No specific treatment is given. Special care should be taken to prevent Adams-Stokes episodes. An artificial pacemaker may be required.

Genetics. Most cases seem to be sporadic. Consanguinity of the parents has been noted.

Leigh's Syndrome

Subacute necrotizing encephalopathy (SNE) is a more homogeneous clinical entity. It was described by Leigh (quoted by Lowenthal, 1981) in 1951, occurring in a child who died at seven months, and it is defined partly by clinical and partly by biochemical criteria.

Onset, Symptoms, Signs, and Prognosis. The symptoms are usually progressive psychomotor slowing, ataxia and nystagmus, muscular hypotonia, peripheral neuropathy, and occasional signs of cranial nerve affection (among which the second and third are most commonly hit). Epileptic seizures may occur. The course is characterized by exacerbations and remissions. Hyperpyrexia to a varying degree is seen during exacerbations. The evolution of the disease may take months or even years (Lowenthal, 1981).

Commonly, this is a disease of infants and children. However, juvenile cases are reported regularly, and even several adult cases have been reported. In juvenile and adult cases the subacute progression is often preceeded by a chronic phase in which different central or peripheral neurological signs—usually one or more of the above given—will occur.

In the terminal stages there is deterioration to coma and marked respiratory depression.

Pathology. The microscopic lesions are found mainly in the basal ganglia (striatum, thalamus, and hypothalamus) and the brain stem (Sipe, 1973; Lowenthal, 1981). Lesions in the cerebellum and the spinal cord are not uncommon (David, Mamunes, and Rosenblum, 1976). Earlier authors hold that the corpora mamillaria are generally spared, in contrast to what is found in Wernicke's encephalopathy, whereas others point to a general sparing of the substantia nigra in the latter condition as different from the findings in Leigh's syndrome (David, Mamunes, and Rosenblum, 1976). The prevailing opinion is that the dendrites and myelin are affected earlier than cell somata or axons proper, and there is an early loosening of the neuropil.

Biochemistry. It was an early finding that the concentrations of alanine, pyruvate, and lactate were elevated in serum, urine, and CSF in Leigh's syndrome. This incriminates the enzyme pyruvate dehydrogenase, and in fact this enzyme is often found to function improperly in patients with SNE, measured in liver (Gruskin et al., 1973; Grobe, Bassewitz, and Dominick, 1975). In one patient, however, pyruvate dehydrogenase activity was found to be normal in a liver biopsy early in the disease, whereas activity was absent at necropsy of the same patient. Cooper, Itokawa, and Pincus (1969) found an inhibitor of thiamine pyrophosphate adenosine triphosphate (ATP) phosphotransferase, an enzyme catalyzing the production of thiamine triphosphate (TTP) from thiamine pyrophosphate (TPP). This inhibitor is also present in fibroblasts from patients—but not from heterozygotes, although it is excreted in the urine of heterozygotes. The relevance of this finding awaits clarification since the relative functions of TTP and TPP in the CNS are not clear. The similarity in the clinicopathologic picture between Leigh's and Wernicke's syndromes points toward common pathogenetic elements, making the observed abnormalities in the metabolism of thiamine interesting. Recently, Sorbi and Blass (1982) have shown an abnormal activation of the pyruvate dehydrogenase complex by dichloracetate in fibroblasts from Leigh's syndrome patients. There has also been demonstrated cytochrome *c* oxidase deficiency in an autopsy-proven patient with Leigh's syndrome (Willems, Monnens, and Trijbels, 1977).

Diagnosis. SNE is suspected on clinical grounds, especially in young patients. A lactic acidosis confirms the suspicion. CSF total protein level is often increased. The EEG is not very informative, but the EMG often shows slowing of neural conduction velocity. Detection of the phosphotransferase inhibitor mentioned above may prove useful, but false positives and false negatives have been reported (Pincus and Solitare, 1976).

Therapy. Thiamine in pharmacological doses has been reported to relieve clinical signs, and reduce the amount of phosphotransferase inhibitor excreted in the urine. However, spontaneous remissions occur, and the effect is difficult to evaluate. Lipoic acid, an important coenzyme of the pyruvate dehydrogenase complex, has been reported to have effect in some cases (Lowenthal, 1981).

Genetics. The infantile and childhood forms are usually inherited in an autosomal recessive mode. Some of the juvenile patients also seem to fit this pattern, whereas most juvenile and all adult forms are sporadics.

EMERGING PATHOGENETIC CLASSIFICATION OF SOME HEREDITARY ATAXIAS WITHIN MITOCHONDRIAL MULTISYSTEM DISORDERS

It should come as no surprise that an increasing number of dysfunctional and degenerative disorders of muscle and nervous tissue now turn out to be related to impairment of normal mitochondrial function. Neurons and muscle cells are among the highest energy consumers within the body's cell populations, and naturally an impairment of cellular "power supply" will exert profound influence on the performance of skeletal, ocular, and cardiac muscle, peripheral nerves, and central nervous structures. In Kearns-Sayre syndrome and Leigh's syndrome, as in Wernicke's encephalopathy and possibly LCCA, preliminary biochemical data are pointing toward defects in the mitochondrial pyruvate translocation and decarboxylation, Krebs cycle and the electron transport chain, all of which are important sequences in the formation of ATP from glucose and oxygen in nervous tissue. Ptosis and paresis of extraocular muscles (ocular "myopathy") are no rarities in the spinocerebellar ataxias (Skre, 1975c). Several hereditary myopathies show signs of dysfunction of the CNS as well, among which ataxia and nystagmus are quite common, and signs of cerebellar dysfunction in myopathies seem to be particularly common where the primary defect is pinpointed to mitochondrial energy metabolism. Morgan-Hughes (1982) and colleagues (Morgan-Hughes et al., 1982) review these disorders, and Stumpf (1979) provides comprehensive overview of the widespread nature of mitochondrial disorders.

In this light we shall examine the case of Friedreich's ataxia somewhat more closely.

An effort to interconnect the various paths of chemical traits described in the section on Friedreich's ataxia leads us to the central energy metabolism used in nervous tissue: glycolysis followed by the mitochondrial citrate circle and the electron transport chain. Barbeau (1982) postulates an energy deprivation state in mitochondria as the focal point of a model for understanding the pathogenesis of FA.

Figure 5.16 shows that a slowing down of acetyl-CoA production from pyruvate due to reduction of activity in the pyruvate dehydrogenase complex (PDHC) leads to decrease of ATP production along the electron transport chain (ETC) (for details, see Figure 5.25). In the nerve cell this will have disastrous consequences in that anabolic activities will lack energy, as will the membrane ionic pumps, leading to decreased ability to participate in intercellular communication by depolarization/repolarization and transmitter production and release activities. Under cellular stress conditions, this lack of energy may lead to cell death. One may presume that reduced access to acetyl-CoA as fuel will force the citrate cycle to exploit other resources, such as alpha-keto-glutarate formed by transamination from glutamate and oxalacetate formed from aspartate. In fac , experiments indicate that developing hypoglycemia correlates with declining cerebral levels of glutamate and GABA and rising levels of aspartate, whereas recovery leads to subnormal levels of aspartate. The reason for this presumably lies in the fact that the transamination creating aspartate also produces alpha-keto-glutarate, whereas glutamate is consumed (see Wilkinson and Prockop, 1976). Lack of acetyl-CoA may also lead to reduced access to acetylcholine and to decreased de novo production of fatty acids, which may impair membrane and myeline maintenance.

However, the PDHC activity defect does not seem to be the primary defect in FA, since the activity is only moderately reduced and reduction is not consistently found in different patients and different tissues (Barbeau, 1982). The report of Stumpf and collaborators (1982) that mitochondrial malic enzyme (ME_m) activity is reduced by some 90 percent in some FA patients may be of significance because ME_m is known to be implied in some way in the regulation of the citrate cycle (Lee and Davis, 1974; Andres, Satrustegui, and Machada, 1980; Davis, Spydevold, and Bremer, 1980). Thus PDHC activity fluctuations could be due to the regulatory effect of ME_m or other regulatory mechanisms. Since ME_m also participates in regulating the NADP/NADPH ratio, fatty acid synthesis may also be affected. The ME_m deficiency in FA patients will have to be confirmed by other centers.

Barbeau (1982) argues for a primary defect in the metabolism of linoleic acid (18 : 2), which is an important source for the coenzyme lipoic acid of the PDHC E_3 component (Figure 5.26) (Barbeau, 1979, 1982). This defect could be located either at the level of the intestinal mucosa (absorption of fatty acids, chylomicron formation), at the level of lipoprotein formation or metabolism, or at the level of exchange of fatty acids between cell membranes

Results from authors using experimental models of neurotoxins leading to neuropathy seem to be of interest here. If some specific antimetabolite or blocking agent could mimic disease, much could be deducted about the fundamental mechanisms involved in the disease process.

Triorthocresylphosphate seems to create a systemic degeneration similar to that seen in FA (Cavanagh, 1964). Intoxications with this organic phosphorus compound have been endemic in western countries for almost a century. Certain episodes, such as the "Jamaica ginger paralysis" in the United States in the 1930s and the Moroccan cooking oil contamination with hydraulic fluid in the 1950s have taken epidemic orders of magnitude (Morgan, 1982). Using this experimental model, Cavanagh shows that pathologic changes first occur in distal parts of the axon, proximal involvement being a later event.

The action of *3-acetylpyridine* has been described earlier. Its interference with NAD metabolism gives the experimental model using this substance special importance in the explanation of defects in cellular energy metabolism.

Toxic *hexacarbons* were used by Spencer and Schaumburg (1977), who demonstrated not only peripheral neuropathy in experimental animals but also degeneration of major spinal tracts, indicating that the "dying-back process" occurs simultaneously in certain vulnerable fibers, regardless of their localization in the peripheral or central nervous system. The authors proposed the term "central peripheral distal axonopathy" to identify disorders with this pattern.

Veronesi et al. (1978) were able to show that fiber diameter was more important than fiber length: damage first occurred in larger fibers. This finding supports Brown's (1962) comments pointing out that there is selective destruction of coarse and a relative preservation of fine fibers in peripheral nerves in FA patients. It is further known that the energy demands of the axon are especially high at the nodes of Ranvier, since this is a portion of the nerve cell with high ionic pump activity. In axonopathies, material from protein synthesis and other anabolic activity near the perikaryon under transport to more distal portions is known to accumulate at the nodes of Ranvier. Thus Spencer and collaborators (1979) postulate "energy depletion" of the axon as a possible etiologic factor in the development of axonopathies, leading to block in energy-dependent axon transport and ionic pump mechanisms. In this way the postulated mitochondrial energy depletion may explain slowing of sensory nerve conduction (in large, myelinated fibers) in FA.

FIGURE 5.25. Details of the pyruvate dehydrogenase complex where lipoic acid is needed as a cofactor. The pyruvate dehydrogenase complex converts pyruvate into acetyl-CoA, which is the fuel of the Krebs cycle in mitochondria. A feedback mechanism regulates the system, as acetyl-CoA inhibits the activator system E_3. (Skre, 1983. By permission of Churchill Livingstone, Edinburgh).

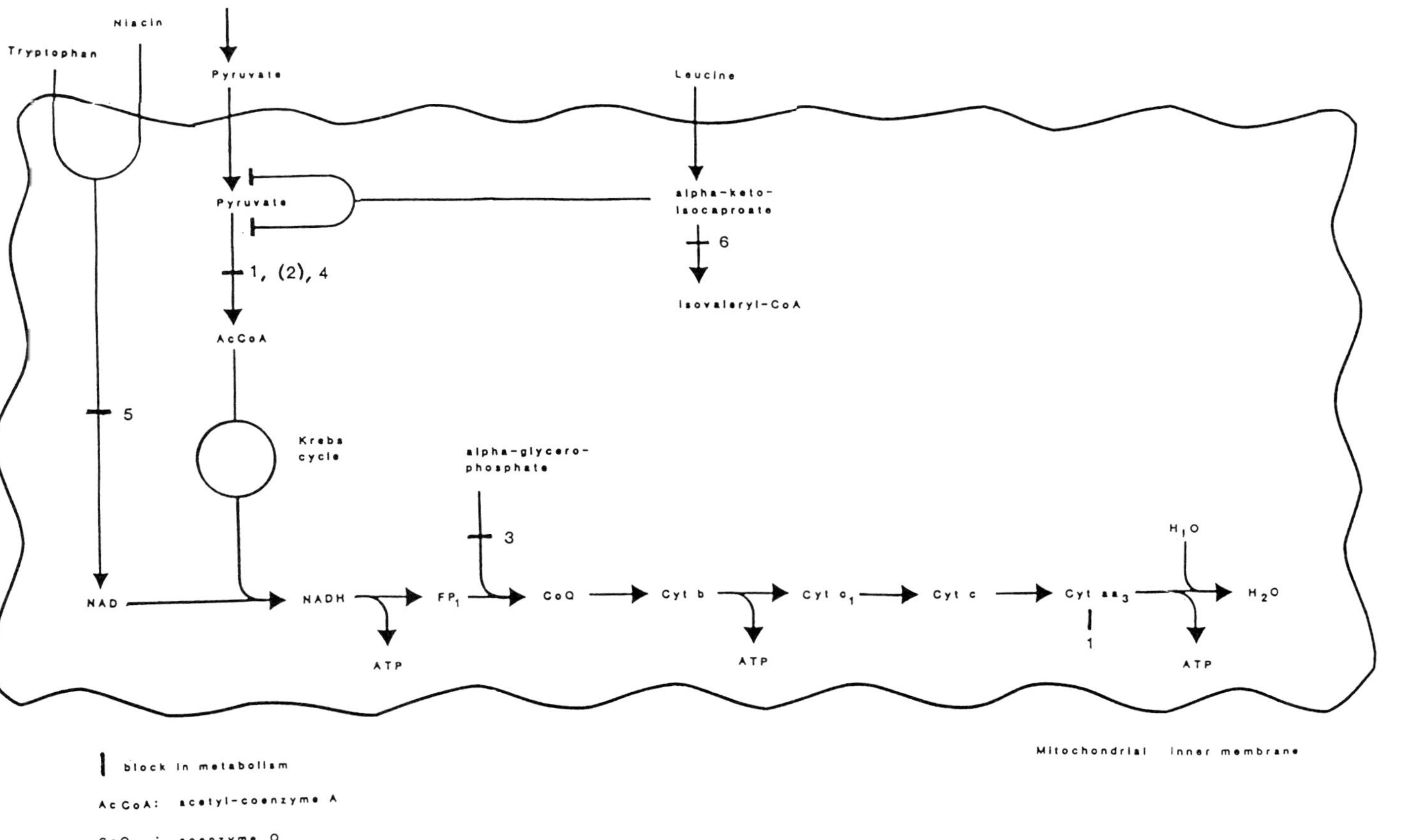
Tryptophan
Niacin
Pyruvate
Leucine
Pyruvate
alpha-keto-
isocaproate
6
Isovaleryl-CoA
1, (2), 4
AcCoA
5
Krebs
cycle
alpha-glycero-
phosphate
3
NAD
NADH
ATP
FP1
CoQ
Cyt b
ATP
Cyt a1
Cyt c
Cyt aa3
1
H1O
H2O
ATP
Mitochondrial inner membrane
block in metabolism
AcCoA: acetyl-coenzyme A
CoQ : coenzyme Q

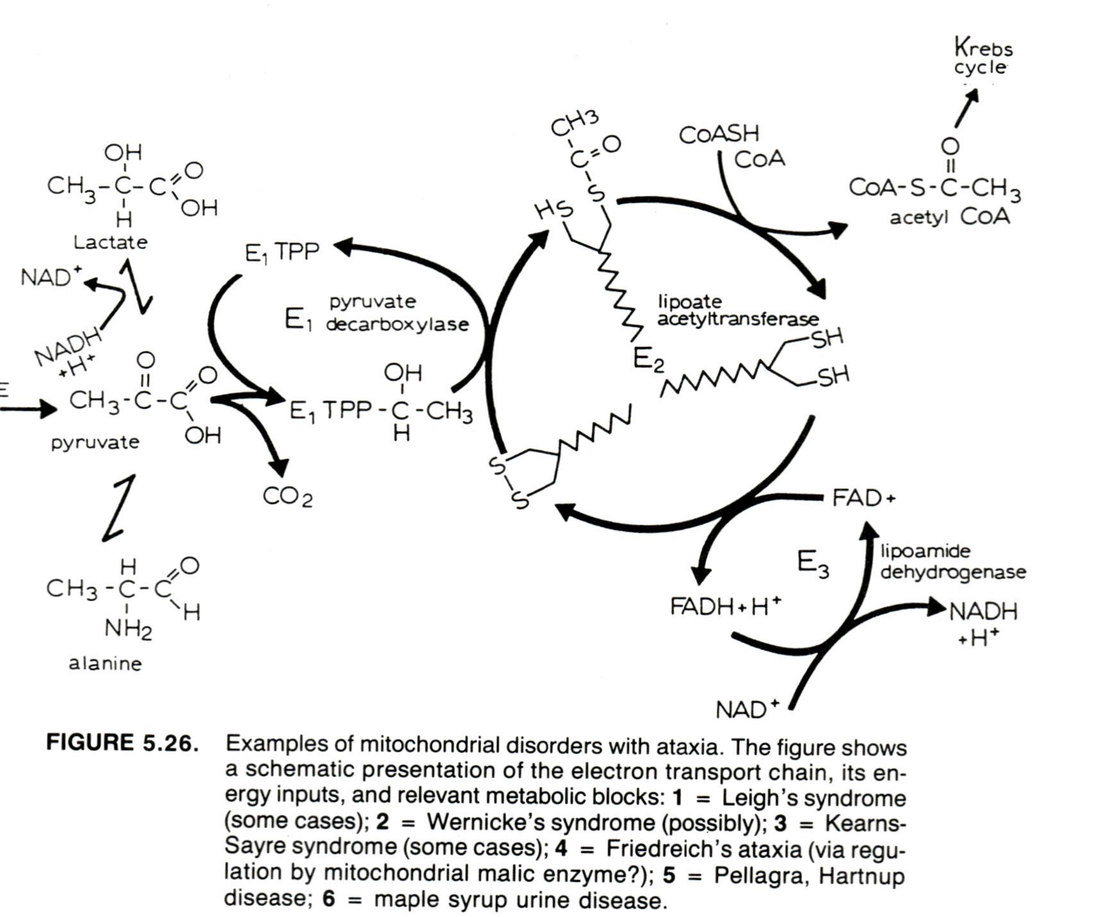

FIGURE 5.26. Examples of mitochondrial disorders with ataxia. The figure shows a schematic presentation of the electron transport chain, its energy inputs, and relevant metabolic blocks: **1** = Leigh's syndrome (some cases); **2** = Wernicke's syndrome (possibly); **3** = Kearns-Sayre syndrome (some cases); **4** = Friedreich's ataxia (via regulation by mitochondrial malic enzyme?); **5** = Pellagra, Hartnup disease; **6** = maple syrup urine disease.

and lipoproteins. The described reduction of unsaturated fatty acids in HDL-esterified cholesterol, especially linoleic acid, in patients with FA (v.s.) could render the cell membranes more rigid. "Rigid membrane" mechanisms may be responsible for the neuropathies in Refsum's and Bassen-Kornzweig syndromes (Cooper, 1977) in which lipid abnormalities are found. Barbeau (1982) argues that the excess of very long chain fatty acids (C_{26}) observed in adrenomyeloneuropathy (Igarashi et al., 1976; O'Neill, Moser, and Marmion, 1980) and phytanic acid (a 20-carbon methylated acid) in Refsum's disease, combined with low serum levels of linoleic acid in these diseases, may point to common elements in the pathogenetic mechanisms of these diseases with that of FA. This may be mediated through a rigid membrane or an energy depletion mechanism or a combination of both with other factors.

In view of emerging data incriminating the central energy metabolism in mitochondria as being the "source of evil" in FA and perhaps other ataxias, we shall try to examine the explanatory power of this hypothesis.

Ataxia

Signs of dysfunction of cerebellum and its connections have always been the prime clinical feature upon which a diagnosis of FA rests. The CNS is particularly prone to damage in energy deprivation states, since it lacks the ability to survive under anaerobic and oligaerobic conditions. Apart from being sensitive to O_2 deprivation, it is equally dependent on glucose as fuel (Bachelard, 1976). In fact, the pathogenetic mechanisms of brain damage by anoxia and hypoglycemia are quite similar (Adams, 1975; Wilkinson and Prockop, 1976). It can readily be presumed that interference with energy production from glucose will have disastrous consequences, especially for cells operating at high levels of metabolic turnover, which seems to be the case with the Purkinje cells of the cerebellum. Further, we have seen that reduced access to acetyl-CoA as fuel leads to derangement of cellular levels of aspartate and glutamate, which may lead to a "metabolic block" of the activities of nerve cells dependant on these amino acids as transmitters. This is particularly interesting because the experimental ataxia produced by 3-acetylpyridine (v.i.), which actually is known to hamper energy metabolism, is partially reversed after intraventricular injection of aspartate and glutamate (De Michele et al., 1980).

Dying-Back Axonopathies

It is recognized that neuropathy in FA is due primarily to axon degeneration (Hughes, Brownell, and Hewer, 1968; Peyronnard et al., 1976) rather than to demyelization. Sensory defects first occur distal in the lower extremities, indicating that longer axons are more prone to damage.

Myopathia/Cardiomyopathy

It is well known that defects in mitochondrial energy metabolism may lead to muscular weakness and myopathia (Morgan-Hughes, 1982). With the rapid energy turnover in myocard, it is not surprising that a possible energy deficit for aerobic conditions may lead to myocard hypertrophy, followed by cell necrosis, inflammation, and fibrosis. However, direct evidence of energy depletion in skeletal and heart muscle in FA is still lacking. Direct examination of muscle cells needs to be undertaken.

Azari and collaborators (1980), examining an experimental myopathy of hamsters, find an increase in calcium concentration in the left ventricle. Taurine given prophylactically corrects calcium accumulation and, to a certain extent, cardiac hypertrophy. This is interesting in view of the findings of Sanchez-Casis, Cote, and Barbeau (1976) in FA patients (v.s.), especially since calcium ions play an important role for both cellular and intracellular membranes during physiologic muscular resting and working conditions. Further experiments point toward an interesting interplay between calcium ions and the amino acids taurine and glutamate (Remtulla, Katz, and Appelgarth, 1979; Rassin, Sturman, and Gaull, 1980).

Glucose Intolerance

As we have seen, Draper, Shapcott, Larose, Stankova, Levesque, and Lemieux (1979) have shown a high incidence of glucose intolerance with hyperinsulinic response in FA patients in the absence of overt diabetes; in addition, they have shown pathologic glucose responses in obligate heterozygotes. Normal siblings had entirely normal responses. This finding has been confirmed by Tolis and collaborators (1980). Draper's results also indicate that insulin receptors are normal in nondiabetic FA patients. Taken together, these results may indicate that the defect in glucose metabolism is located somehow to the level of glucose entry into the cell—the mechanisms responsible for this being some other than the hormonal effect of insulin. Further, it has been demonstrated (Barbeau, 1975; Barbeau et al., 1976) that pyruvate is abnormally handled in response to glucose loading.

This phenomenon may be explained as follows: Interference with cellular handling of pyruvate, as represented by a reduction of PDHC activity, will result in an accumulation of metabolites proximal to pyruvate in glycolysis (see Figure 5.16), in cellular accumulation of glucose, and in further accumulation of glucose extracellularly. In an effort to do away with the hyperglycemia thus created, the beta cells in the pancreas pour out insulin, creating the hyperinsulinic hyperglycemia. In this way the "diabetes" of FA patients may be incorporated in and explained by the "energy depletion hypothesis." As yet, this theory must still be regarded as vague hypothesis; further investigations in different FA patient populations are needed to test its validity.

There are numerous other syndromes that include signs of cerebellar dysfunction in which heredity plays a role in the pathogenesis where one suspects involvement in mitochondria. Suffice it here to mention two: *pellagra*, with its photodermatitis; diarrhea, peripheral neuropathia; ataxia; and mental, basal ganglia, and pyramidal symptoms; and its counterpart, *Hartnup disease*, being more clearly genetic in etiology with its suspected autosomal recessive mode of inheritance (Still, 1976; Jepson, 1978). *Maple syrup urine disease* is another disorder of amino acid metabolism that in its mild or intermittent forms may show examples of cerebellar dysfunction (Rosenberg, 1977; Tanaka and Rosenberg, 1983), clearly expressed in a case report from India (Kannan and Subramanyam, 1977). Some of these disorders show vitamin responsiveness (Rosenberg, 1977; Tanaka and Rosenberg, 1983), suggesting an incomplete activation or incorporation of vitamin-derived cofactors into essential enzymes, again being suggestive of the combined effect of hereditary and environmental factors in these disorders.

From a genetic point of view, the mitochondrial disorders are interesting in that nuclear as well as mitochondrial genes may be involved. Cytoplasmic inheritance (defect in the circular double-stranded mitochondrial DNA) is suspected in Leber's optic atrophy (Stumpf, 1979). Being mainly derived from maternal sources (through the ovum), some role of mitochondria may be sought for in the Schut-Haymaker family of OPCA, which shows earlier onset and more severe manifestations in the offspring of maternal probands (Schut and Böök, 1953).

GENETIC LINKAGE AS A SOURCE OF CLINICAL VARIATION IN HEREDITARY ATAXIAS

Hereditary ataxias is a group of disorders in which significant clinical variation may occur. When such variation occurs in a Mendelian disorder, the following explanations are traditionally applied:

1. Environmental differences, nutritional differences (before or after birth), or effects of unrelated diseases;
2. Varying expressivity, sex, and age effects;
3. Presence of modifying genes or different "genetic background";
4. Pleiotropism; and
5. Genetically different disorders (different genes)—"new syndrome."

One or the other of the first three mechanisms would usually be invoked as an explanation when the clinical variation observed is of a minor degree, whereas items 4 and 5 would be considered when variation is substantial. Doubtless to say, any of these traditional interpretations may be relevant in given cases. However, the possibility that in some instances genetic linkage may be the cause for clinical variation in inherited disorders has received some attention (Berg and Skre, 1976; Skre and Berg, 1976; Skre and Berg, 1977; Berg, 1978a; Berg, 1978b). The study by two of us (HS and KB) of hereditary ataxias has provided several examples of clinical variation possibly reflecting genetic linkage, and it seems likely that many more cases of possible linkage may be detected with increasing awareness of this mechanism.

Hereditary Spastic Paraplegia and Retinitis Pigmentosa

One of the first examples of concurrence in the same kindred of two traits known as genetic entities in their own right was the occurrence in two out of five Norwegian kindreds with autosomal recessive spastic paraplegia of retinitis pigmentosa (Skre, 1974d; Skre and Berg, 1974b). The information concerning this clinical variation is limited, as reflected by low lod score values. No individual exhibited a phenotype suggesting that a recombinant event had taken place. At the clinical level, however, the debut of ocular and spastic signs appeared to be completely independent. We interpret these data as merely a suggestion of linkage between recessive hereditary spastic paraplegia and retinitis pigmentosa.

Tyrosinase-Negative Albinism and Cerebellar Ataxia

In a highly inbred kindred, from a relatively isolated area in Norway, four persons in a sibship of nine had ataxia as well as total albinism, whereas the remaining five members of the sibship had neither disorder (Skre and Berg, 1974a). The ataxia was clinically of a cerebellar type with little spinal involvement and resembled Holmes's cerebello-olivary form (Holmes, 1907b). The disease started in all four patients in the fifth decade, as expected for this type, and had a progressive course. Nystagmus and dysarthria were found together with muscular hypotonia and cerebellar signs. There was a complete absence of pigmentation in skin, hair, and eyes (Figure 5.27). The albinism was of a tyrosinase-negative type, known to be inherited as an autosomal recessive trait.

A paternal aunt and her son were reportedly albinos, and the parents of the paternal aunt were also related (see Figure 5.28). The fact that albinism segregated without cerebellar ataxia in a different part of the kindred led us to consider the possibility that the combination of the two traits did not form a "new syndrome" but, rather, reflected genetic linkage.

Lod scores for linkage between cerebellar ataxia and total albinism obtained from the informative sibship are given in Table 5.17. The lod score of 2.437 at recombination fraction zero strongly suggested linkage, but it is insufficient to prove linkage formally. The phenotype of the paternal aunt

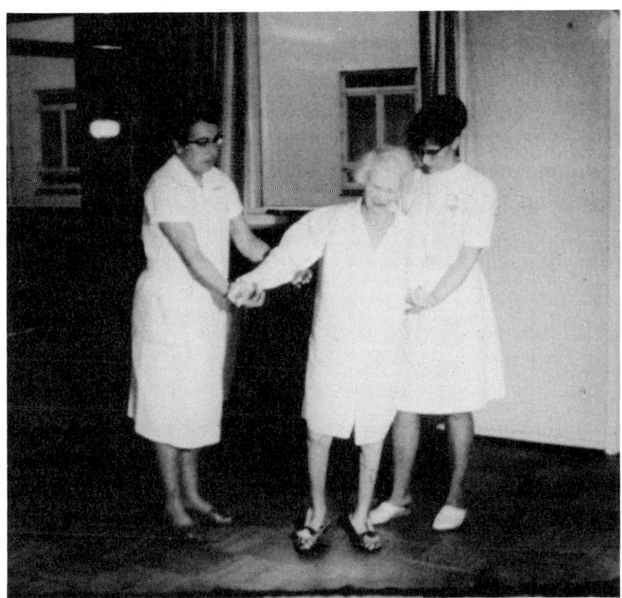

FIGURE 5.27. Patient with cerebellar ataxia (type 2B, see paragraph on spinocerebellar ataxias in Norway) and albinism (proband in pedigree 54, Figure 5.28).

who had albinism but not cerebellar ataxia may reflect a recombinant event, or one of her albinism genes may stem from a source different from that of those present in the homozygous state in the four children affected with both traits. No distinction between these two possibilities can be made. If a recombinant event is assumed, this could have taken place in any one of several different ancestors. However, if it is assumed that the parents of the doubly affected persons were heterozygous for total albinism and cerebellar ataxia because of common ancestry, the two genes had traveled together, at least corresponding to four nonrecombinant events among the ancestors of the proband sibship. Accordingly, the lod score analysis was repeated after adding information on four nonrecombinant events and one recombinant event to the information extracted from the proband kinship. A recombinant event necessarily makes a lod score of $-\infty$ for a recombination fraction zero. However, highly suggestive lod scores (2.30 and 2.25, respectively) remained for recombination fractions 0.05 and 0.10. Thus the case for linkage was not significantly weakened by taking the additional information into consideration. The curve for relative probabilities of various recombination fractions based on the expanded set of data is shown in Figure 5.29.

Total albinism has very rarely been reported to occur together with other syndromes and has not been seen in ataxia of a cerebellar type, to our knowl-

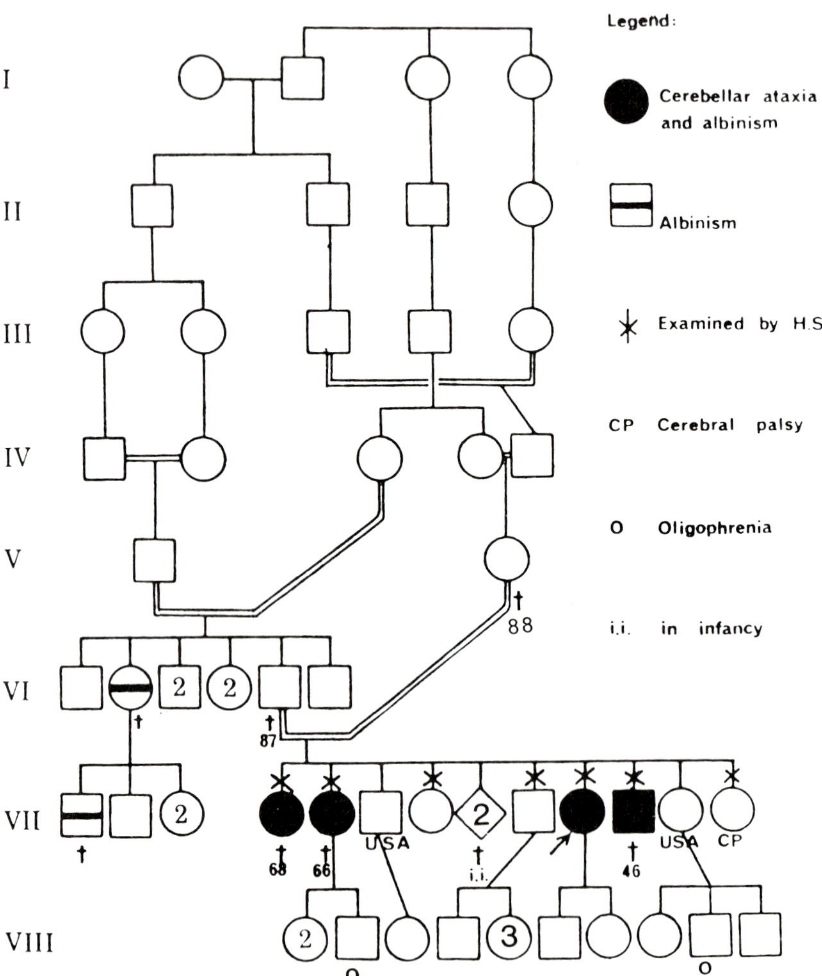

FIGURE 5.28. Pedigree chart of kindred 54. (Skre and Berg. © 1974 Munksgaard International Publishers Ltd., Copenhagen, Denmark.)

TABLE 5.17. Lod Score for Linkage between Cerebellar Ataxia and Total Albinism from the Informative Sibship (VII, 5–15) in Kindred 54

	Recombination Fraction					
	0.00	0.05	0.10	0.20	0.30	0.40
Score: z_3	2.426	2.177	1.927	1.377	0.802	0.242
Correction: e_3	0.011	0.009	0.007	0.002	0.001	0
Total score	2.437	2.186	1.934	1.379	0.803	0.242
Relative probability	274	154	86	24	6	2

Source: Skre and Berg (1974a).

edge. We concluded from this study that it appeared more likely that the concurrence of cerebellar ataxia and tyrosinase-negative albinism was caused by linkage between two independent traits than by a pleiotropic effect of one gene or represented a previously unrecognized syndrome (Skre and Berg, 1974a).

In the mouse, the non-alpha-globin loci are linked to albinism. In man the non-alpha-globin region is known to be on the short arm of chromosome 11. Because of the considerable degree of conservatism in chromosome evolution, there is a distinct possibility that tyrosinase-negative albinism is also linked to the non-alpha-globin region in man. Therefore, if the distribution of total albinism and cerebellar ataxia in the kindred we have studied truly reflects relatively close linkage, it is an intriguing possibility that the locus for the cerebellar ataxia is on chromosome 11.

Marinesco-Sjögren Syndrome and Hypergonadotropic Hypogonadism

In two Norwegian kindreds (Figures 5.18 and 5.30) Marinesco-Sjögren syndrome occurred together with hypergonadotrophic hypogonadism in 9 out of 10 affected individuals. The last patient had Marinesco-Sjögren syndrome without manifestations of hypogonadism (Skre and Berg, 1977). One patient in kindred 87, (Figure 5.17**A**) had normal ovaries at autopsy (Case XII,6).

Although physical retardation is typical of Marinesco-Sjögren syndrome, hypogonadism is not considered to be part of the syndrome. In addition to the two kindreds where Marinesco-Sjögren syndrome and hypogonadism occurred together, we have examined a third family with two affected sisters in whom the manifestations of Marinesco-Sjögren syndrome were very similar to those of the patients in the two previous kindreds, but the sisters evidenced no sign of infantilism or hypergonadotropic hypogonadism (Table 5.18). The

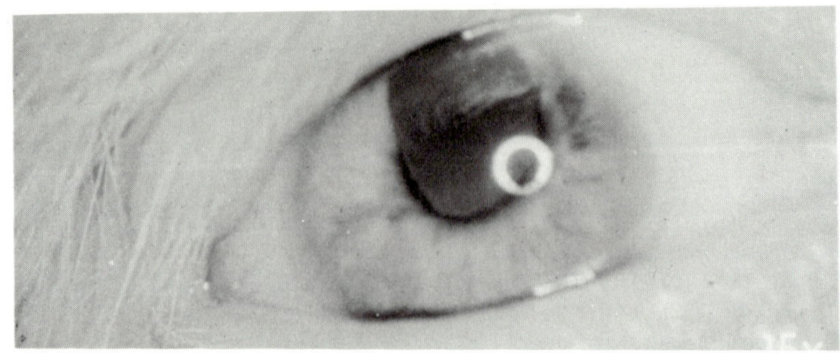

FIGURE 5.29. (**A**) Cerebellar ataxia and total albinism. (**B**) Relative probabilities of various recombination fractions, given four nonrecombinant events (Cases IV,3; IV,4; V,4; V,2; VI,7;) and one recombinant event (Case VI,2).

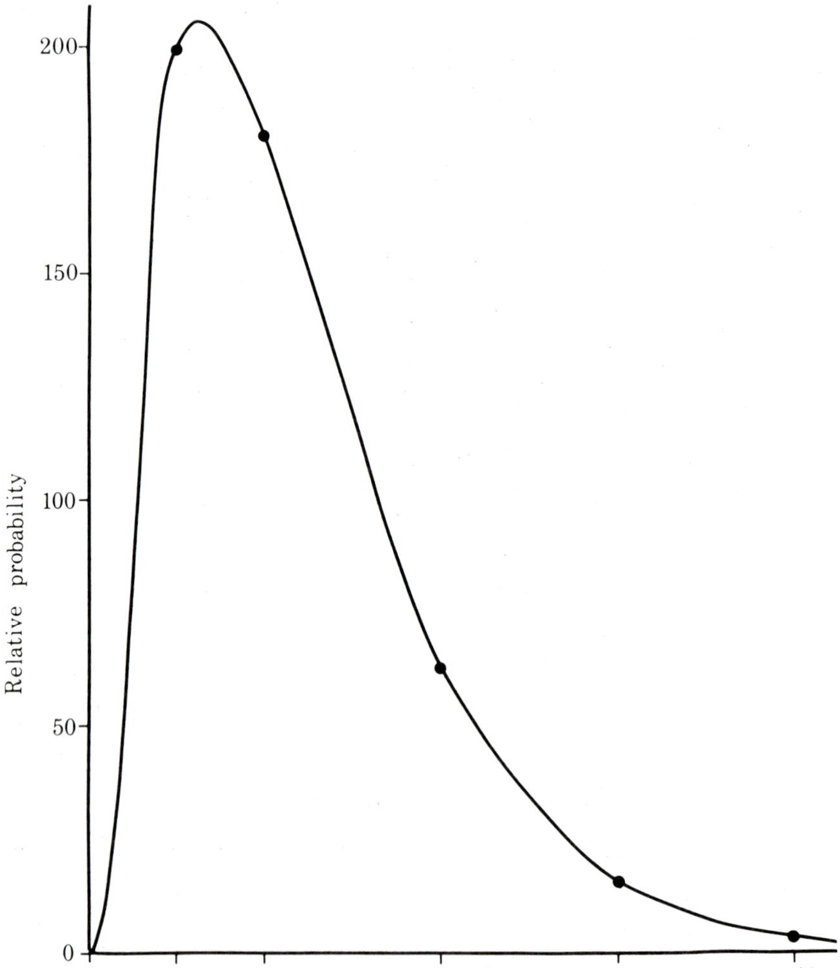

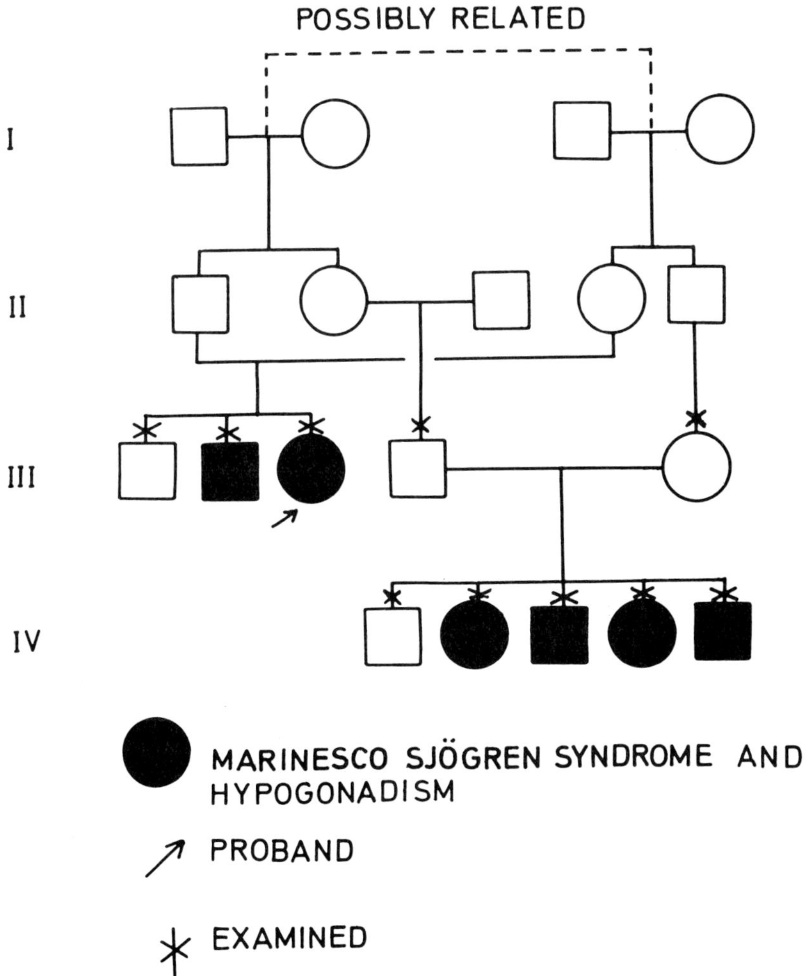

FIGURE 5.30. Pedigree chart of kindred H. (Modified from Berg and Skre, 1976.)

comparative study among affected persons from the three kindreds led to the conclusion that the same neurological disorder was segregating in all families and that hypergonadotropic hypogonadism is not an obligatory part of the syndrome. This conclusion is in agreement with data reported in the literature.

Although hypogonadotropic hypogonadism could occur secondarily to neurological disorders of several kinds, it is difficult to visualize how hypergonadotropic hypogonadism could be secondary to such disorders. Hypergonadotropic hypogonadism is known to occur as a distinct, albeit rare, autosomal recessive condition.

On the assumption that the concurrence of Marinesco-Sjögren syndrome and hypergonadotropic hypogonadism was caused by genetic linkage, lod score analysis was conducted. Table 5.19 shows the lod score for this linkage relationship at different values of the recombination fraction, from the four

TABLE 5.18. Findings in Three Kindreds with Marinesco-Sjögren Syndrome

Sign	Kindred 87	Kindred H	Kindred 66
Ataxia	4/4	6/6	2/2
Cataract	4/4	6/6	1/2
Mental deficiency	3/3	5/6	2/2
Deformities	3/3	6/6	2/2
Infantilism	3/4	6/6	0/2
HHG*	3/4	6/6	0/2

*HHG = hypergonadotropic hypogonadism
Note: Number with sign/total number examined.
Source: Skre and Berg (1976).

TABLE 5.19. Lod Scores for Linkage between the Marinesco-Sjögren Syndrome and Hypergonadotropic Hypogonadism in Four Informative Sibships from Two Kindreds

		Recombination Fraction					
Kindred	Sibship	0.00	0.05	0.10	0.20	0.30	0.40
87	XII, 1-6	1.087	0.941	0.795	0.508	0.251	0.066
87	XI, 1-6	$-\infty$	−0.472	−0.242	−0.076	−0.021	−0.004
H	III, 1-3	0.790	0.669	0.553	0.340	0.163	0.043
H	IV, 1-5	2.114	1.878	1.641	1.158	0.660	0.208
Total score		$-\infty$	3.016	2.747	1.930	1.053	0.313
Relative probability		0	1,038	559	85	11	2

Source: Skre and Berg (1977).

informative sibships. The lod score exceeded 3 at the recombination fraction 0.05. The curve of relative probability for different values of the recombination fraction (Figure 5.31) had its peak at 6 percent recombination, making this the best estimate of the frequency of recombination between Marinesco-Sjögren syndrome and hypergonadotropic hypogonadism. We concluded that if Marinesco-Sjögren syndrome and hypergonadotropic hypogonadism are considered as distinct entities, the lod score analysis formally proves linkage between their loci (Skre and Berg, 1977).

When Is Genetic Linkage a Reasonable Interpretation?

It is likely that genetic linkage may underlie some occurrences of unusual traits in Mendelian disorders, and in the case of Marinesco-Sjögren syndrome and hypergonadotropic hypogonadism formal lod score evidence for linkage is available. Keep in mind, however, that several other interpretations may be applicable, as suggested above. The following criteria may be useful for situations where genetic linkage seems to explain clinical variation (Berg and Skre, 1976; Skre and Berg, 1976; Berg, 1978a):

1. Two disorders, each known as a Mendelian trait in its own right, concur in a relatively small number of families.
2. The degree of manifestation of each disorder is reasonably constant in any affected person, so that each disorder behaves as an all-or-none phenomenon.
3. No biochemical or other link between the two disorders is known or seems plausible.
4. There is definite evidence of recombination, proving that the two entities are separable.

If conditions 1, 2, and 3 are fulfilled, and the lod score reaches the height conventionally accepted as proof of linkage, evidence for recombination would strengthen the case for linkage significantly. Thus the case would be particularly strong if at least one individual had only one of the two traits in question.

All of these requirements were fulfilled in the case of Marinesco-Sjögren syndrome and hypergonadotropic hypogonadism. So this concurrence may be a particularly suitable example of striking clinical variation caused by genetic linkage between independent traits. The case of cerebellar ataxia and tyrosinase-negative albinism also go a long way to fulfill these criteria, but since the person who had only one of the traits was not in the same sibship as the doubly affected individuals, and since the total lod score did not reach the conventional level for considering linkage proven, this case is weaker.

It can reasonably be assumed that other examples of clinical variation

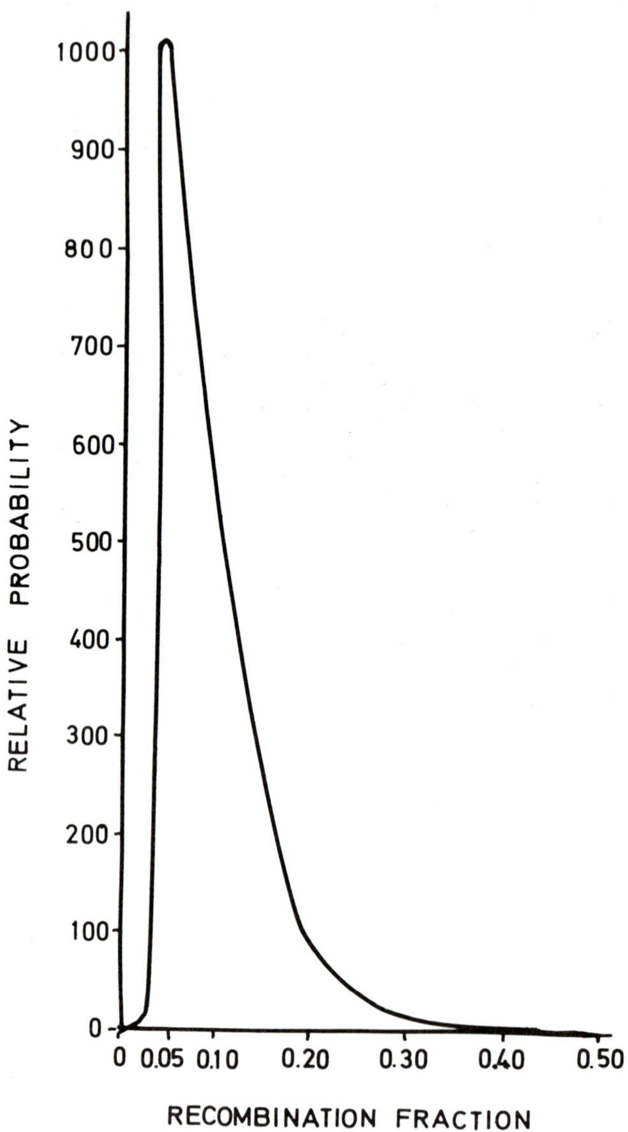

FIGURE 5.31. Marinesco-Sjögren syndrome and hypergonadotrophic hypogonadism. Relative probabilities of various recombination fractions. (Modified from Berg and Skre, 1976.)

caused by genetic linkage will be uncovered. It would be most useful if clinicians encountering unexpected, striking variation in disorders known to be monogenic would bear in mind the possibility that the atypical findings could be caused by genetic linkage. Such kindreds should be reported so that in the future it will be possible to establish the relative importance of genetic linkage as a source of clinical variation in inherited disorders.

GENETIC COUNSELING IN HEREDITARY ATAXIAS

In the cases of monogenic hereditary ataxias, genetic risk estimates are based on Mendelian ratios. The importance of a thorough neurological examination to classify the disorder correctly, and of family investigation to establish the mode of inheritance in the kindred of any given patient, cannot be overemphasized.

Several of the hereditary ataxias are diseases of adult life having their debut after decisions concerning family planning have been made. Although many of these diseases are compatible with normal life well into middle age, some of them, such as olivopontocerebellar atrophy, have a more serious course.

The risk to offspring of patients with autosomal recessive types is generally low despite the fact that all offspring will receive the mutant gene. This is so because the frequency of the carrier state in the general population is very low (in Norway, 1 and 5 per 1,000 for spinocerebellar ataxia type 2A and 2B, respectively). However, marriage with a relative carries a much higher risk because of the possibility that both partners have received the rare disease gene from a common ancestor. Therefore, it is important to emphasize the risk for affected offspring when a member of a family with a rare recessive condition marries a relative, such as a first or second cousin. In agreement with Mendelian ratios, the risk for affected offspring of marriage between a homozygous affected and a carrier for a recessive disorder is 0.5.

Healthy brothers and sisters of a person with an autosomal recessive disorder have a 2 : 3 risk of being a carrier for that disorder. Again, the risk to offspring would be negligible unless marriage is to a related person. Risk to offspring of two heterozygous carriers is 0.25.

The neurological scoring system introduced by Skre (1972, 1974a; see paragraph on phenotypic variation and the appendix) makes it possible to detect persons with minor neurological signs. When this system was supplied to sibships of probands with certain types of hereditary ataxias, a significant number of brothers and sisters to persons with autosomal recessive ataxia exhibited signs of unspecific neuropathy. Unspecific neuropathy in such sibships may, at least in some cases, reflect the carrier state for their recessive neurological disorder (Skre, 1974a, 1974d, 1975b). Thus an extremely careful neurological examination employing the scoring system may be helpful in identifying those healthy sibs that carry the disease gene. In some (rare) cases, unspecific neuropathy could presumably reflect early stages of disease in persons homozygous for the disease gene. Therefore, neurological examination of brothers and sisters of patients with recessive ataxias can be useful for genetic counseling purposes, but it must be emphasized that even the most elaborate scoring system does not provide a definitive answer to the question of the genotype of the individual. At the very most, half of the carriers may be detected in this way (Table 5.20).

Persons who have an autosomal dominant form of hereditary ataxia have a 50 percent risk of passing on the disease gene to a given son or daughter. In this case also, the disease often becomes manifest only after the patient has reproduced. The dilemma in the genetic counseling situation often is that a healthy consultand whose mother or father has an autosomal dominant hereditary ataxia wants to know the risk for passing on the disease to offspring. Young consultands must be given a risk of 0.25 for any one child (his or her own 0.5 risk of having received the gene multiplied by the 0.5 risk of passing on an autosomal gene to a given child). As mentioned above, careful family study is necessary. If there are many affected persons in the kindred, and

TABLE 5.20. Hereditary Ataxias

	Number of Persons Examined			
Diagnosis	With Main Disorder	With Un	Healthy	Total
A: Distribution of individuals in sibships of Norewegian probands with some hereditary ataxias according to presence or absence of main disorder and unspecific neuropathy (Un), respectively				
Autosomal dominant inheritance				
Spinocerebellar ataxia, type I	7	6*	4	17
Familial spastic paraplegia	5	2	7	14
Autosomal recessive inheritance				
Spinocerebellar ataxia, type IIA	4	4	8	16
Spinocerebellar ataxia, type IIB	6	6	13	25
Friedreich's ataxia	3	12	7	22
Familial spastic paraplegia, recessive	4	9	7	20
Marinesco-Sjögren syndrome	8	1	4	13
B: Distribution of probands' parents (only disorders with autosomal recessive inheritance)				
Spinocerebellar ataxia, type IIA	—	3	3	6
Spinocerebellar ataxia, type IIB	—	2	2	4
Friedreich's ataxia	—	5	5	10
Familial spastic paraplegia	—	2	2	4
Marinesco-Sjögren syndrome	—	2	4	6

*Three cases were probably "formes frustes."
Source: Skre (unpublished).

therefore extensive data available on age at onset of the disease in the family, this information may be useful in genetic counseling situations.

A closely linkėd genetic marker could be a very useful tool in diagnosing the presence of a disease gene prior to disease onset or, for autosomal recessive conditions, the presence of a disease gene in the heterozygous carrier state. The linkage detected between olivopontocerebellar ataxia type I and HLA is too loose to be very useful for genetic counseling purposes. The recent detection of several restriction fragment length polymorphisms in DNA holds promise that a great number of new genetic markers will become available over the next several years. There is reason for cautious optimism that markers even for these relatively rare disorders may become available. Such markers will also be very useful for prenatal diagnosis of the disorder in question.

In all cases genetic counseling should be provided in a nondirective manner, leaving all the decisions to be made to the consultand on a truly informed basis. Great attention should be paid to the way in which genetic counseling is offered, particularly in cases where the consultand or the spouse is at risk himself or herself of developing a serious neurological disorder. In cases of serious dominant neurological disorders, genetic counseling should, in a tactful manner and through relatives who are consultands, be offered to family members who are unaware of their risk situation.

APPENDIX

The procedures utilized for rating neurological signs in the Norwegian investigation are described in the section on phenotypic variation.

Table A5.1 presents the general instructions for neurological testing and the rating of signs. A form used during such a testing/scoring procedure is shown in Table A5.2. A third table—Table A5.3—provides prevalence figures for individual scores in the general adult population used in the correcting procedure (corrections for age and sex).

TABLE A5.1. Procedure for a Rated Testing System Based on the Neurological Exam

Subtest
Number

1 Mental status

Feeblemindedness: Feeblemindedness relies mostly upon performance during the examination (cooperation and comprehension). Motor impersistence is regarded as a sign of mental deficit. Disordered behavior as a result of mental illness has to be considered and corrected for. Scored as 1.

Neurosis and psychosis: Diagnosis of mental disorder is recorded, at present or in the past. Scored as 1.

(*continued*)

TABLE A5.1. Procedure for a Rated Testing System Based on the Neurological Exam (*continued*)

Dementia: Dementia should only be considered if loss of memory or concentration/orientation not due to the mental illness is observed. Scored as 3.

Epilepsy: Epilepsy is mostly unequivocal. Scored as 3.

2 Oculoauditive signs

Most of the findings under this heading are unequivocal.

Optic atrophy (degenerative): Optic atrophy is one without a history of multiple sclerosis, vascular disease, or papilledema, developing insidiously. Scored as 3.

Major refraction error: A major refraction error is one requiring 4D correction or more. Scored as 1.

Acquired strabismus: Acquired strabismus is one developing after childhood and presenting with double vision. Scored as 3. Strabismus with amblyopia is scored as 1.

Neural hearing loss: Neural hearing loss is reduced hearing (try whispering or watch with a normal Rinne test; if unilateral, with Weber's, to the normal side). Scored as 1.

3 Deformities[a]

Hammer toes: A slight finding equal to a score of 1 means that only two or more of the lesser toes are deformed or pes transverso planus is present. A reading of 2 also means that the big toe is deformed. A rating of 3 means marked changes.

Pes cavus: Pes cavus is rated according to the general instructions.

Camptodactyly: If there are contractures in one or two fingers, it is a moderate finding, giving a score of 1. If there are contractures in more than two fingers, this is given a score of 2. If there are marked changes, it receives a score of 3.

Excavated palms: This term refers to increased hollowness of the palm. It is rated according to general instructions.

Scoliosis: A moderate scoliosis is rated as 1. Torsion of the vertebrae gives a rating of 2. Marked kyphoscoliosis is rated as 3.

Anomalies in digits: Anomalies in digits other than those mentioned above receive a score of 1.

Hallux valgus: Hallux valgus is rated as a moderate sign regardless of the degree of change. Score 1 if unilateral; score 2 if bilateral.

Talipes equinovarus: Rate this according to the general instructions.

Luxation or dysplasia: Luxation or dysplasia of the hip joint is rated as a moderate sign regardless of the degree of findings (see *hallux valgus*).

4 Neuromuscular affection

In general, testing of muscle strength follows the procedure utilized by physiotherapists in which normal strength is rated 6, whereas complete paralysis is rated as 0. Muscle performance rated as 3 means, in this system, patient has strength enough to compensate for weight of the limb tested. A moderate

TABLE A5.1. Procedure for a Rated Testing System Based on the Neurological Exam (*continued*)

finding means strength loss less than a rating of 3, whereas a marked finding means loss of strength at below 3.

"Fat bottle calves": This is a sign of atrophy seen preferably in peroneal muscle atrophy and is located to the distal part of the leg, which appears very thin with a sharp delineation to the normal proximal part of the leg. Rated as 3.

Paresis/atrophy of short muscle in feet: Such findings are regarded as a moderate sign regardless of the changes (see *hallux valgus*). The short muscles are tested by observing the short extensor muscle to the toes and by testing extension in the distal joints. The short muscles are also tested by flexion in the basal joints in the toes and by fanning of toes.

Short muscles in hands: These are tested by inspection of the thenar and hypothenar prominences in the hand, by testing ab—and adduction in the fingers, and by testing flexion of fingers in the basal joints (of fingers 2 to 5). Thumb adduction and opposition are tested separately.

Muscles in legs: These are tested by dorsi and plantar flexion in ankle and by supination and pronation in the ankle. These movements are best tested while patient is walking.

Muscles in forearms: Testing of muscles in the forearm is done by flexion/extension at the wrist and by testing the patient's grip.

Muscles in thighs: These are tested by extension or flexion at the knee.

Muscles in arms: These are tested by extension and flexion at the elbow.

Muscles in hip girdle: These are tested best while the patient is lying on the examining table, and the main movements, extension, flexion, adduction and abduction are tested. A good test of abduction in the hip is to ask the patient to hop on one foot.

Muscles in shoulder girdle: These are best tested while the patient is seated, and the main movements should be tested as for the hip girdle.

Muscles of the trunk: Only flexion and extension are tested. A positive finding is given a score of 3 regardless of its degree.

Paresis/atrophy of the muscles of bulbar region: Cranial motor nerves 5, 7, 9, 10, 11, and 12 are involved. A positive finding is given a score of 3.

Fasciculations: Fasciculations are looked for while the patient is resting on the examining table. A positive finding is given a score of 3.

5 **Sensory affection**

In general, differences between two points in the strength of perception of stimuli should be asked for. Leading questions should be avoided.

Loss of touch: A marked finding is a total loss of perception; relative loss of perception is a moderate one:
a. Below knees.
b. Below elbows.

Loss of pain, temperature sense:
a. Below knees.
b. Below elbows.

(*continued*)

TABLE A5.1. Procedure for a Rated Testing System Based on the Neurological Exam (*continued*)

Loss of superficial sensation:
a. Proximal portions, lower extremities (LE).
b. Proximal portions, upper extremities (UE).

Loss of position sense in the big toes (and ankles): This is regarded as a marked sign regardless of the degree of changes. Scored as 3 when unilateral; as 4 when bilateral.

Loss of vibration sense, 256 Hz tuning fork:
a. In big toes (base). Only total loss is recorded, rated as a moderate sign (scored as 1 or 2).
b. In foot (on dorsum corresponding to the base of first metatarsal bone). Only total loss is recorded, rated as moderate sign (scored as 1 or 2).
c. Inner malleoli.
d. Tibial tuberosities.
e. In dorsum of hands.
f. In sternum.

Sign of transverse lesion, any quality: Testing of vibration sense in the anterior iliac spines is included here.

6 **Hyporeflexia[b] of tendon and plantar reflexes**:
a. In LE. A tendon reflex of + is regarded a moderate sign. If both reflexes in a leg are weaker than this, (+) or 0, it is regarded a severe sign.
b. In UE. A tendon reflex of (+) is regarded as a moderate sign. If two or all reflexes are abolished in an arm, it is regarded as a severe sign.
c. Abolished plantar response not resulting from paresis is regarded a moderate sign only (scored as 1 or 2).

7 **Ataxic signs**

Fixation nystagmus: Often it presents most prominent in deviant eye fixation, with a rapid phase in the deviant direction. Given a score of 3 when present.

Ataxic ocular movements: These are seen when subject follows examiner's finger (moving in circles before the patient) as jerky deviations and stops. Given a score of 1 when present.

Dysarthria, nonparetic: This scans speech and difficulties in articulation. Given a score of 3 when present.

Intention and/or static tremor: This is best seen when subject keeps his or her hand in front of himself or herself—fingers slightly apart, arms stretched, eyes closed. Intention tremor is best seen during the finger-nose test or when subject attempts to drink a full glass of water. When present given a score of 1.

Bradykinesia: This includes any bradykinesia of nonparetic origin. Should be scored according to number of extremities involved, from 1 to 4.

Disturbance of coordination:
a. In LE. Subject is asked to place the heel directly on the other knee with eyes closed and then to slide the heel along the shin. Rated as usual. The test can also be used as a test for dysmetria or asynergia ("decomposition des mouvements").

TABLE A5.1. Procedure for a Rated Testing System Based on the Neurological Exam (*continued*)

---

b. In UE. Subject is asked to point with the tip of the index finger to the tip of the nose and then to point the tips of the two index fingers against each other in front of himself or herself with semiflexed elbows, beginning far laterally. This is done with eyes open and then with eyes shut. Failures to perform are rated as usual. (Buttoning and unbuttoning might be used as an additional test.)

Diadochokinesia: Subject is asked to pronate-supinate his or hand in rapid succession. If this is not done smoothly, but jerky at turns, it is regarded as a moderate sign regardless of the degree of disordered movements.

Other cerebellar signs: Among these are tests for asynergia, dysmetria, hypotonia, asthenia, and rebound phenomenon. All these examinations are in fact done during the performance of tests on motor function and coordination in general. Rebound phenomenon is tested for in examining strength of arm flexion. Subject is asked to pull his or her hand while wrist is held firmly by the examiner. The hand is then released suddenly. Too much rebounding despite the subject's knowledge of a sudden release is a positive sign.

Romberg's sign: The patient is asked to stand with feet close together, first with eyes open, then with eyes shut. Coarse reeling without movements of feet is a moderate sign; corrective stepping is a severe sign. Rated as score 1 and 3, respectively.

8 **Pyramidal and spastic signs**

Spastic paresis and/or bradykinesia: Paresis in a limb with spasticity or signs of pyramidal tract affection and without muscle wasting is regarded as a spastic paresis. Bradykinesia is then invariably present but may even be present without paresis. Such signs are recorded in:
a. LE. Scores as in general.
b. UE. Scores as in general.

Spasticity: "Clasp knife phenomenon" is tested in (a) LE by abrupt passive flexion of subject's knee from complete extended position and in (b) UE by abrupt passive extension (and in some instances flexion) of subject's elbow or wrist. Do not confuse with rigidity, which is found as a plastic or cogwheel resistance at all angles and movements in the joints. These signs are scored independently in LE and UE according to the general rating rules.

Spastic gait: The subject walks without the normal smoothness of movements, often with stiff knees and by shuffling the toes along the ground. A positive sign gives a score of 3.

Exaggerated tendon reflexes[c] *in*:
a. LE. A reflex response of + + +(+) is regarded as a moderate sign. If both patellar and Achilles' reflexes are clonic, it is regarded as a severe sign (if in the same limb). Scores according to general rules.
b. UE. A response of + + + is regarded as a moderate sign. Clonic or subclonic [+ + +(+)] is regarded as a severe sign if present in at least two test points in the same limb. Scores according to general rules.

Babinski's sign or inverted plantar responses: A response of A, OA, or OA? from the soles, or a Chaddock's sign, is regarded as moderate. If a response of A is received from all test points, it is regarded as a severe sign. A plantar

(*continued*)

TABLE A5.1. Procedure for a Rated Testing System Based on the Neurological Exam (*continued*)

response of O? by lateral stroke is regarded as slight inversion if the medial stroke elicits no reaction and the abdominal reflexes are reduced on the actual side and/or the tendon reflexes are exaggerated. Scores according to general rules. Withdrawal reaction in LE is regarded as a severe sign.

9 Basal ganglia signs

a. Rigidity. Score rating according to number of limbs involved.
b. Parkinsonian tremor, tremor at rest. Score rating as in foregoing.
c. Essential or senile tremor. Score rating as in foregoing, but head/trunk gets a 3.
d. Immobility of face. Score rating as in (a).
e. Oligokinesia. Score rating as in (a).
f. Abnormal synkinetic movements. Score rating as in foregoing (face: 3).
g. Choreatic or athetotic movements. Score rating as in foregoing.
h. Torsion dystonia. Score 3.

[a]From subtest 3 on, signs are given scores from 1 to 4. A score of 1 or 2 means a moderate sign, 1 signaling unilateral findings, 2 signaling bilateral findings. A score of 3 and 4 means marked findings unilaterally or bilaterally, respectively. This reading is adhered to unless special instructions are given.

[b]Comments to (a) and (b): Generally, + + means a normal response. If a reflex is + in an UE and +(+) in a LE in the presence of signs of spasticity in the same limb, it is also regarded as a sign of moderate hyporeflexia. Similar conclusions are drawn if the contralateral, symmetric reflexes are + or brisker, or if another reflex in the same limb is + or brisker.

[c]Comments to (a) and (b): If at least one reflex is + +(+) in an UE or + + + in a LE, it is regarded as a moderate hyperreflexia if other spastic signs and/or an inverted plantar response and/or reduced abdominal reflexes are present on the same side, or if the response is at least one + brisker than the contralateral one.

Source: Walton JN. Brain's Diseases of the Nervous System. Oxford University Press, Oxford, New York, Toronto. 1977, pp. 53–57.

Corrected score after weighting

TABLE A5.2. Clinical Examination of the Nervous System

HISTORY

Name | Sex | Profession

Subtest Number		Score		
1	*Status mentalis*			
	Feeblemindedness (IQ)	1		
	Neurosis (Type)	1		
	Psychosis/personality deviation (Type)	1		
	Dementia (moderate/severe)		3	
	Epilepsia (Type)		3	
	Sum test			1 →
2	*Oculoauditive signs*			
	Tapetoretinal degeneration/hemeralopia/degeneration of macula		3	

TABLE A5.2. Clinical Examination of the Nervous System (*continued*)

Subtest Number		Score
	Optic atrophy (degenerative)	3
	Retrobulbar neuritis/sequelae	1
	Major refraction error	1
	Secondary/senile cataract	1
	Congenital cataract	3
	Congenital strabismus (and amblyopia)	1
	Anisocoria, other pupillary anomaly	3
	Acquired strabismus	3
	Ocular myopathy, gaze paresis, exophthalmos, ptosis	3
	Congenital surdity	3
	Neural hearing loss	1
	Other hearing loss	
	Sum test	2 →
3	*Deformities*	
	Hammer toes, slight/moderate/marked	1 2 3
	Talipes cavus	1 2 3 4
	Camptodactyly, slight/moderate/marked	1 2 3
	Excavated palms	1 2 3 4
	Scoliosis, slight/moderate/marked	1 2 3
	Anomalies in digits	1
	Hallux valgus	1 2
	Talipes equinovarus	1 2 3 4
	Dysplasia (luxation) of hip joint	1 2
	Dysraphic signs	
	Sum test	3 →
4	*Neuromuscular affection*	
	"Fat bottle calves"	3
	Paresis/atrophy of short muscle in feet	1 2
	Paresis/atrophy of short muscle in hands	1 2 3 4
	Paresis/atrophy of muscles in legs	1 2 3 4
	Paresis/atrophy of muscles in forearms	1 2 3 4
	Paresis/atrophy of muscles in thighs	1 2 3 4
	Paresis/atrophy of muscles in arms	1 2 3 4
	Paresis/atrophy of muscles in hip girdle	1 2 3 4
	Paresis/atrophy of muscles in shoulder girdle	1 2 3 4
	Paresis/atrophy of muscles of trunk	3
	Paresis/atrophy of muscles of bulbar region	3
	Fasciculations	3
	Sum test	4 →
5a	*Loss of superficial sensation*	
	Touch, below knees	1 2 3 4
	Touch, below elbows	1 2 3 4
	Pain/temperature, below knees	1 2 3 4
	Pain/temperature, below elbows	1 2 3 4
	All qualities, proximal parts	3 4
	All qualities, lower extremities	

(*continued*)

TABLE A5.2. Clinical Examination of the Nervous System (*continued*)

Subtest Number		Score			
	All qualities, proximal parts			3	4
	All qualities, upper extremities				
5b	*Loss of proprioception (vibration/position senses)*				
	Position in toes/ankles			3	4
	Total loss of vibration sense in toes (256 Hz)	1	2		
	Same, in ankles (metatarsals)	1	2		
	Partial/total loss of vibration sense in inner malleoli (256 Hz)	1	2	3	4
	Same, in tibial tuberosities	1	2	3	4
	Same, in hands	1	2	3	4
	Same, in sternum			3	
5c	*Some sign of transverse lesion*			3	
	Sum test 5 →				
6	*Hyporeflexia*				
	In upper extremities	1	2	3	4
	In lower extremities	1	2	3	4
	Plantar reflexes abolished	1	2		
	Sum test 6 →				
7	*Ataxia*				
	Nonparetic nystagmus			3	
	Jerky eye movements	1			
	Nonparetic dysarthria			3	
	Static/intention tremor	1	2		
	Cerebellar bradykinesia	1	2	3	4
	Incoordination, lower extremities	1	2	3	4
	Incoordination, upper extremities	1	2	3	4
	Dys (a) diadochokinesia	1	2		
	Other cerebellar signs	1	2	3	4
	Romberg's test, unsteady/positive	1		3	
	Ataxic gait			3	
	Sum test 7 →				
8	*Pyramidal tract signs and spasticity*				
	Bradykinesia/clumsiness in feet	1	2	3	4
	Same, in hands	1	2	3	4
	Spasticity in legs	1	2	3	4
	Spasticity in arms	1	2	3	4
	Spastic gait			3	
	Hyperreflexia, lower extremities	1	2	3	4
	Hyperreflexia, upper extremities	1	2	3	4
	Inverted plantar responses	1	2	3	4
	Shortening reflexes			3	4
	Sum test 8 →				
9	*Basal ganglia signs*				
	Rigidity	1	2	3	4
	Parkinsonian tremor (resting)	1	2	3	4

TABLE A5.2. Clinical Examination of the Nervous System (*continued*)

Subtest Number		Score
	Essential (senile) tremor	1 2 3 4
	Immobile face	3
	Oligokinesia	1 2 3 4
	Synkinesis	1 2 3 4
	Chorea/athetosis	1 2 3 4
	Tortion dystonia	3
	Sum test 9	→
	Total corrected score	→

TABLE A5.3. Prevalence of Neurological Signs in a Randomized Population Sample (in percent)

Subtest and Signs	Maximum Test Score	Age Group 1		Age Group 2		Age Group 3		Age Group 4		Age Group 5		Mean Percent Score
		M	F	M	F	M	F	M	F	M	F	
1 *CNS dysfunction*												
Feebleminded	1											1,3
Neurotic	1											2,1
Psychotic, and the like	1											0,3
Dementia	3	0	0	0	0	0	0	0	0	2,6	2,9	
Epilepsy	3											0,3
2 *Special sense organs*												
Tapetoretinal degeneration	3											0
Optic atrophy, and the like	3	0	0	0	2,5	0	0	0	1,4	3,8	5,9	
Error of refraction	1											3
Cataract	3	0	0	0	0	0	0	0	1,8	0,9	2,9	
Strabismus	3	0,9	0	0	0	0,9	1	0	0,9	0	8,8	
Ocular "myopathy"	3	0	0	2,6	2,6	0	0	2,6	2,7	5,1	5,7	
VIIIth nerve affected	3	0,9	0	0	0	0,9	0	3,5	2,7	3,4	1	
3 *Deformities*												
Hammer toes	3	3,6	4,6	7,2	6,7	9	15,2	14	18,9	12,8	20	
Talipes cavus	4	0,7	0	0	0	1,3	1,4	1,9	1,4	0	0	
Camptodactyly	3	0	0	5,4	2,5	6,1	2,8	11,4	9,9	17,1	11,4	
Excavated palms	4	0	0	0	0	0	0	0,7	1,4	0,6	2,9	
Scoliosis	3											0,7
Anomalies in digits	1											0,5
Hallux valgus	2	0	0	2,7	2,5	2,7	4,3	2,6	13,5	2,6	15,7	
Pes equinovarus	4											0,6
Luxatio coxae congentia	2											0,7

4 *Neuromuscular affections*												
“Fat bottle calf”	3											0
In toes	2	2,8	9,5	13,5	22,5	12,2	40	52,6	73	53,8	66,2	
In hands	4	2,1	0	1,4	2,5	0,7	1,4	2,6	2,7	5,8	6,3	
In legs	4	0	0	4,7	0	0	0	1,3	1,4	2,6	2,1	
In forearms, thighs, arms, and limb girdles, each	4											0
In trunk	3											0
Of bulbar motor nerves 5 and 9–12	3											0
Fasciculations	3											0
5 *Sensibility loss*												
Touch, below knees	4	2,8	9,5	6,8	6,3	7,9	7,1	3,9	2,7	9,6	5,7	
Do., below elbows	4	9,7	13,5	6,8	10	10,5	5,7	5,3	5,4	9	1,4	
Pain temperature, below knees	4	2,8	8,1	10,8	2,6	7,9	7,1	9,2	6,1	10,3	5,7	
Do., below elbows	4	7	13,5	8,1	7,5	13,2	7,1	9,2	8,1	9	1,4	
Superficial, proximal LE and UE, each	4											0
Position sense, toes and ankles	4	0	2,7	0	0	0	0	2,7	2,7	19,2	5,7	
Vibration, total loss in toes	2	9,7	14,9	22,2	12,5	45,6	32,9	59,2	68,9	86,6	71,4	
Do., in ankles	2	0	8,1	9,5	5	24,3	15,7	30,3	36,5	71,8	50	
Vibration, in malleoli	4	0,7	0	2	0,6	18,2	5,7	13,2	25	43,6	27,1	

(continued)

TABLE A5.3. Prevalence of Neurological Signs in a Randomized Population Sample (in percent) (*continued*)

Subtest and Signs	Maximum Test Score	Age Group 1		Age Group 2		Age Group 3		Age Group 4		Age Group 5		Mean Percent Score
		M	F	M	F	M	F	M	F	M	F	
Do., in tibial tuberosity	4	0	0	0	0	0	2,8	7,9	6,6	26,3	5,7	
Vibration, in hands	4	0	0	0	0	1,4	0	0	0	7,1	0	
Do., in the sternum	3											0
All qualities, transverse lesion	3	0	0	0	0	0	0	5,3	2,7	15,4	4,3	
6 *Hyporeflexia*												
In UE	4	0	0	0,7	0	0,7	2,9	4,6	0,7	10,3	0	
In LE	4	2,8	2,7	3,4	1,3	2	5,7	15,1	0	16	12,1	
Abolished plantar response, not paretic	2	0	1,4	2,7	1,3	0	0	0	5,4	6,4	11,1	
7 *Ataxia, and the like*												
Fixation nystagmus	3											0,3
Ataxic ocular movements	1											0,5
Nonparetic dysarthria	3											0
Intention/static tremor	2	8,3	13,5	5,4	7,4	21,6	14,3	26,4	25,7	56,2	54	
Nonparetic bradykinesia	4	0	1,4	0	0	0	0	1,3	5,4	3,8	2,9	
Incoordination, in LE	4	0	0	0	0	0	0	2,6	0	0,4	4,3	
Do., in UE	4	0	0	0	0	0	0	1,3	0	1,3	1,4	
Dysdiadochokinesia	2	0	0	2,7	0	5,4	10	10,5	5,4	17,9	8,6	
Other cerebellar signs	4	0	0	0	0	0	0	0	0	0	0,7	
Romberg's test	3	0	0	0	0	2,6	0	2,6	2,7	5,1	5,7	
Ataxic gait	3	0	0	0	0	0	0	0	0	5,1	0	
8 *Pyramidal and spastic signs*												

	Paresis/bradykinesia, distal LE	4	0	0	1,4	0	0	1,4	0	1,4	0	0	
	Do., distal UE	4											0
	Spasticity, LE	4	0	0	0	1,9	0	0	0	0	1,3	0	
	Do., UE	4											0
	Spastic gait	3											0
	Exaggerated tendon reflexes, LE	4	0	0	0	1,3	0	0,7	0	1,4	0	0	
	Do., UE	4	0	0	0	0	0	0,7	0	0	0	0	
	Babinski's and Chaddock's reflexes	4	0	0	0	1,9	0	4,3	0,7	3,4	1,9	2,1	
	Flexion reflexes	4											0
9	*Basal ganglia signs*												
	Rigidity	4	0	0	0	0	0	0	0	1,4	5,1	4,3	
	Tremor at rest	4	0	0	0	0	0	0	0	0	1,3	0	
	Senile tremor	4	0	0	0	0	0	0	0	0	2,6	2,9	
	Facial mask	3	0	0	0	0	0	0	0	0	2,6	0	
	Other oligo- or hyperkinesis, 4 tests, each	4											0
	Number investigated		36	37	38	40	37	35	38	37	39	35	

Note: When a mean score is used in this table, no specific trend in the age/sex distribution of scores is observed. The mean score applies equally to all age/sex categories.

Source: Skre (1972).

Correction of the observed scores was a weighting based on the figures in Table A5.3 corresponding to the actual test for the age group and sex in question:

$$\text{Corrected score} = \frac{\text{observed score}}{\text{prevalence figure}}.$$

No correction was done if the prevalence was ≤ 1. A sum score was obtained by adding all corrected score figures. A subtest score was obtained by adding all corrected score figures within a subtest (compare Tables A5.1 and A5.2).

We thank Munksgaard International Publishers (Copenhagen), Masson Editeur (Paris), and Churchill Livingstone (Edinburgh) for permission to reproduce material from published papers. We also would like to acknowledge Dr. Lene Pedersen, Rigshospitalet, University of Copenhagen, for permission to include Figure 5.9 and Professor Paul Trouillas, Hospital de l'Antiquaille, Lyon, for permission to include Figure 5.10.

REFERENCES

Adams JH. (1975) Hypoxic brain damage. British Journal of Anesthesiology. 1975: 47: 121–129.

Aggerbeck LP, McMahon JP, Scanu AM. (1974) Hypobetalipoproteinemia: clinical and biochemical description of a new kindred with "Friedreich's ataxia." Neurology (Minneapolis). 1974: 24: 1051–1063.

Aguilar MJ, Kamoshita S, Landing BH, Boder E. (1968) Pathological observations in ataxia telangiectasia. A report on five cases. Journal of Neuropathology and Experimental Neurology. 1968: 27: 659–676.

Alter M, Talbert OR. (1962) Cerebellar ataxia, congenital cataracts, and retarded somatic and mental maturation. Report of cases of Marinesco-Sjögren syndrome. Neurology (Minneapolis). 1962: 12: 836–847.

Amman AJ, Cain WA, Ishizaka K, Hong R, Good RA. (1969) Immunoglobulin E deficiency in ataxia telangiectasia. New England Journal of Medicine. 1969: 281: 469–472.

Andersen B. (1965) Marinesco-Sjögren syndrome: spinocerebellar ataxia, congenital cataract, somatic and mental retardation. Developmental Medicine Child Neurology. 1965: 7: 249–257.

Andersson WA, Flumerfelt BA. (1980) A light and electron microscopic study of the effects of 3-acetylpyridine intoxication of the inferior olivary complex and cerebellar cortex. Journal of Comparative Neurology. 1980: 190: 157–174.

André van Leeuven, M. (1948) Hérédoataxies par dégénérescence spino-ponto-cérébelleuse. Les manifestations optiques. Revue Oto-neuro-ophthalmologique. 1948: 20: 43–108.

Andres A, Satrustegui J, Machada A. (1980) Development of NADPH-producing pathways in rat heart. Biochemical Journal. 1980: 186: 799–803.

Azari J, Brumbaugh P, Barbeau A, Huxtable R. (1980) Taurine decreases lesion severity in the heart of cardiomyopathic hamsters. Canadian Journal of Neurological Sciences. 1980: 7(4): 435–440.

Azari J, Reisine T, Barbeau A, Yamamura HI, Huxtable R. (1979) The Syrian golden hamster: a model for the cardiomyopathy of Friedreich's ataxia. Canadian Journal of Neurological Sciences. 1979: 6: 223–226.

Azizi E, Zaidman JL, Eschar J, Szeinberg A. (1978) Abetalipoproteinemia treated with parenteral and oral vitamins A and E, and with MCT. Acta Paediatrica Scandinavica. 1978: 67: 797–801.

Bachelard HS. (1976) Carbohydrate and energy metabolism of the central nervous system: biochemical approach. In: Vinken PG, Bruyn GW (eds). Handbook of clinical neurology, vol 27. North-Holland/Elsevier, Amsterdam, Oxford, New York, ch 1, pp 1–26.

Bank WJ, Pizer L, Pfendner W. (1978) Glycine metabolism and spinal cord disorders. In: Kark RAP, Rosenberg RN, Schut LJ (eds). Advances in neurology, vol. 21: The inherited ataxias. Raven Press, New York, pp 267–278.

Bar R, Lewis RS, Rechler MM, Harrison LC, Siebert C, Podskalny J, Roth V, Muggeo M. (1978) Extreme insulin resistance in ataxia telangiectasia. Defect in affinity of insulin receptors. New England Journal of Medicine. 1978: 298: 1164–1171.

Baraitser M. (1982) Cerebellar syndromes. In: Fraser Roberts JA, Carter CD, Motulsky AG (eds). The genetics of neurological disorders. Oxford monographs on medical genetics, Oxford University Press, Oxford, New York, Toronto, pp 125–154.

Barbeau A. (1975) Preliminary studies on pyruvate metabolism in Friedreich's ataxia. Transactions. American Neurological Association. 1975: 100: 164–165.

Barbeau A. (1976) Quebec cooperative study of Friedreich's ataxia. Design of investigation. Canadian Journal of Neurological Sciences. 1976: 3: 271–274.

Barbeau A. (1978a) Emerging treatments: replacement therapy with choline or lecithin in neurological diseases. Canadian Journal of Neurological Sciences. 1978a: 5: 157–160.

Barbeau A. (1978b) Friedreich's ataxia 1978: an overview. Canadian Journal of Neurological Sciences 1978b: 5: 161–165.

Barbeau A. (1979) Friedreich's ataxia 1979: an overview. Canadian Journal of Neurological Sciences. 1979: 6: 311–319.

Barbeau A. (1982) Pathophysiology of Friedreich's ataxia. In: Matthews WB, Glaser HG (eds). Recent advances in clinical neurology. Churchill Livingstone, Edinburgh, pp 125–145.

Barbeau A, Melancon S, Huxtable RJ, Lemieux B. (1981) Taurine and Friedreich's ataxia: an update. Advances of experimental medicine and biology, 1981 vol 139: 389–399.

Barbeau A, de Siegle M, Breton G, Coallier R, Bouchard JP. (1976) Friedreich's ataxia: preliminary results of some genealogic research. Canadian Journal of Neurological Sciences. 1976: 3: 303–306.

Bassen FA, Kornzweig AL. (1950) Malformation of the erythrocytes in a case of atypical retinitis pigmentosa. Blood. 1950: 5: 381–387.

Becker PE. (1966a) Krankheiten mit hauptsachlicher Beteiligung des spino-cerebellaren Systems (Erbliche Ataxien). In: Humangenetik, vol V/1. Thieme, Stuttgart, pp 245–313.

Becker PE. (1966b) Spastische Spinalparalysen. In: Humangenetik, vol V/1. Thieme, Stuttgart, pp 314–330.

Becker PE, Sabuncu N, Hopf HC. (1971) Dominant erblicher Typ von "cerebellarer Ataxie." Zeitschrift für Neurologie. 1971: 119: 116–139.

Behan PO. (1983) Immunological mechanisms in Refsum's disease. Acta Neurologica Scandinavica. 1983: 67: 191.

Behan WM, Maia M. (1974) Strümpell's familial spastic paraplegia: genetics and neuropathology. Journal of Neurology, Neurosurgery and Psychiatry. 1974: 37: 8–20.

Bell J, Carmichael EA. (1939) On hereditary ataxia and spastic paraplegia. Treasury of Human Inheritance. 1939: 4: 141–281.

Berg K. (1978a) Genetic linkage as a cause of clinical variation in inherited disorders. In: Hamerton JL, Klinger HP, McKusick VA, Evans J (eds). Human Gene Mapping 4, Birth Defects: Original Article Series Vol. XIV, No. 4, The National Foundation, March of Dimes.

Berg K. (1978b) Genetic linkage as a source of clinical variation in inherited disorders. Abstract presented at the Sigrid Juselius Symposium, Åland, August 13–16, 1978.

Berg K. (1979) Inherited lipoprotein variation and atherosclerotic disease. In: Schanu AM, Wissler RW, Getz GS (eds). The biochemistry of atherosclerosis. Marcel Dekker, New York, pp 419–490.

Berg K, Skre H. (1976) Possible linkage between the Marinesco-Sjögren syndrome and hypergonadotropic hypogonadism. Birth Defects: Original Article Series. 1976: 12: 271–274.

Bickerstaff ER. (1950) Hereditary spastic paraplegia. Journal of Neurology, Neurosurgery and Psychiatry. 1950: 13: 134–145.

Billimoria JD, Clemens ME, Gibberd FB, Whitelaw MN. (1982) Metabolism of phytanic acid in Refsum's disease. Lancet. 1982: I: 194–196.

Blass JP. (1979) Disorders of pyruvate metabolism. Neurology (Minneapolis). 1979: 29: 280–286.

Bloom FE, Hoffer BV, Siggins GR. (1971) Studies on norepinephrine containing afferents to Purkinje cells of rat cerebellum. I. Localisation of the fibres and their synapses. Brain Research. 1971: 25: 501–521.

Bobillier P, Seguin S, Petitjean F, Salvert D, Touret M, Jouvet M. (1976) The raphe nuclei of the cat brain stem: a topographical atlas of their afferent projections as revealed by autoradiography. Brain Research. 1976: 113: 449–486.

Boder E, Sedgwick RP. (1963) Ataxia telangiectasia: a review of 101 cases. Little Club Clinics in Developmental Medicine. 1963: 8: 110–118.

Boller F, Segarra JM. (1969) Spino-pontine degeneration. European Neurology. 1969: 2: 356–373.

Boller F, Segarra JM. (1975) Spino-pontine degeneration. In: Vinken PJ, Bruyn GW (eds). Handbook of clinical neurology, vol 21. North-Holland/Elsevier, Amsterdam, Oxford, New York, pp 389–402.

Borud O, Torp KH, Purkiss P. (1979) Increased urinary oligosaccharides in Marinesco-Sjögren syndrome. Paper read at 3rd European Congress of Clinical Chemistry, Brighton, June 3–8, 1979.

Bouchard JP, Barbeau A, Bouchard R, Paquet M, Bouchars RW. (1979) A cluster of Friedreich's ataxia in Rimovski, Québec. Canadian Journal of Neurological Sciences. 1979: 6: 205–208.

Bowden DH, Danis PG, Sommers SC. (1963) Ataxia-telangiectasia: case with lesions of ovaries and adenohypophysis. Journal of Neuropathology and Experimental Neurology. 1963: 22: 549–554.

Brown JR. (1962) Diseases of the cerebellum. In: Baker AB (ed). Clinical neurology, vol 3. Harper & Row, New York, pp 1406–1455.

Brown S. (1892) On hereditary ataxy with a series of twenty-one cases. Brain. 1892: 15: 250–282.

Butterfield DA, Leung PK, Markesberry A. (1979) Evidence for an altered physical state of membrane proteins in erythrocytes in Friedreich's ataxia. Canadian Journal of Neurological Sciences. 1979: 6: 295–298.

Butterworth RF, Shapcott D, Melancon S, Breton G, Geoffroy G, Lemieux B, Barbeau A. (1976) Clinical laboratory findings in Friedreich's ataxia. Canadian Journal of Neurological Sciences. 1976: 3: 355–360.

Cammermeyer J. (1946) Om de anatomiske funn i to tilfeller av Dr. S Refsums materiale av "et tidligere ikke beskrevet familiaert syndrom". Nordisk Medicine. 1946: 26: 617.

Cammermeyer J. (1975) Refsum's disease. Neuropathological aspects. In: Vinken PJ, Bruyn GW (eds). Handbook of clinical neurology, vol 21. North-Holland/Elsevier, Amsterdam, Oxford, New York, ch 11, pp 231–261.

Carpenter S, Schumacher GA. (1966) Familial infantile cerebellar atrophy associated with retinal degeneration. Archives of Neurology (Chicago). 1966: 14: 82–94.

Carter HR, Sukavaja C. (1956) Familial cerebello-olivary degeneration with late development of rigidity and dementia. Neurology (Minneapolis). 1956: 6: 876–884.

Cavanagh JB. (1964) The significance of the "dying back" process in experimental and human neurological disease. In: Richter GW, Epstein MA (eds). International review of experimental pathology, vol 3. Academic Press, New York, London, pp 219–267.

Chamberlain S, Robinson N, Walker J, Smith C, Benton S, Kennard C, Swash M, Kilkenny B, Bradbury S. (1980) Effect of lecithin on disability and plasma free-choline levels in Friedreich's ataxia. Journal of Neurology, Neurosurgery and Psychiatry. 1980: 43: 843–845.

Chan-Palay V. (1975) Fine structure of labelled axons in the cerebellar cortex and nuclei of rodents and primates after intraventricular infusions with tritiated serotonin. Anatomy Embryology. 1975: 148: 235–265.

Chan-Palay V. (1977) Cerebellar dentate nucleus organisation, cytology and transmitters. Springer-Verlag, Berlin, pp 390–463.

Chan-Palay V. (1979) Indolaminergic neurons and their processes in the normal rat brain and in chronic diet-induced thiamine deficiency, demonstrated by ^{3}H-serotonin. Journal Comparative Neurology. 1979: 176: 467–494.

Chen K-M, Brody JA, Kurland LT. (1968) Patterns of neurologic diseases on Guam. 1. Epidemiologic aspects. Archives of Neurology (Chicago). 1968: 19: 573–578.

Coërs C, Woolf AL. (1959) The innervation of muscle. A biopsy study. Blackwell, Oxford, pp 65–70.

Cooper JR, Itokawa Y, Pincus JH. (1969) Thiamine triphosphate deficiency in subacute necrotizing encephalomyelopathy. Science. 1969: 164: 74–75.

Cooper R. (1977) Abnormalities of cell membrane fluidity in the pathogenesis of disease. New England Journal of Medicine. 1977: 297: 371–377.

Coquerelle TM, Weibezahn KF. (1981) Rejoining of DNA double-strand breaks in human fibroblasts and its impairment in one ataxia telangiectasia and two Fanconi strains. Journal of Supramolecular Structure and Cellular Biochemistry. 1981: 17: 369–376.

Cote M, Bureau M, Léger C, Martin J, Gattiker H, Cimon M, Larose A, Lemieux B. (1979) Evolution of cardiopulmonary involvement in Friedreich's ataxia. Canadian Journal of Neurological Sciences. 1979: 6: 151–157.
Coutinho P, Andrade C. (1978) Autosomal dominant system degeneration in Portuguese families of the Azores islands: a new genetic disorder involving cerebellar, pyramidal, extrapyramidal and spinal cord motor functions. Neurology (Minneapolis). 1978: 28: 703–709.
Cross HE, McKusick VA. (1967a) The Mast syndrome. A recessively inherited form of presenile dementia with motor disturbances. Archives of Neurology (Chicago). 1967a: 16: 1–13.
Cross HE, McKusick VA. (1967b) The Troyer syndrome. A recessive form of spastic paraplegia with distal muscle wasting. Archives of Neurology (Chicago). 1967b: 16: 473–485.
Curtius F, Störring FK, Schönberg K. (1935) Über Friedreich'scher Ataxie und Status dysraphicus (Zugleich ein Beitrag zu den Beziehungen zwischen Friedreich'scher Ataxie und Diabetes mellitus). Zeitschrift den gesamte Neurologie und Psychiatrie. 1935: 153: 719–743.
David RB, Mamunes P, Rosenblum WI. (1976) Necrotizing encephalomyelopathy (Leigh). In: Vinken PJ, Bruyn GW (eds). Handbook of clinical neurology, vol 29. North-Holland/Elsevier, Amsterdam, Oxford, New York, ch 14, pp 349–363.
Davignon J, Huang YS, Wolf JP, Barbeau A. (1979) Fatty acid profile of major lipid classes in plasma lipoproteins of patients with Friedreich's ataxia. Demonstration of a low linoleic content most evident in the cholesteryl-ester fraction. Canadian Journal of Neurological Sciences. 1979: 6: 275–283.
Davis EJ, Spydevold Ø, Bremer J. (1980) Pyruvate carboxylase and propionyl-CoA carboxylase as anapleurotic enzymes in skeletal muscle mitochondria. European Journal of Biochemistry. 1980: 110: 225–262.
De Michele G, Jolicoeur JB, Rondeau DB, Butterworth RF, Barbeau A. (1980) Effects of glutamate and aspartate on ataxic gait induced by 3-acetylpyridine in rats. Canadian Journal of Neurological Sciences. 1980: 7(4): 451–454.
DiMauro S, Schotland DL, Bonilla E, Lee CP, Gambetti P, Rowland LD. (1973) Progressive ophthalmoplegia, glycogen storage and abnormal mitochondria. Archives of Neurology (Chicago). 1973: 29: 170–179.
Dimitrijevic MR, Lenman J, Prevec T, Wheathy K. (1982) A study of posterior column function in familial spastic paraplegia. Journal of Neurology, Neurosurgery and Psychiatry. 1982: 45: 46–49.
Drachman DA. (1968) Ophthalmoplegia plus; the neurodegenerative disorders associated with progressive external opthalmoplegia. Archives of Neurology. 1968: 18: 654–674.
Drachman DA. (1975) Opthalmoplegia plus; a classification of the disorders associated with progressive external opthalmoplegia. In: Vinken PJ, Bruyn GW (eds). Handbook of clinical neurology, vol 22. North-Holland/Elsevier, Amsterdam, Oxford, New York, ch 9, pp 203–216.
Draper R, Shapcott D, Larose A, Stankova J, Levesque F, Lemieux B. (1979) Glucose tolerance and erythrocyte insulin receptors in Friedreich's ataxia. Canadian Journal of Neurological Sciences. 1979: 6: 233–239.
Draper R, Shapcott D, Langlois M, Lemieux B. (1980) Glycosylated hemoglobin in Friedreich's ataxia. Canadian Journal of Neurological Sciences. 1980: 7: 405–407.
Draper R, Huang YS, Schapco HD, Lemieux B, Brennan N, Barbeau A, Davignon J. (1979) Erythrocyte membrane lipids in Friedreich's ataxia. Canadian Journal of Neurological Sciences. 1979: 6: 291–294.
Eadie MJ. (1975a) Cerebello-olivary atrophy (Holmes type) In: Vinken PJ and Bruyn GW. Handbook of clinical neurology, vol 21. North-Holland/Elsevier, Amsterdam, Oxford and New York, ch 19, pp 403–414.
Eadie MJ. (1975b) Hereditary spastic ataxia. In: Vinken PJ, Bruyn GW (eds). Handbook of clinical neurology, vol 21. North-Holland/Elsevier, Amsterdam, Oxford, New York, ch 15, pp 365–375.
Eadie MJ. (1975c) Olivo-ponto-cerebellar atrophy (Dejerine-Thomas). In: Vinken PJ, Bruyn GW (eds). Handbook of clinical neurology, vol 22. North-Holland/Elsevier, Amsterdam, Oxford, New York, ch 20, pp 415–431.
Eadie MJ. (1975d) Olivo-ponto-cerebellar atrophy (Mentzel type). In: Vinken PJ, Bruyn GW (eds). Handbook of clinical neurology, vol 21. North-Holland/Elsevier, Amsterdam, Oxford, New York, pp 433–449.
Eadie MJ. (1975e) Olivo-ponto-cerebellar atrophy (variants). In: Vinken PJ, Bruyn GW (eds).

Handbook of clinical neurology, vol 21. North-Holland/Elsevier, Amsterdam, Oxford, New York, ch 22, pp. 451-457.
Ebels EJ. (1974) Underlying illness in Wernicke's encephalopathy. Analysis of possible causes of underdiagnosis. European Neurology. 1974: 12: 226-228.
Eccles JC. (1966) Functional organisation of the cerebellum in relation to its role in motor control. In: Granit R (ed). Muscular afferents and motor control. Nobel symposium, Lindingö, 1965. Södergran, Stockholm: 19-36.
Eccles J. (1977) Cerebellar function in the control of movement. In: Rose PC (ed). Physiological aspects of clinical neurology. Blackwell Scientific Publications, pp. 157-178.
Edwards JH. (1960) The simulation of Mendelism. Acta Genetica (Basel). 1960: 10: 63-70.
Eldjarn L, Stokke O, Try K. (1977) Biochemical aspects of Refsum's disease and principles for dietary treatment. In: Vinken PJ, Bruyn GW (eds). Handbook of clinical neurology, vol 27. North-Holland/Elsevier, Amsterdam, Oxford, New York, ch 23, pp 519-541.
Elias E, Muller DPR, Scott J. (1981) Association of spinocerebellar disorders with cystic fibrosis or chronic childhood cholestasis and very low serum vitamin E. Lancet. 1981: 1319-1321.
Falconer DS. (1965) The inheritance of liability to certain diseases, estimated from the incidence among relatives. Annals of Human Genetics. 1965: 29: 51-76.
Ferguson FR, Critchley M. (1929) A clinical study of an heredo-familial disease resembling disseminated sclerosis. Brain. 1929: 52: 203-225.
Fickler A. (1911) Klinische und Patologisch-anatomische Beiträge zu dem Erkrankungen des Kleinhirns. Deutschen Zeitschrift für Nervenheilkunde.
Filla A, Butterworth RF, Geoffroy G, Lemieux B, Barbeau A. (1978) Platelet taurine uptake in spinocerebellar degeneration. Canadian Journal of Neurological Sciences. 1978: 5: 119-123.
Ford MD, Houldsworthy J, Lavin MF. (1981) DNA-repair synthesis in ataxia telangiectasia lymphoblastoid cells. Mutation Research. 1981: 84: 419-427.
Fraser D. (1880) Defect of the cerebellum occurring in a brother and sister. Glasgow Medical Journal 1880: 13: 199-210.
Fredrickson DS, Goldstein JL, Brown MS. (1978) Familial hyperlipoproteinemias. In. Stanbury JB, Wyngaarden JB, Fredrickson DS (eds). The metabolic basis of inherited disease. 4th ed. McGraw-Hill, New York, ch 30, pp 604-655.
Friedreich N. (1863) Ueber degenerative Atrophie der spinalen Hinterstränge. Virchows Archiv für pathologische Anatomie und Phisilogie und für klinische Medizin. 1863: 26: 391-419.
Friedreich N. (1876) Ueber degenerative Atrophie der spinalen Hinterstränge. Virchows Archiv für pathologische Anatomie und Phisilogie und für klinische Medizin. 1876: 68: 145-245.
Fryer DG, Winckelman AC, Ways PO, Swanson AG. (1971) Refsum's disease. A clinical and pathological report. Neurology (Minneapolis). 1971: 21: 162-167.
Garrod AE. (1908) Inborn errors of metabolism. Lancet. 1908: II: 1.
Gautier JC, Landat PH, Rosa A, Gray F, Lhermitte F. (1973) Maladie de Refsum. Test de charge en phytol chez un descendant. Presse medicale. 1973: 2: 2029-2032.
Geoffroy G, Barbeau A, Breton G, Lemieux B, Aube M, Léger C, Bouchard JP. (1976) Clinical description and roentgenologic evaluation of patients with Friedreich's ataxia. Canadian Journal of Neurological Sciences. 1976: 3: 279-286.
Gilbert GJ, McEntee III WJ, Glaser GH. (1963) Familial myoclonus and ataxia. Pathophysiologic implications. Neurology (Minneapolis). 1963: 13: 365-372.
Gilman S, Horenstein S. (1964) Familial amyotropic dystonic paraplegia. Brain. 1964: 87: 51-66.
Goldman H, Scriver CR. (1967) A transport system in mammalian kidney with a preference for beta-amino compounds. Pediatric Research. 1967: 1: 212-213.
Gowers WR. (1902) A lecture on abiotrophy. Lancet. 1902: 1: 1003-1007.
Greenfield JG. (1954) The spino-cerebellar degenerations. Blackwell, Oxford.
Grobe H, Bassewitz DBv, Dominick HC. (1975) Subacute necrotizing encephalomyelopathy. Clinical, ultrastructural, biochemical and therapeutic studies in an infant. Acta Paediatrica Scandinavica. 1975: 64: 755-762.
Gruskin AB, Patel MS, Linshaw M, Ettrenger R, Huff D, Grower W. (1973) Renal function studies and kidney pyruvate carboxylase in subacute necrotizing encephalomyelopathy (Leigh's syndrome). Pediatric Research. 1973: 7: 832.
Guidotti A, Biggio G, Costa E. (1975) 3-acetylpyridine: a tool to inhibit the tremor and the increase of cGMP content in cerebellar cortex elicited by harmaline. Brain Research 1975: 96: 201-205.
Hall PV, Smith JE, Lane J, Mote T, Campell R. (1970) Glycine and experimental spinal spas-

ticity. Neurology (Minneapolis). 1970: 29: 262-267.
Hallet M. (1979) Physiology and pathophysiology of voluntary movement. In: Tyler HR, Dawson DM (eds). Current neurology. Houghton-Mifflin Medical Division, Boston, ch 21 pp 351-376.
Harding AE. (1981a) Friedreich's ataxia: a clinical and genetic study of 90 families with an analysis of early diagnostic criteria and intrafamilial clustering of clinical features. Brain. 1981a: 104: 589-620.
Harding AE. (1981b) Hereditary "pure" spastic paraplegia: a clinical and genetic study of 22 families. Journal of Neurology, Neurosurgery and Psychiatry. 1981b: 44: 871-883.
Harding AE. (1982) The clinical features and classification of the late onset autosomal dominant cerebellar ataxias. A study of 11 families, including descendants of "the Drew family of Walworth." Brain. 1982: 105: 1-28.
Harding AE. (1983) Classification of the hereditary ataxias and paraplegias. Lancet. 1983: I: 1151-1155.
Headley PM, Lodge D, Duggan AN. (1976) Drug induced rhythmical activity in the inferior olivary complex of the rat. Brain Research. 1976: 101: 461-478.
Herbert PN, Gotto AM, Fredrickson DS. (1978) Familial lipoprotein deficiency. In: Stanbury JB, Wyngaarden JB, Fredrickson DS (eds). The metabolic basis of inherited disease. 4th ed. McGraw-Hill, New York, ch 28, pp 533-569.
Herndon JH, Steinberg D, Uhlendorf BW. (1969) Refsum's disease. Defective oxidation of phytanic acid in tissue cultures derived from homozygotes and heterozygotes. New England Journal of Medicine. 1969: 281: 1034-1038.
Herndon RM. (1978) Selective vulnerability in the cerebellum. In: Kark RAP, Rosenberg RN, Schut LJ (eds). Advances in neurology, vol 21: The inherited ataxias. Raven Press, New York, ch 23, pp 319-330.
Hill W, Dysart D. (1975) Hereditary periodic ataxias. In: Vinken PJ, Bruyn GW (eds). Handbook of clinical neurology, vol 21. North-Holland/Elsevier, Amsterdam, Oxford, New York, ch 31, pp 563-571.
Hodskins MB, Yakovlev PI. (1930) Anatomo-clinical observations on myoclonus in epileptics and on related symptom complexes. American Journal of Psychiatry. 1930: 86: 827-848.
Hoffman DG, Sladek Jr. JR (1973) The distribution of catecholamines within the inferior olivary complex of the gerbil and rabbit. Journal Comparative Neurology. 1973: 151: 101-112.
Hökfelt T, Fuxe K. (1969) Cerebellar monoamine terminals, a new type of afferent fibres to the cortex cerebelli. Experimental Brain Research. 1969: 9: 63-72.
Holmes G. (1907a) An attempt to classify cerebellar disease, with a note on Marie's hereditary cerebellar ataxia. Brain. 1907a: 30: 545-567.
Holmes G. (1907b) A form of familial degeneration of the cerebellum. Brain. 1907b: 30: 466-489.
Hughes JI, Brownell B, Hewer RL. (1968) The peripheral sensory pathway in Friedreich's ataxia. Canadian Journal of Neurological Sciences. 1968: 5: 83-91.
Huxtable R, Azari J, Reisine T, Johnson P, Yamamura H, Barbeau A. (1979) Regional distribution of amino acids in Friedreich's ataxia brains. Canadian Journal of Neurological Sciences. 1979: 6: 255-262.
Igarashi M, Schlumberg HH, Powers J, Kishimoto Y, Kilondy E, Suzuki K. (1976) Fatty acid abnormality in adrenoleukodystrophy. Journal of Neurochemistry. 1976: 26: 851-860.
Illingworth DR, Connor WE, Miller RG. (1980) Abetalipoproteinemia. Report of two cases and review of therapy. Archives of Neurology (Chicago). 1980: 37: 659-662.
Ionescu M. (1970) Tulburâri endocrine si de la metabolism la a bolnavâ cu sindrom Marinesco-Sjögren. Neurologia, Psihiatria, Neurochirurgia. 1970: 15: 145-157.
Ito M. (1978) Recent advances in cerebellar physiology and pathology. In: Kark RAP, Rosenberg RN, Schut LJ (eds). Advances in neurology, vol 21: The inherited ataxias. Raven Press, New York, ch 4, pp 59-84.
Jackson JF, Currier RD, Terasaki PI, Morton NE. (1977) Spinocerebellar ataxia and HLA linkage. New England Journal of Medicine. 1977: 296: 1138-1141.
Jampell RS, Falls HF. (1958) Atypical retinitis pigmentosa, acanthocytosis and heredo-degenerative neuromuscular disease. Archives of Ophthalmology. 1958: 59: 818-820.
Jellum E, Kluge T, Börresen HC, Stokke O, Eldjarn L. (1970) Pyroglutamic aciduria: a new inborn error of metabolism. Scandinavian Journal of Clinical and Laboratory Investigation. 1970: 26: 327-335.
Jepson JB. (1978) Hartnup disease. In: Stanbury JB, Wyngaarden JB, and Fredrickson DS.

(eds). The metabolic basis of inherited disease. 4th ed. McGraw-Hill, New York, ch 66, pp 1563–1577.

Johnson WG. (1981) The clinical spectrum of hexosaminidase deficiency diseases. Neurology (Minneapolis). 1981: 31: 1453–1456.

Jones J, Baraitser M, Halliday AM. (1980) Peripheral and central somato-sensory nerve conduction defects in Friedreich's ataxia. Journal of Neurology, Neurosurgery, and Psychiatry. 1980: 43: 495–503.

Kannan V, Subramanyam K. (1977) Maple syrup urine disease. Thiamine responsive branched chain aminoaciduria: a unique case in an adult. Journal of the Association of Physicians of India. 1977: 25: 163–166.

Kark RAP, Blass JP, Engel WK. (1974) Pyruvate oxidation in neuromuscular disease. Evidence for a genetic defect in two families with the clinical syndrome of Friedreich's ataxia. Neurology (Minneapolis). 1974: 24: 964–971.

Kark RAP, Blass JP, Spence MA. (1977) Physostigmine in familial ataxias. Neurology (Minneapolis). 1977: 27: 70–72.

Kark RAP, Rodriguez-Budelli M, Blass JP. (1978) Evidence for a primary defect of lipoamide dehydrogenase in Friedreich's ataxia. (1978) In: Kark RAP, Rosenberg RN, Schut LJ (eds). Advances in neurology, vol 21: The inherited ataxias. Raven Press, New York, ch 11, pp 163–180.

Kaufman DB, Miller HC. (1977) Ataxia telangiectasia: an autoimmune disease associated with cytotoxic antibody to brain and thymus. Clinical Immunology and Immunopathology. 1977: 7: 288–299.

Kearns TP, Sayre GP. (1958) Retinitis pigmentosa, external ophthalmoplegia. A study of two cases. Archives of Ophthalmology. 1958: 60: 280–289.

Kiran O, Yalaz K, Taysi K, Say B. (1973) Immunological studies in ataxia-telangiectasia. Clinical Genetics. 1973: 4: 40–45.

Kjellin KG. (1959) Familial spastic paraplegia with amyotrophy, oligophrenia, and central retinal degeneration. Archives of Neurology (Chicago). 1959: 1: 133–140.

Kjellin KG. (1975) Hereditary spastic paraplegia and retinal degeneration. In: Vinken PJ, Bruyn GW (eds). Handbook of clinical neurology, vol 21. North-Holland/Elsevier, Amsterdam, Oxford, New York, ch 20: pp 467–473.

Klein D. (1937) Familienkundliche, körperliche und psychopatologische Untersuchungen über eine Friedreich-Familie. Schweizerische Archiv für Neurologie und Psychiatrie. 1937: 39: 89–116, 320–329.

Klenk E, Kahlke W. (1963) Über das Vorkommen der 3,7,11,15-Tetramethylhexadecansäure Phytansäure) in den cholesterinestern and anderen Lipoidfractionen der Organe bei einem Krankheits-Fall unbekannter Genese. (Verdacht auf Heredopathia atactica polyneuritiformis (Refsum-Syndrom). Hoppe Seylers Zeitschrift für Physiologische Chemie. 1963: 333: 133.

Klippel M, Durante G. (1892) Contribution à l'étude des affections nerveuses familiales et héréditaires. Revista Médica (Paris). 1892: 12: 745–785.

Kondo K, Hirota K. (1980) Friedreich's ataxia in Japan: clinical and genetic analyses in 52 cases. Brain & Development. 1980: 2: 365–371.

Kondo K, Hirota K, Katagiri T. (1981) Genetic and clinical patterns of heritable cerebellar ataxias in adults: II. Clinical manifestations. Journal of Medical Genetics. 1981: 18: 276–284.

Kondo K, Sobue I. (1980) Genetic and clinical patterns of heritable cerebellar ataxias in adults: I. Genetic analysis. Journal of Medical Genetics. 1980: 17: 416–423.

Konigsmark BW, Weiner LP. (1970) The olivopontocerebellar atrophies: a review. Medicine (Baltimore). 1970: 49: 227–241.

Lancet, Editorial. (1979) Wernicke's preventable encephalopathy. Lancet. 1979: 1: 1122–1123.

Landau WM, Gitt JJ. (1951) Hereditary spastic paraplegia and hereditary ataxia. A family demonstrating a variety of phenotypic manifestations. Archives of Neurology and Psychiatry (Chicago). 1951: 66: 346–354.

Lee S-H, Davis EJ. (1974) Carboxylation and decarboxylation reactions: anapleurotic flux and removal of citrate cycle intermediates in skeletal muscle. Journal of Biochemistry. 1974: 254: 420–430.

Lemieux B, Barbeau A, Berionade V, Shapcott D, Breton G, Geoffroy G, Melancon S. (1976) Amino acid metabolism in Friedreich's ataxia. Canadian Journal of Neurological Sciences. 1976: 3: 373–378.

Livingstone IB, Mastaglia FL. (1979) Choline chloride in the treatment of ataxia. British Medical

Journal. 1979: 2: 939.
Livingstone IB, Mastaglia FL, Edis R, Howe JW. (1981) Visual involvement in Friedreich's ataxia and hereditary spastic ataxia. A clinical and visual evoked response study. Archives of Neurology (Chicago). 1981: 38: 75–79.
Livingston IB, Mastaglia FL, Pennington RJI, Skilbeck C. (1981) Choline chloride in the treatment of cerebellar and spinocerebellar ataxia. Journal of the Neurological Sciences. 1981: 50: 161–174.
Llinas RR, Simpson JI. (1981) Cerebellar control of movement. In: Towe AL, Luschei ES (eds). Handbook of behavioural neurobiology, vol 5: Motor coordination. Plenum Press, New York, pp 231–302.
Lorrain, M. (1898) Contribution a l'étude de la paraplégie spasmodique familiale. Thése de Paris No. 216.
Louis-Bar D. (1941) Sur un syndrome progressif comprenant des telangiectasies capillaires cutanées et conjunctivales symmétriques, á disposition naevoide et des troubles cérébelleux. Confinia Neurologica (Basel). 1941: 4: 32–42.
Lowenthal A. (1981) Subacute necrotizing encephalopathy. In: Vinken PJ, Bruyn GW (eds). Handbook of clinical neurology, vol 42. North-Holland/Elsevier, Amsterdam, Oxford, New York, pp 625–626.
Lundberg A, Lilja LG, Try K. (1972) Heredopathia atactica polyneuritiformis (Refsum's disease). Experiences of dietary treatment and plasmapheresis. European Neurology. 1972: 8: 309–324.
Mahloudji M. (1963) Hereditary spastic ataxia simulating disseminated sclerosis. Journal of Neurology, Neurosurgery and Psychiatry. 1963: 26: 511–513.
Mahloudji M. (1975) Marinesco-Sjögren syndrome. In: Vinken PJ, Bruyn GW (eds). Handbook of clinical neurology, vol 21. North-Holland/Elsevier, Amsterdam, Oxford, New York, ch 30, pp. 550–561.
Malin JP. (1976) Beteiligung des peripher-motorischen Neurons bei hereditärer spastische Spinalparalyse. Zeitschrift für Elektroenzephalographie, Elektromyographie und verwandte Gebiete. 1976: 7: 140–145.
Mancall EL. (1975) Late (acquired) cortical cerebellar atrophy. In: Vinken PJ, Bruyn GW (eds). Handbook of clinical neurology, vol 21. North-Holland/Elsevier, Amsterdam, Oxford, New York, ch 25, pp 477–508.
Marie P. (1893) Sur l'hérédoataxie cérébelleuse. Semiology Medicale (Paris). 1893: 13: 444–447.
Marie P, Foix C, Alajouanine T. (1922) De l'atrophie cérébelleuse tardive a prédominance corticale. (Atrophie parenchymateuse primitive des lamelles du cervelet, atrophie paleocérébelleuse primitive.) Revue Neurologique (Paris). 1922: II: 849.
Marinesco G, Draganesco ST, Vasiliu D. (1931) Nouvelle maladie familiale caractérisée par une cataracte congénitale et un arrêt du développement somatoneurophysique. Encéphale. 1931: 26: 97–109.
Marsden CD, Obesco JA, Lang AE. (1982) Adrenoleukomyeloneuropathy presenting as spinocerebellar degeneration. Neurology (Minneapolis). 1982: 32: 1031–1032.
Marstein S, Jellum E, Halpern B, Eldjarn L. (1976) Biochemical studies of erythrocytes in a patient with pyroglutamic aciduria (5-oxoprolinemia). New England Journal of Medicine. 1976: 295: 406–412.
McFarlin DE, Strober W, Barlow M, Waldmann T. (1971) The immunological deficiency state in ataxia-telangiectasia. In: Rowland LP (ed): Research publications of ARNMD, Vol. XLIX. Baltimore, Md., Williams and Wilkins 1971: 275–292.
McLeod JG, Evans WA. (1981) Peripheral neuropathy in spinocerebellar degenerations. Muscle & Nerve. 1981: 4: 51–61.
McNutt W, Klingman WO, Harlan HT. (1960) Pes cavus and hereditary ataxia. Texas Reports on Biology and Medicine. 1960: 18: 222–232.
McReynolds MEW, Dabbous MK, Hanissian AS, Duenas D, Kimbrell R. (1976) Abnormal collagen in ataxia telangiectasia. American Journal of Diseased Children. 1976: 130: 305–307.
Meister A. (1978) Relation between ataxia and defects of the gamma-glutamyl cycle. In: Kark RAP, Rosenberg RN, Schut LJ (eds). Advances in neurology, vol 21: The inherited ataxias. Raven Press, New York, ch 20, pp 289–302.
Melancon SB, Orignon B, Potier M, Dallaire L. (1979) Taurine and beta-alanine uptake in cultured human skin fibroblasts from patients with Friedreich's ataxia. Canadian Journal of Neurological Sciences. 1979: 6: 251–253.
Mentzel P. (1891) Beitrag zur Kenntnis der hereditären Ataxie und Kleinhirnatrophie. Archiv für Psychiatrie und Nerven Krankheiten. 1891: 22: 160–190.

Misiti J, Waldmann TA. (1981) A defect in antigen-specific antibody production due to a lack of T helper cells in a patient with ataxia-telangiectasia. Transactions of the Association of American Physicians. 1981: 94: 349-356.

Morgan JP. (1982) The Jamaica ginger paralysis. Journal of the American Medical Association. 1982: 248: 1864-1867.

Morgan-Hughes JA. (1982) Defects of the energy pathways of skeletal muscle. In: Matthews WB, Glacer GH (eds). Recent advances in clinical neurology. Churchill Livingstone, Edinburgh, pp 1-46.

Morgan-Hughes JA, Hayes DJ, Clark JB, Landon DN, Swash M, Stark RJ, Rudge P. (1982) Mitochondrial encephalomyopathies. Biochemical studies in two cases revealing defects in the respiratory chain. Brain. 1982: 105: 553-582.

Morton NE, Lalouel J-M, Jackson JF, Currier RD, Yee S. (1980) Linkage studies in spinocerebellar ataxia. American Journal of Medical Genetics. 1980: 6: 251-257.

Muller DPR, Harries JT, Lloyd JK. (1970) Vitamin E therapy in abetalipoproteinemia. Archives of Diseases of Children. 1970: 45: 715.

Muller DPR, Lloyd JK. (1983) Vitamin E and neurological function. Lancet. 1983: I: 225-228.

Muller DPR, Lloyd JK, Bird AC. (1977) Long-term management of abetalipoproteinemia. Archives of Diseases of Children. 1977: 52: 209-214.

Nadi NS, Kanter D, McBride WJ, Aprison MH. (1977) Effects of 3-acetylpyridine on several putative neurotransmitter amino acids in the cerebellum and medulla of the rat. Journal of Neurochemistry. 1977: 28: 661-662.

Nino HE, Noreen HJ, Dubey DP, Resch JA, Namboodiri K, Elston RC, Yunis EJ. (1980) A family with hereditary ataxia: HLA typing. Neurology (Minneapolis). 1980: 30: 12-20.

Nonne M. (1891) Über eine eigentümliche familiäre Ekrankungsform des Zentralnervensystems. Archiv für Psychiatrie und Nervenkrankheiten. 1891: 22: 283-316.

Nordhagen E, Grøndahl J. (1964) Heredopathia atactica polyneuritiformis (Refsum's disease). Acta Ophthalmologica. 1964: 42: 629-633.

Ogawa N, Kuroda H, Ota Z, Yamamoto M, Otsuki S. (1982) Cerebrospinal fluid gammaaminobutyric acid variations in cerebellar ataxia. Lancet. 1982: II: 215

Olson, W, Engel WK, Walsh GO, Einagler R. (1972) Oculo-cranio-somatic neuromuscular disease with ragged-red fibres. Histochemical and ultrastructural changes in limb muscles of a group of patients with idiopathic progressive external ophthalmoplegia. Archives of Neurology. 1972: 26: 193-211.

O'Neill BP, Moser HW, Marmion LC. (1980) The adrenoleukomyeloneuropathy (ALMN) complex: Elevated C_{26} fatty acid in cultured fibroblasts and correlation with disease expression in three generations of a kindred. Neurology (Minneapolis). 1980: 30: 352.

Oppenheimer DR. (1979) Brain lesions in Friedreich's ataxia. Canadian Journal of Neurological Sciences. 1979: 6: 173-176.

Ørbeck AL, Quelprud T. (1954) Chorea progressiva hereditaria (Setesdalsrykkja). University Press, Oslo.

Ottersen OP, Storm-Mathisen J. (1984) In: Björklund A, Hökfelt T (eds). Handbook of chemical neuroanatomy, vol 3, (in press).

Paterson MC, Smith BP, Lohmann PHM, Anderson AK, Fishman L. (1976) Defective excision repair of gamma-ray-damaged DNA in human (ataxia-telangiectasia) fibroblasts. Nature. 1976: 260: 444.

Pedersen L. (1980) Hereditary ataxia in a large Danish pedigree. Clinical Genetics. 1980: 17: 385-393.

Pedersen L, Gyldensted C. (1978) Computerized tomography in hereditary ataxias. Acta Neurologica Scandinavica. 1978: 58: 81-88.

Pedersen L, Paltz P, Ryder LP, Lamm LU, Dissing J. (1980) A linkage study of hereditary ataxias and related disorders. Human Genetics. 1980: 54: 371-383.

Pedersen L, Trojaborg W. (1981) Visual, auditory and somatosensory pathway involvement in hereditary cerebellar ataxia, Friedreich's ataxia and familial spastic paraplegia. Electroencephalography and Clinical Neurophysiology. 1981: 52: 283-297.

Pentland B, Martyn CN, Steer CR, Christie JE. (1981) Lecithin treatment in Friedreich's ataxia. British Medical Journal. 1981: 282: 1197-1198.

Perlman SL, Kark RAP, Schlecter B, Budelli MR, Youkeles L. (1980) Preliminary studies of oral physostigmine in the longterm management of inherited ataxias. Neurology (Minneapolis). 1980: 30: 303-314.

Perry TL, Currier RD, Hansen S, MacLean J. (1977) Aspartate-taurine imbalance in dominantly

inherited olivopontocerebellar atrophy. Neurology (Minneapolis). 1977: 27: 257–261.
Perry TL, Hansen S, Currier RD, Berry K. (1978) Abnormalities in neurotransmitter amino acids in dominantly inherited cerebellar disorders. In: Kark RAP, Rosenberg RN, Schut LJ (eds). Advances in neurology, vol 21: The inherited ataxias. Raven Press, New York, ch 21, pp 303–314.
Peterson RDA, Cooper MD, Good RA. (1966) Lymphoid tissue abnormalities associated with ataxia-telangiectasia. American Journal of Medicine. 1966: 41: 342–359.
Peyronnard JM, Lapontine L, Bouchard JP, Lamontagne A, Lemieux B, Barbeau A. (1976) Nerve conduction studies and electromyography in Friedreich's ataxia. Canadian Journal of Neurological Sciences. 1976: 3: 313–317.
Pincus JH, Solitare JR. (1976) Thiamine triphosphate levels and histopathology: correlation in Leigh's disease. Archives of Neurology (Chicago). 1976: 33: 759–765.
Plaitakis A, Berl S, Yahr M. (1982) Abnormal glutamate metabolism in an adult-onset degenerative disorder. Science. 1982: 216: 193–196.
Pratt RTC. (1967) The genetics of neurological disorders. Oxford University Press, London, ch 4, pp 41–42.
Purkiss P, Baraitser M, Borud O, Chalmers RA. (1981) Biochemical and clinical studies of Friedreich's ataxia. Journal of Neurology, Neurosurgery and Psychiatry. 1981: 44: 574–582.
Rassin DK, Sturman JA, Gaull GE. (1980) Taurine and glutamate: developmental interrelationships in monkey brain. American Society of Neurochemistry Proceedings. 1980: 11: 70.
Refsum S. (1946) Heredopathia atactica polyneuritiformis. A familial syndrome not hitherto described. Acta Psychiatrica of Neurologica Scandinavica (Copenhagen). 1946: suppl. 38.
Refsum S. (1975) Heredopathia atactica polyneuritiformis. Phytanic acid storage disease (Refsum's disease). In: Vinken PJ, Bruyn GW (eds). Handbook of clinical neurology, vol 21. North-Holland/Elsevier, Amsterdam, Oxford, New York, ch 10, pp 181–229.
Refsum S, Mohr J. (1977) Genetic aspects of neurology. In: Baker AB (ed). Clinical neurology, vol 3. Harper & Row, New York, ch 47, p 29.
Refsum S, Scillicorn SA. (1954) Amyotrophic familial spastic paraplegia. Neurology (Minneapolis). 1954: 4: 40–47.
Remtulla MA, Katz S, Appelgarth DA. (1979) Effect of taurine and passive ion transport in rat brain synaptosomes. Canadian Federation of Biological Societies Proceedings. 1979: 22: 80.
Richind KE, Boder E, Teplitz RL. (1982) Fetal proteins in ataxia-telangiectasia. Journal of the American Medical Association. 1982: 248: 1346–1347.
Robinson N. (1966) Chemical changes in the spinal cord in Friedreich's ataxia and motor neuron disease. Journal of Neurology, Neurosurgery and Psychiatry. 1966: 31: 330–333.
Rosenberg L. (1977) Vitamin-responsive aminoacidopathies. Pediatrician. 1977: 6: 387–395.
Rosenberg RN. (1980) Genetic diseases of the nervous system. In: Dietschy JM (ed). The science and practice of clinical medicine, vol 5. Grune & Stratton, New York, pp 165–242.
Rosenberg RN, Nyhan WL, Continho P, Bay C. (1978) Joseph's disease: an autosomal dominant neurological disease in the Portuguese of the United States and the Azores islands. In: Kark RAP, Rosenberg RN, Schut LJ (eds). Advances in neurology, vol 21: The inherited ataxias. Raven Press, New York, ch 3, pp 33–57.
Rosenberg RN, Schotland D, Lovelace R, Rowland L. (1968) Progressive ophthalmoplegia. Archives of Neurology (Chicago). 1968: 19: 362–376.
Rowland LP. (1975) Progressive external ophthalmoplegia. In: Vinken PJ, Bruyn GW (eds). Handbook of clinical neurology, vol 22. North-Holland/Elsevier, Amsterdam, Oxford, New York, ch 8, pp 177–202.
Salam M. (1975) Metabolic ataxias. In: Vinken PJ, Bruyn GW (eds). Handbook of clinical neurology, vol 21. North-Holland/Elsevier, Amsterdam, Oxford, New York, ch 32, pp 580–582.
Salisachs P, Codina M, Findley LJ, Martinez-Lage JM. (1982) The significance of widely different motor conduction velocities in Refsum's disease. Muscle & Nerve. 1982: 5: 255.
Salt HB, Wolff OH, Lloyd JK, Fosbrooke AS, Cameron AH, Hubble DV. (1960) On having no beta-lipoprotein: a syndrome comprising a-beta-lipoproteinemia, acanthocytosis and steatorrhea. Lancet. 1960: II: 325–329.
Sanchez-Casis G, Cote M, Barbeau A. (1976) Pathology of the heart in Friedreich's ataxia and report of one case. Canadian Journal of Neurological Sciences. 1976: 3: 349–354.
Schenck L, Adachi M, Briet P, Wolintz A. (1973) Ophthalmoplegia plus. Journal of Neurological Sciences. 1973: 19: 37–44.

Schoenberg BS. (1978) Epidemiology of the inherited ataxias. In: Kark RAP, Rosenberg RN, Schut LJ (eds). Advances in neurology, vol 21: The inherited ataxias. Raven Press, New York, ch 2, pp 15–32.

Schut JW. (1950) Hereditary ataxia: clinical study through six generations. Archives of Neurology and Psychiatry (Chicago). 1950: 63: 535–568.

Schut JW, Böök JA. (1953) Hereditary ataxia. Difference between progeny of male and female members and a definition of certain signs useful in detecting disease prior to onset of clinical symptoms. Archives of Neurology (Chicago). 1953: 70: 169–179.

Schut JW, Haymaker W. (1951) Hereditary ataxia: a pathologic study of five cases of common ancestry. Journal of Neuropathology and Clinical Neurology. 1951: 1: 183–213.

Scotto JM, Hadchouel M, Odievre M, Laudat M-H, Saudubray J-M, Dulac O, Beucler I, Beaune P. (1982) Infantile phytanic acid storage disease, a possible variant of Refsum's disease: three cases, including ultrastructural studies of the liver. Journal of Inherited Metabolic Diseases. 1982: 5: 83–90.

Serratrice G, Gastaut JL, Dubois-Gambarelli D. (1973) Amyotrophie neurogéne périphérique au cours du syndrome Marinesco-Sjögren. Révue Neurologique (Paris). 1973: 128: 432–441.

Shinnar S, Maciewicz RJ, Shofer RJ. (1975) A raphe projection to the cerebellar cortex. Brain Research. 1975: 97: 139–143.

Sipe JC. (1973) Leigh's syndrome: the adult form of subacute necrotizing encephalomyelopathy with predilection for the brainstem. Neurology (Minneapolis). 1973: 23: 1030–1038.

Sjögren T. (1943) Klinische und erbbiologische Untersuchungen über die Heredoataxien. Acta Psychiatrica Neurologica Scandinavica. suppl. 27.

Sjögren T. (1950) Hereditary congenital spinocerebellar ataxia combined with congenital cataract and oligophrenia. Confinia Neurologica (Basel). 1950: 10: 293–308.

Sjögren T, Larsson T. (1957) Oligophrenia in combination with congenital ichthyosis and spastic disorders: a clinical and genetic study. Acta Psychiatrica Scandinavica. 1957: 32: suppl 113: 1–112.

Skre H. (1963) Heredoataxia in western Norway. Some experiences from a preliminary investigation. Journal de Génétique Humaine. 1963: 13: 86–107.

Skre H. (1972) Neurological signs in a normal population. Acta Neurologica Scandinavica. 1972: 48: 575–606.

Skre H. (1974a) Application of a quantitative scoring system in the investigation of some hereditary neurological disorders. Clinical Genetics. 1974a: 5: 163–172.

Skre H. (1974b) Genetic and clinical aspects of Charcot-Marie-Tooth's disease. Clinical Genetics. 1974b: 6: 98–118.

Skre H. (1974c) Hereditary spastic paraplegia in western Norway. Clinical Genetics. 1974c: 6: 165–183.

Skre H. (1974d) Spinocerebellar ataxia in western Norway. Clinical Genetics. 1974d: 265–288.

Skre H. (1975a) Friedreich's ataxia in western Norway. Clinical Genetics. 1975a: 7: 209–218.

Skre H. (1975b) The significance of "unspecific neuropathy" in hereditary ataxias and related disorders. Clinical Genetics. 1975b: 7: 287–298.

Skre H. (1975c) A study of certain traits accompanying some inherited neurological disorders. Clinical Genetics. 1975c: 8: 117–135.

Skre H. (1980) Epidemiology of spinocerebellar degenerations in western Norway: hereditary ataxias. In: Sobue I (ed). Spinocerebellar degenerations. University of Tokyo Press, Tokyo, pp 103–120.

Skre H. (1983) Spinocerebellar disorders. In: Emery AEH, Rimoin DL (eds). Principles and practice of medical genetics, vol 1. Churchill Livingstone, Edinburgh, pp 267–283.

Skre H, Bassöe HH, Berg K, Frövig AG. (1976) Cerebellar ataxia and hypergonadotropic hypogonadism in two kindreds. Chance concurrence, pleiotropism or linkage? Clinical Genetics. 1976: 9: 234–244.

Skre H, Berg K. (1974a) Cerebellar ataxia and total albinism: a kindred suggesting pleiotropism or linkage. Clinical Genetics. 1974a: 7: 196–204.

Skre H, Berg K. (1974b) The pleiotropism – linkage problem in clinical medicine as exemplified by the occurrence of oculo-cutaneous syndromes in heredo-degenerative diseases. European Society of Clinical Investigation Abstracts. 1974b: 4: 365.

Skre H, Berg K. (1976) Genetic linkage as a cause of clinical variation in hereditary disorders. In: Armendares S, Lisker R (eds). Excerpta Medica, International Congress Series No 397, V International Congress of Human Genetics, Abstracts, 100.

Skre H, Berg K. (1977) Linkage studies on the Marinesco-Sjögren syndrome and hypergonadotropic hypogonadism. Clinical Genetics. 1977: 11: 57–66.
Skre H, Eldjarn L, Stokke O, Try K. (1969). A new case of heredopathia atactica polyneuritiformis (Refsum's disease). In: Excerpta Medica International Congress Series No 175, Proceedings of the Second International Congress of Neuro-Genetics and Neuro-Ophthalmology, Montreal, September 1967, Vol 1, pp 710–722.
Skre H, Løken AC. (1970) Myoclonus epilepsy and subacute presenile dementia in heredo-ataxia. Acta Neurologica Scandinavica. 1970: 46: 18–42.
Solitaire GB, Lopez VF. (1967) Louis-Bar's syndrome (ataxia-telangiectasia). Neuropathological observations. Neurology (Minneapolis). 1967: 17: 23–31.
Sorbi S, Blass JP. (1982) Abnormal activation of pyruvate dehydrogenase in Leigh disease fibroblasts. Neurology (Minneapolis). 1982: 32: 555–558.
Spencer PS, Sabri MI, Schaumberg HH, Moore CL. (1979) Does a defect in energy metabolism in the nerve fibre underlie axonal degeneration in polyneuropathies? Annals of Neurology. 1979: 5: 501–507.
Spencer PS, Schaumburg HH. (1977) Ultrastructural studies of the dying back process. IV. Different vulnerability of PNS and CNS fibres in experimental central-peripheral distal axonopathies. Journal of Neuropathology and Experimental Neurology. 1977: 36: 300–320.
Spillaine JD. (1940) Familial pes cavus and absent tendon jerks: its relationship with Friedreich's disease and peroneal muscular atrophy. Brain. 1940: 63: 275–290.
Stephens J, Hoover ML, Denst J. (1958) On familial ataxia, neuronal amyotrophy and their association with progressive external ophthalmoplegia. Brain. 1958: 81: 556–566.
Stern C. (1960) Principles of human genetics. Freeman, San Francisco, London, pp 322–325.
Stern P, Bokanjic R. (1974) Glycine therapy in seven cases of spasticity. Pharmacology. 1974: 12: 117–119.
Still CN. (1976) Nicotinic acid and nicotinamide deficiency: pellagra and related disorders of the nervous system. In: Vinken PJ, Bruyn GW (eds). Handbook of clinical neurology, vol 28. North-Holland/Elsevier, Amsterdam, Oxford, New York, ch 3, pp 49–104.
Strümpell A. (1880)Beiträge zur Pathologie des Rückenmarks. Archiv für Psychiatrie und Nervenkrankheiten. 1880: 10: 676–717.
Stumpf DA. (1979) Mitochondrial multisystem disorders: clinical, biochemical and morphologic features. In: Tyler HR, Dawson DM (eds). Current neurology, vol 2. Houghton-Mifflin Professional Publishers, Medical Division, Boston, pp 117–149.
Stumpf DA, Parks JK, Eguren LA, Haas R. (1982) Friedreich ataxia: III. Mitochondrial malic enzyme deficiency. Neurology (Minneapolis). 1982: 32: 221–227.
Sutherland JM. (1975) Familial spastic paraplegia. In: Vinken PJ, Bruyn GW (eds). Handbook of clinical neurology, vol 22. North-Holland/Elsevier, Amsterdam, Oxford, New York, ch 17, pp 421–443.
Tanaka K, Rosenberg LE. (1983) Disorders of branched chain amino acid and organic acid metabolism. In: Stanbury JB, Wyngaarden JB, Fredrickson DS (eds). The metabolic basis of inherited disease. 5th ed. McGraw-Hill, New York, pp 440–473.
Taniguchi R, Konigsmark BW. (1971) Dominant spino-pontine atrophy. Report of a family through three generations. Brain. 1971: 94: 349–358.
Thieffry S, Arthuis M, Aicardi J, Lyon G. (1961) L'ataxie-télangiectasie. Révue Neurologique (Paris). 1961: 105: 390–405.
Thorpe FT. (1935) Familial degeneration of the cerebellum in association with epilepsy. A report of two cases, one with pathological findings. Brain. 1935: 58: 97–114.
Todorov A. (1965) Le syndrome de Marinesco-Sjögren. Premiére étude anatomo-clinique. Journal de Génétique Humaine (Basel). 1965: 14: 197–234.
Tolis G, Metha A, Anderman E, Harvey C, Barbeau A. (1980) Friedreich's ataxia and glucose tolerance: I. The effect of ingested glucose on serum glucose and insuline values in homozygotes, obligate heterozygotes and potential carriers of the Friedreich's ataxia gene. Canadian Journal of Neurological Sciences. 7, 1980: 4: vol 3.
Trouillas P, Garde A, Robert JM, Renaud B, Adeleine P, Bard J, Brudon F. (1982) Régression du syndrome cérébelleuse sous administration a long terme de 5-HTP ou de l'association 5 HTP-benserazide. 26 observations quantifiées et traitées par ordinateur. Révue Neurologique (Paris). 1982: 138: 415–435.
Tyrer JH. (1975) Friedreich's ataxia. In: Vinken PJ, Bruyn GW (eds). Handbook of clinical neurology, vol 22. North-Holland/Elsevier, Amsterdam, Oxford, New York, ch 14, pp 319–364.

Tyrer JH, Sutherland JM. (1961) The primary spino-cerebellar atrophies and their associated defects, with a study of the foot deformity. Brain. 1961: 84: 289–300.
van Bogaert L. (1949) Les maladies systematisées. In: Lemierre A, Lenormant C, Paginez P, Savy P, Fiessiger N, de Gennes L, Ravina A (eds). Traité de médicine, vol 16. Masson & Cie, Paris, pp 142–200.
van Rossum J, Veenema H, Went LN. (1981) Linkage investigations in two families with hereditary ataxia. Journal of Neurology, Neurosurgery, and Psychiatry. 1981: 44: 516–522.
Veronesi B, Peterson ER, Di Vincenso G, Spencer PS, Schaumberg HH. (1978) A tissue culture model of distal (dying-back) axonopathy – its use in determining primary neurotoxic compounds. Journal of Neuropathology and Experimental Neurology. 1978: 37: 703.
Victor M. (1976) The Wernicke-Korsakoff syndrome. In: Vinken PJ, Bruyn GW (eds). Handbook of clinical neurology, vol 28. North-Holland/Elsevier, Amsterdam, Oxford, New York, ch 9, pp 243–270.
Victor M, Adams RD, Mancall EL. (1959) A restricted form of cerebellar cortical degeneration occurring in alcoholic patients. Archives of Neurology (Chicago). 1959: 1: 579–688.
Walker JL, Chamberlain S, Robinson N. (1980) Lipids and lipoproteins in Friedreich's ataxia. Journal of Neurology, Neurosurgery and Psychiatry. 1980: 43: 111–117.
Walton JN, Gardner-Medwin FD. (1981) Ocular myopathy or progressive external ophthalmoplegia. In: Walton JN (ed). Disorders of voluntary muscle. 4th ed. Churchill Livingstone, Edinburgh, ch 14, pp 506–508.
Wiener JB. (1962) Statistical principles in experimental design. McGraw-Hill, New York, 1962.
Weiner LP, Konigsmark BW. (1971) Hereditary diseases of the cerebellar parenchyma. Birth defects: Original Article series. 1971: 7: 192–196.
Weiner LP, Konigsmark BW, Stoll J, Magladery JW. (1967) Hereditary olivopontocerebellar atrophy with retinal degeneration. Report of a family through six generations. Archives of Neurology (Chicago). 1967: 16: 364–376.
Wernicke C. (1881) Lehrbuch der Gehirnkrankheiten für Ärtze und Studierende. Theodor Fischer, Kassel, 2: 229–242.
Wiklund L, Björklund A, Sjölund B. (1977) The indolaminergic innervation of the inferior olive. I. Convergence with direct spinal afferents in the area projecting to the cerebellar anterior lobe. Brain Research. 1977: 131: 1–21.
Wilkinson DS, Prockop LD. (1976) Hypoglycemia: effect on the central nervous system. In: Vinken PJ, Bruyn GW (eds). Handbook of clinical neurology, vol 27. North-Holland/Elsevier, Amsterdam, Oxford, New York, ch 4, pp 53–78.
Willems JL, Monnens LAH, Trijbels JMF. (1977) Leigh's encephalomyelopathy in a patient with cytochrome c oxidase deficiency in muscle tissue. Pediatrics. 1977: 60: 850–857.
Winkler C. (1923) A case of olivo-pontine cerebellar atrophy and our conception of neo- and paleocerebellum. Schweitzerische Archiv für Neurologie und Psychiatrie. 1923: 13: 684–702.
Yakura H, Wakisaka A, Fujimoto S, Itakura K. (1974) Hereditary ataxia and HLA genotypes. New England Journal of Medicine. 1974: 291: 154–155.

6 Genetically Controlled Resistance to Virus Infections of the Central Nervous System

Raymond P. Roos

Department of Neurology, University of Chicago Medical Center, Chicago, Illinois

---

Acknowledgment: The secretarial assistance of Zayda Stewart and Peggy Peach is gratefully acknowledged.

INTRODUCTION

Host genetic factors are critical in determining the outcome of virus infection. The fact that there are separate groups of plant viruses and animal viruses, as well as viruses that infect only certain species of animals, is a reflection of the inherent genetic susceptibility or resistance of the host. In this chapter we will deal with genetic factors that guide resistance within an animal species to virus infections primarily involving the central nervous system; a review of the pathogenesis of viral infections of the central nervous system was published by Johnson in 1982.

Genetic factors may play an extremely important role in viral diseases beyond determining host susceptibility. For example, in the case of certain retrovirus infections, the genetic factor is the integrated endogenous virus itself. With other virus infections, genetic factors can change the disease phenotype. Lactic dehydrogenase virus produces a subclinical, persistent infection in most mouse strains, but under appropriate conditions infection of C58 mice results in a chronic paralytic disease (Brinton, 1981a). In a similar fashion the extent and distribution of amyloid plaques in scrapie depend on the particular mouse strain used (Bruce, Dickinson, and Fraser, 1976).

The complex nature of resistance of humans to virus infections has prompted investigators to focus on infections of inbred mouse strains. So this review will concentrate on studies in mice, but wherever possible reference to human infections will be made. The relevance of mouse studies to human disease is supported by the genetic homologies between mouse and human genes. For example, the mouse major histocompatibility locus H-2 bears structural and organizational similarities to the human major histocompatibility locus HLA. Although much has been learned through these genetic studies, we must be aware of several cautions in analyzing the data (from mouse studies as well as from human ones):

1. A variety of nongenetic factors influence the outcome of virus infection—for example, the route of inoculation and the age of the animal. The effects of the genes determining the host response to herpes simplex type 1 virus intraperitoneal (IP) infection are far less, if at all, apparent following intracerebral (IC) inoculation (Lopez, 1980).
2. Host genetic factors tend to be relative. The effect of retrovirus *FV-1* gene can be overridden by a large virus inoculum or immunosuppression (Jolicoeur, 1979). Mouse hepatitis type 3 virus (MHV-3) involves mouse strains that are resistant, susceptible, and semisusceptible (Dupuy, Lafforet-Cresteil, and Dupuy, 1980).
3. Genetic factors may vary depending on the particular type or strain of a virus. Presumably, a different passage history of a virus can lead to poorly understood molecular changes in a virus that alter its response to resistance gene products.
4. Genetic factors influencing infection by the same virus can vary depending

on the animal species studied. Resistance to JHM strain of mouse hepatitis virus type 4 (MHV-4) is controlled by a dominantly inherited gene in rats (Sorensen et al., 1982) but by a recessive gene in mice (Stohlman and Frelinger, 1978; Knobler, Haspel, and Oldstone, 1981).

5. Strains of mice vary in susceptibility to different viruses. A/J mice are resistant to MHV-3 but susceptible to MHV-4 (Knobler, Haspel, and Oldstone, 1981).

These studies have occasionally delineated single genes determining resistance—as in the case of flaviruses or influenza virus—but generally the genes governing resistance have been found to be varied and multiple. The resistance phenotypes may present at varying ages of the host and have varying mechanisms. These observations are not unexpected given the heterogeneity of viruses and the many available sites at which a gene product can interfere with the outcome of a virus infection. Genes may act at the level of the host response (immune response, natural killer cells, macrophages, interferon), of the cell membrane (receptors important for virus adsorption and penetration), and of the cell cytoplasm (factors important for viral replication, translation, assembly, egress).

This chapter describes several viruses and the host genes thought to play a role in their diseases. Table 6.1 summarizes the host genetic factors—and their proposed mechanisms—pertaining to the viruses discussed in this review. The subacute spongiform encephalopathies will be discussed in an effort to elucidate how these diseases, which display an apparent dominant inheritance pattern, can be transmissible. The chapter will not deal with inherited immunodeficiency diseases or with inflammatory disorders of the central nervous system that are of questionable viral etiology, such as multiple sclerosis. The final section will describe mechanisms of genes that determine host resistance.

HOST GENETIC INFLUENCES ON VIRUS INFECTIONS

Retrovirus

The retroviruses are a group of oncogenic ribonucleic acid (RNA) viruses that have attracted interest because of the recent effort to delineate genes important in cancer formation. Their clinical pathological manifestations are varied and have given us a glimpse of the staggering complexity of host genetic factors important in influencing disease susceptibility and the outcome of infection. The retrovirus perhaps most relevant to human neurological disease is the murine neurotropic retrovirus, which is responsible for spongiform polioencephalomyelopathy in mice.

Murine neurotropic virus was first isolated in several populations of wild mice (*Mus musculus*) in southern California (Gardner et al., 1973) that exhibit a chronic progressive paralytic disease and lymphoma. The murine

TABLE 6.1. Some Host Genetic Factors Affecting Virus Infections in Experimental Animals

Virus	Inheritance Pattern of Gene(s) Determining Resistance	Proposed Mechanism(s)	Reference
Retrovirus*	Dominant (Fv-1)*	Intracellular interference with viral DNA replication and integration	Jolicoeur, 1979
Togavirus			
Flavivirus	Dominant	Intracellular interference with replication cycle	Brinton, 1981a
LDV	Dominant (Fv-1)	See above regarding retrovirus	Pease, Abrams, and Murphy, 1982
Coronavirus			
MHV-2	Recessive	Lymphocyte macrophage interactions	Bang and Cody, 1980
MHV-3	Variable (see text)	?	Levy-Leblond, Oth, and Dupuy, 1979
MHV-4 (JHM)	Recessive; dominant and recessive	Neuronal block in virus replication after penetration	Knobler, Haspel, and Oldstone, 1981 Stohlman and Frelinger, 1978
Picornavirus			
TMEV	Dose dependent (see text) (H-2 linked, non-H-2 linked)	?	Lipton and Melvold, 1984
EMC (M)	Dominant	Control of virus adsorption and ?intracellular virus replication by β cell	Onodera et al., 1978

Orthomyxovirus			
Influenza virus	Dominant	IFN	Haller, 1981
Paramyxovirus			
Measles virus	Dominant	?	Rager-Zisman, Neighbour, and Bloom, 1980
Herpesvirus			
HSV-1	Dominant	?NK;?IFN	Lopez, 1981
HSV-2	Dominant (X-linked)	?	Mogensen, 1980
MCMV	Dominant (H-2–linked)	?NK	Grundy, Mackenzie, and Stanley, 1981
MDV	Dominant (MHC-linked, nonlinked)	Immune response (MHC-linked); control of number of target cells	Gallatin and Longenecker, 1979

LDV = lactic dehydrogenase virus
MHV-2, MHV-3, MHV-4 (JHM) = mouse hepatitis virus type 2, 3, 4, (JHM strain)
TMEV = Theiler's murine encephalomyelitis viruses
EMC(M) = M variant of encephalomyocarditis virus
HSV-1, HSV-2 = herpes simplex virus type 1, 2
MCMV = murine cytomegalovirus
MDV = Marek's disease virus
MHC = major histocompatibility complex
H-2 = MHC of mouse
IFN = interferon
NK = natural killer cell
*For a more complete list of genes affecting retroviruses, see Pincus, 1980.

neurotropic virus implicated in the paralytic disease is actually distinct from a lymphomagenic virus also isolated from these mice. Infection with the neurotropic virus, but not with the lymphomagenic virus, produces primarily paralytic disease. The neurotropic virus also has lymphomagenic properties of its own since inoculation of neonatal NIH mice with a neurotropic isolate will produce a 25 percent incidence of lymphoma (Rasheed, Toth, and Gardner, 1977).

The neurotropic virus produces a noninflammatory vacuolization of neurons and glia with reactive astrocytosis predominantly in the spinal cord, brain stem, and dentate nucleus (Brooks, Swarz, and Johnson, 1980) following an incubation period of as long as a year (Andrews and Gardner, 1974). The pathology resembles that seen in the wobbler mouse mutant, scrapie in animals (see later in this chapter), and human motor neuron disease. The latter observation has provoked interest in the possible involvement of retroviruses in human neurological degenerative diseases of unknown etiology.

The murine neurotropic virus is a member of the subgroup of leukemia viruses and can be classified as ecotropic, signifying that it replicates primarily in murine cells. In contrast, xenotropic viruses do not replicate in murine cells but can replicate in other species, whereas amphotropic viruses replicate in murine cells as well as in cells from other species. Because none of the three classes of murine leukemia viruses replicate in hamster cells, Oie et al. (1978) and Ruddle et al. (1978) effectively used hamster × mouse somatic cell hybrids to delineate the chromosome coding for permissiveness to ecotropic virus infection. Chromosome 5 and cellular binding of ^{125}I-gp71—the retroviral major envelope glycoprotein essential for adsorption—segregated in all somatic cell hybrids. These results indicated that chromosome 5 codes for the receptor for ecotropic (Rauscher leukemia) virus binding. Studies have shown that the murine receptor for amphotropic virus is distinct and coded for by chromosome 8 (Gazdar et al., 1977).

Murine leukemia viruses can be divided into an N-tropic, B-tropic, or NB-tropic groups (reviewed by Jolicoeur, 1979). N-tropic viruses infect certain strains of mice and grow 100- to 1,000-fold more efficiently than B-tropic virus in cells of these strains of mice (N-type mice, such as NIH Swiss mice); conversely, B-tropic viruses grow better in other strains (B-type, such as Balb/c mice). Mouse strains behave as if they were governed by a gene denoted *Fv-1* with two main alleles; N-type homozygote mice are designated $Fv\text{-}1^{n/n}$, whereas B-type homozygotes are $Fv\text{-}1^{b/b}$. The F_1 hybrid, ($Fv\text{-}1^{n/b}$), resists both N- and B-tropic virus, indicating that *Fv-1* restricts the growth (that is, prevents growth to high titer) of the virus of opposite tropism: that is, $Fv\text{-}1^{n/n}$ mice restrict the growth of B-tropic virus, $Fv\text{-}1^{b/b}$ mice restrict the growth of N-tropic virus, and $Fv\text{-}1^{n/b}$ displays codominance, restricting the growth of both N- and B-tropic virus. The restriction of virus growth is not absolute since resistant mouse strains can be made to replicate virus more efficiently if inoculated at a high multiplicity of infection or if treated with immunosuppressors. The *Fv-1* gene has been mapped to chromosome 4 by linkage studies and somatic cell genetics.

On the basis of studies in inbred mouse strains, Gardner et al. (1976) has classified the murine neurotropic virus as N-tropic; that is, *Fv-1*$^{n/n}$ mouse strains most efficiently support the growth of the virus. Mohan and Pal (1982) have shown that murine neurotropic virus binds to spinal cord cells of all the mouse strains studied, indicating that the *Fv-1* restriction is not at the level of viral adsorption.

Much work has been directed toward an elucidation of the precise site of action of the *Fv-1* gene during the retrovirus replication cycle. Retroviral RNA is normally copied into deoxyribonucleic acid (DNA) by reverse transcriptase, and then a second complementary strand is synthesized to make full-length double-stranded DNA. The linear form of DNA is the precursor of a closed circular species that is eventually integrated into the cellular genome. Cells of the *Fv-1* restrictive host exhibit a decrease in viral integrated DNA. Circular forms of DNA are definitely decreased in amount, but the amount of the linear form varies and is more difficult to evaluate. The pattern of viral forms seen in the restrictive host is similar to that seen in cells following treatment with aphidicolin, an inhibitor of eukaryotic DNA polymerase α. This mimicry suggests that the action of *Fv-1* may be at the level of DNA polymerase α (Chinsky and Soeiro, 1982), but the exact site of the block is unclear and these speculations must be confirmed more directly.

The *Fv-1* gene product interacts with a viral protein to restrict virus production. By constructing recombinant viral DNA between cloned, infectious N- and B-tropic viral genomes, DesGroseillers and Jolicoeur (1983) delineated the viral sequences which determine host range of the virus. They found that only two consecutive amino acids of the p30 viral protein differentiate an N-tropic form from a B-tropic sequence, and are necessary and sufficient to combine with the *Fv-1* gene product to confer a given host range to the virus.

It seems appropriate to mention briefly some of the other genes controlling murine leukemia virus disease expression because they are relatively well defined and because they illustrate the large number of genes that can influence disease caused by a group of viruses (reviewed by Steeves and Lilly, 1977; Pincus, 1980); admittedly, the effects of these genes on murine neurotropic virus remain unexplored. The following paragraph describes some of the genetic influences on murine leukemia virus disease expression.

Many inbred mouse strains that spontaneously produce high levels of ecotropic murine leukemia virus carry the actual genomes of endogenous murine leukemia viruses (provirus) in the host chromosome (Chattopadhyay et al., 1975). Varying levels of virus expression correspond to the number of copies of the virus within the host genome; that is, the virus inducibility locus contains viral structural genes. The endogenous provirus behaves like a stable genetic element, and virus expression depends on host and viral factor interactions. These loci are located on different chromosomes, and the locations vary among mouse strains (Quint et al., 1982). Loci reported for ecotropic virus inducibility include: Akr/N strain maps to chromosomes 7 and 16; Balb/c and C3H/He mice to chromosome 5; C57B1/6J to chromosome 8; and DBA/2J to chromosome 9 (Jenkins et al., 1981). Besides the endogenous proviruses, there are a variety of other genes influencing virus or disease express-sion. *Fv-2*, a dominant susceptibility gene located on chromosome 9, is the major determinant of induction of leukemia by Friend leukemia virus (a rapidly transforming acute murine leukemia virus that is a complex of repli-cation-defective spleen focus forming virus and its helper leukemia virus) (Kumar, Goldschmidt, Eastcott, and Bennett, 1978). The gene's effect may be mediated by natural killer cell activity (Kumar and Bennett, 1981) and changes in the size of the target cell population. *Fv-3* is a dominant assistance gene that regulates immunosuppression by Friend leukemia virus (Kumar, Resnick, Eastcott, and Bennett, 1978). *Fv-4* is a dominant gene for resistance to a variety of N-tropic and NB-tropic murine leukemia viruses, presumably overriding the effect of *Fv-1* (Kai et al., 1976). *Rgv-1* and *Rfv-1* map closely to the H-2 complex on chromosome 17 and confer relative resistance against a variety of retroviruses (Lilly, 1966, 1968). *Rgv-1* maps closely to the major histocompatibility complex (MHC) locus of the mouse (H-2)-linked murine immune response genes, suggesting an immune mechanism of resistance in which an association of certain H-2 molecules with viral proteins may allow clearance by cytotoxic lymphocytes.

In summary, there are a large number of genes affecting resistance to murine leukemia virus. Genes exist that represent the actual viral genome, control the expression of this endogenous virus, restrict viral replication (for example, *Fv-1*), affect the immune response (for example, *Rgv-1*), and af-fect the number of target cells. Many of these genes, numbering over 20, have not been localized to a particular chromosome.

Togavirus

Togaviruses are a large group of enveloped viruses containing single-stranded RNA of positive polarity (infectious). The term *envelope* refers to a lipoprotein material covering that is derived from the cell membrane as the virus matures by budding. It should also be mentioned that RNA viruses con-tain either positive-stranded RNA, which can be translated directly into a pro-tein, or negative-stranded RNA, which must first be used as a template to make usable messenger RNA (positive-stranded) in order to be translated. This section will deal with the flavivirus subgroup of togaviruses and an unclassified togavirus member, the lactic dehydrogenase virus (LDV). Flavi-viruses are primarily arthropod-borne viruses and show serological cross-reactivity. They are a cause of subclinical infections, acute meningoenceph-alitis, and a variety of systemic abnormalities in man. A prototype of the subgroup, from which the name *flavivirus* is derived, is the yellow fever virus. Flaviviruses are of special interest because of the extensive work that has been performed concerning the role of host genetic factors in transmission of flaviviruses to mice.

LDV has been known for many years as a cause of a chronic, persistent infection in mice manifest by viremia with infectious immune complexes, elevated levels of lactic dehydrogenase in serum, and no signs of clinical disease (Riley, 1974). Recent interest in LDV has arisen because of its identification as the cause of murine immune polioencephalomyelitis (Martinez et al., 1980).

Flavivirus

Work with inbred mouse strains as well as wild mice established the presence of an autosomal dominant gene regulating resistance to infection by several representative flaviviruses, such as Banzi virus and West Nile virus (WNV) (reviewed by Brinton, 1981a). The creation of congenic strains of resistant C3H/RV and susceptible C3H/He mice has been useful in delineating factors important in resistance. Over 10^5 times more Banzi virus is required to produce death in C3H/RV mice than in C3H/He mice following intraperitoneal inoculation (Jacoby and Bhatt, 1976).

Although the phenotypic expression of the gene is modified by the virus route of inoculation and host age and immune status, these factors do not absolutely control the action of the gene, and their function is considered distinct from the mechanism of the resistant gene product. For example, IC inoculation tends to make the C3H/RV mice more susceptible to disease; however, slight differences remain between IC inoculated C3H/RV and C3H/He mice in survival time, 50 percent lethal dose (LD_{50}), brain virus titers, and histopathology (Jacoby and Bhatt, 1976). Immunosuppression converts an asymptomatic infection of a resistant mouse to a fatal one, but differences in brain virus titers and the tempo of disease between resistant and susceptible mice remain (Bhatt and Jacoby, 1976).

The finding that the resistance gene is expressed in vitro in a variety of cell types suggests a cellular mechanism of action (Brinton, 1981a). The block in virus replication is not in adsorption or penetration since virus antigen develops in cells of resistant animals following a high multiplicity of infection. Resistant cells have a lower level of viral RNA and protein synthesis, indicating that the restriction to virus growth is at a relatively early stage.

Evidence suggests that defective interfering (DI) particles are responsible for WNV restriction (Brinton, 1981a). DI particles contain deletion mutations of the viral genome and interfere with wild type virus replication. These mutants not infrequently appear in a variety of virus infections after multiple cycles of growth of virus at high multiplicities of infection. Replication of the mutant depends on complementation of their deleted function by genes of coinfecting wild type virus. The mechanism of interference with growth of wild type virus is still unclear. DI particles have been implicated in persistent and nonlytic infections. The production of DI particles is known to be influenced by the host cell.

Brinton noted that serial passage of virus in both susceptible and resistant cultured mouse embryo fibroblasts produced a cycling of the titer of infectious virus, as is frequently seen when DI particles are present. Supernatant fluid, taken from C3H/RV mouse cells following three serial undiluted passages with WNV, interfered with the multiplication of hamster brain-propagated WNV, whereas fluids from infected C3H/He mouse cultures did not; this finding suggested that the number of DI particles produced in cultures is significantly influenced by the host cell and is greater from resistant mice than from susceptible mice. A role for DI particles is supported by reports of a WNV mutant which replicates significantly more efficiently than standard virus in C3H/RV cell cultures and is insensitive to interference by WNV DI particles (Brinton 1981b; Brinton and Fernandez 1983b). Interfering activity was also present in virus passed in brains of resistant mice (Brinton, 1980). Although Smith, Jacoby, and Bhatt (1980) found interfering activity in resistant (C3H/He) mouse brains, but not susceptible (C3H/RV) mouse brains infected with Banzi virus, they were unable to detect interfering activity in serially passed infected macrophage cultures from these strains, emphasizing the importance of specific virus-host relationships.

Brinton (1983) recently has directly examined the progeny virus replicating in resistant C3H/RV and susceptible C3H/HE mouse cell cultures for evidence of DI particles. Virus progeny of C3H/RV cells contained substantially more small, subgenomic-sized RNAs than progeny of C3H/HE cells which previously contained standard virus. The production of significant amounts of small RNAs by C3H/RV cells in association with interfering activity suggests that the small RNAs are DI particles. It will be important to purify the RNA species to determine their specific interfering activities. It will also be important to characterize the virus progeny and delineate the role of DI particles in in vivo infection of the two mouse strains. Host factors, which are known to be important in RNA virus replication, may be under the control of the flavivirus resistance gene and lead to the production of DI particles.

Interferon (IFN) does not seem to play a major role in the intracellular block to flavivirus replication, but it is still involved in the infection and some of these molecular events. The supernatant fluid from resistant mouse cultures has only one unit of IFN (Darnell and Koprowski, 1974). Administration of anti-IFN serum in vivo does not affect mortality of resistant mice, nor does it increase virus titers in their brains (Brinton, 1981a). Nevertheless, IFN is involved in the infection in general and presumably can modify the expression of the resistant gene by exerting an antiviral effect; for example, anti-IFN serum increases the susceptibility of C3H/He mice to WNV infection, and administration of IFN in vitro causes a greater suppression of WNV replication in cultured cells from resistant mice than from susceptible mice.

Lactic Dehydrogenase Virus

Until recently, little was known about the cause of immune polioencephalomyelitis, also known as age-dependent polioencephalomyelitis, an inflammatory motor neuron disease produced after transfer of mouse line Ib leukemic cell suspensions to old C58 (Murphy et al., 1970). In 1980 Martinez et al. identified LDV as the etiologic agent that had inadvertently contaminated the cell suspensions. The continued unraveling of the many idiosyncrasies of this disease has been exciting and instructional, emphasizing the elaborate requirements for successful disease induction.

The conditions that are presently understood to affect polioencephalomyelitis disease induction significantly are neurovirulence of the virus strain, host age, immune status, and genetics (reviewed by Brinton, 1981a). If C58 mice are immunocompetent, they must be over six months of age to develop paralysis after IP inoculation of strain Ib-LDV; younger C58 animals will become susceptible if they are x-irradiated, indicating that the loss of a T-lymphocyte subset is required for susceptibility. Genetic crosses between inbred strains demonstrated that susceptibility to polioencephalomyelitis segregated with the *Fv-1* linkage group (see "Retrovirus") (Pease, Abrams, and Murphy, 1982). Susceptible mouse strains are *Fv-1*$^{n/n}$ and have multiple copies of N-tropic retroviruses, whereas mice with the *Fv-1*^{b} allele or that carry few copies of N-tropic retroviruses in their genome are resistant. Interestingly, mice that are *Fv-1*$^{n/n}$ but that carry only a few copies of N-tropic retroviruses in their genome can be made more susceptible if they are bred to C58 mothers, suggesting that C58 mothers can transmit their own endogenous retroviruses to the offspring and enhance susceptibility. Presumably one needs the presence of both a retrovirus and LDV for disease induction. Pease, Abrams, and Murphy (1982) also noted a maternal resistance effect that in preliminary studies appeared to be linked to the H-2 locus.

Susceptibility to LDH virus infection is even more complicated than the above studies suggest since mouse strains resistant to clinical polioencephalomyelitis can display inflammatory lesions restricted to the white matter (Stroop and Brinton 1983). The development of this encephalomyeloradiculitis does not depend on host age or prior immunosuppression and is not under exclusive control by the H-2 complex.

The pathogenesis of the LDV complex disease is still far from understood. If induction of still unrecognized unconventional human viral infections involves as many variables as does age-dependent paralytic LDV infection, successful experimental transmission of these diseases will be extremely difficult.

Coronavirus

Coronaviruses are enveloped viruses with positive-stranded RNA (that is, can be translated directly into protein) that are found in many species. Human coronaviruses are a common cause of respiratory disease and, in an unconfirmed report, have been cited as a possible cause of multiple sclerosis (Burks et al., 1980). This section will deal with mouse hepatitis virus (MHV), a member of this group, because of the extensive literature concerning the genetic susceptibility of different mouse strains to MHV and because of the neurovirulence of certain MHV strains. MHV occurs naturally as a latent infection and can experimentally produce hepatitis as well as several neurological syndromes. The variety of virus-host relationships seen adds additional complexity to the subject.

In pioneer investigations Bang described genetic resistance to MHV type 2 (MHV-2) (reviewed by Bang and Cody, 1980). C3H mice are not susceptible to IP inoculation of certain strains of MHV-2 and the subsequent hepatitis manifest by the more susceptible mouse strain PRI. The restriction, inherited as a single autosomal recessive gene, is expressed at the level of cultured macrophages; that is, macrophages from resistant strains replicate MHV-2 to lower titer than macrophages from susceptible strains. Macrophages are important in host defenses and are believed to play a role in antigen presentation to certain T-lymphocytes. They have phagocytic and enzymatic properties (esterase, peroxidase) and secrete a multitude of bioactive compounds. The block to MHV-2 replication in macrophages occurs after virus adsorption since this process is efficient in both resistant and susceptible macrophages (Shif and Bang, 1970). The phenotypic expression of the macrophage can be altered by changing the culture media and serum source. The resistance of macrophages also appears to be influenced by lymphokines and lymphocytes themselves (Weiser and Bang, 1976); for example, supernatants from mixed lymphocyte reactions render resistant macrophages more susceptible. Lymphocytes may play a critical role in in vivo resistance as well, since C3H mice can be rendered susceptible by treatment with lymphocytotoxic agents (although macrophage cultures from the treated animals are still resistant). Perhaps lymphocytes alter the phenotypic expression of resistance manifest by the macrophage. IFN may also be involved in the lymphocyte-macrophage interactions.

In the case of MHV type 3 (MHV-3), the pathogenesis of disease is more complicated, partly because there is both acute and chronic infection (Lévy-Leblond, Oth, and Dupuy, 1979). The acute infection is manifest by necrosis of the liver and lymphoid organs. Mice with the chronic disease develop ataxia and progressive paralysis with evidence of chronic chorioependymitis and meningitis or nervous system vasculitis (Virelizier, Dayan, and Allison, 1975). A/J strain mice are resistant to acute disease if older than three weeks of age, whereas mice of susceptible strains (for example, C57B1/6) develop an acute fatal infection, regardless of age. Some F1 hybrid mice from susceptible and resistant parents, as well as certain inbred strains, are *semisusceptible*, a term denoting the induction of acute disease in 10 to 30 percent of adults and the production of chronic disease in survivors. The incidence of acute disease in resistant and susceptible mice in crosses and backcrosses after IP inoculation at six weeks of age suggested that one or two recessive genes determine resistance. When animals were inoculated at three months of age, however, the data suggested a dominant gene for resistance. The differences in gene effects depending on host age may be due to alterations of virus cell receptors or of immune defenses that develop with maturity. The genes that determine susceptibility to the acute viral disease clearly segregate independently of the H-2 complex. Studies with mice of congenic lines, however, suggested that one or two dominant genes linked to the H-2 complex conferred resistance to the chronic disease; this gene may be important in the genesis of this putatively immunopathological disease.

The genetic factors important in MHV-3 resistance are not absolute, and they must interact with host immune defenses. One can induce fatal acute disease in the resistant A/J mice by a variety of immunosuppressive actions. Conversely, susceptible animals may survive acute disease after inoculation with minute amounts of virus or through the administration of immune serum.

There is some controversy as to whether resistance to MHV-3 is expressed in vitro at the level of the macrophage. Virelizier (1981) reports high titers of virus in macrophages of susceptible strains and no virus multiplication in resistant strains. Three other groups, however, are in disagreement, finding little or no correlation between macrophage in vitro permissiveness and in vivo susceptibility of the mouse strain (Dupuy, Lafforet-Cresteil, and Dupuy, 1980; MacNaughton and Patterson, 1980; Taguchi et al., 1981). Dupuy, Lafforet-Cresteil, and Dupuy (1980) found that macrophages from different strains show no differences in viral cytopathic effect when infected in vitro, but macrophages taken from resistant and susceptible mice one to three days after IP injection do show differences in cytopathic effect when cultured and observed in vitro. Taguchi et al. (1981) and Yamada, Taguchi, and Fujiwara (1979) report a varying permissiveness of cultured T-lymphocytes, not macrophages, corresponding to susceptibility of the mouse strain. These studies emphasize the importance of the milieu of the macrophage and possibly its interactions with lymphocytes as a critical feature in resistance. Perhaps the variable purity of macrophage preparations and the changing number of contaminating lymphocytes have been responsible for the conflicting results of in vitro susceptibility studies. The presumed role of macrophages and lymphocytes in resistance gains less importance with a report that MHV-3 resistance is expressed at the level of cultured hepatocytes (Arnheiter, Baechi, and Haller, 1982).

It seems unlikely that IFN controls susceptibility to MHV-3 at the cellular level. IFN had a similar inhibitory effect on replication of MHV-3 in hepatocytes from both resistant and susceptible mouse strains (Arnheiter, Baechi, and Haller, 1982). Treatment of cultures with anti-IFN serum had no effect on virus yields. Strains differing in resistance to MHV-3 have similar levels of circulating IFN (Virelizier, 1981). However, it is possible that there may be important local differences in IFN availability, or there may be genetic differences in sensitivity to IFN between sensitive and resistant mice. IFN clearly exerts a protective effect on the infection since treatment of resistant mice with anti-IFN serum causes death, whereas treatment of susceptible mice accelerates the onset of death (Virelizier and Gresser, 1978).

MHV type 4 (MHV-4) has attracted recent attention because the JHM strain provides an experimental murine model of demyelination. Mice develop an acute or recurrent white matter disease generally thought to represent a direct oligodendroglialytic infection (Weiner, 1973; Herndon et al., 1975). Knobler, Haspel, and Oldstone (1981) found that mouse strains differed by 30- to 75-fold in susceptibility to acute fatal encephalitis following IC inoculation of MHV-3. The dose of virus required to kill 50 percent of SJL/J mice was greater than 100 times the dose for other strains checked (such as Balb/c). Studies of crosses and backcrosses between resistant and susceptible strains, as well as between recombinant imbred strains, suggested that one recessive gene not related to the H-2 complex determined resistance (Knobler, Haspel, and Oldstone, 1981; Knobler, Linthicum, and Cohn, in press). Note that Stohlman and Frelinger (1978) reported discrepant findings and claimed that both a dominant and recessive gene were required for resistance.

Knobler, Haspel, and Oldstone (1981) found that the resistance of SJL/J mice to MHV-4 was reflected in the titers of virus in brain, which were 100 to 1,000 times lower than those of Balb/c within the first two days following inoculation and never reached levels of Balb/c. Knobler and coworkers determined that this resistance was expressed at the level of neuron in vitro by observing the effect of infection of neuronal cell cultures from different strains. SJL/J neurons in culture replicated MHV-4 little, whereas neurons from susceptible strains or from an F_1 hybrid between SJL/J and a susceptible strain replicated MHV-4 to high titer. Similar findings were obtained with experiments with macrophages. The block in viral growth in neurons is not at the receptor level since viral antigen is present in neurons of SJL/J mice in vivo or in vitro after inoculation with a large dose of virus. Presumably, virus adsorbs to the neurons and penetrates the cell, but a block prevents the efficient production of infectious virus.

Similarities exist in the findings of host genetic influences on MHV-2, MHV-3, and MHV-4 infection. In each type one or more genes regulates resistance both in vivo as well as in the primary target tissue in vitro. As Knobler and coworkers note, however, there are remarkable differences regarding the behavior of particular mouse strains challenged with one or the other virus type; for example, A/J mice are resistant to MHV-3 but susceptible to MHV-4.

Picornavirus

Picornaviruses are positive-stranded, small RNA viruses that are a common cause of human disease. These agents produce subclinical infections, enteric disease, and meningoencephalitis. Perhaps the most studied of the diseases is poliomyelitis. Encephalomyocarditis virus is of interest because the M strain produces a diabetic syndrome in mice. Other strains of encephalomyocarditis produce a lethal meningoencephalitis with mild, if any, pancreatic damage. Studies with the diabetogenic variant may provide a foundation for our understanding of host genetic factors affecting susceptibility to the neurotropic strains.

Poliovirus

The great majority of individuals with poliovirus infection have subclinical disease; only 4 to 8 percent have minor illness, and less than 2 percent have neurological symptoms and signs. The frequency of subclinical disease and the epidemic nature of poliovirus infections have prompted direct investigations of human genetic factors affecting poliomyelitis. What determines whether patients develop paralytic disease?

There are a varity of genetic and nongenetic host factors affecting the clinical course following poliovirus infection. Few are well understood. Nongenetic factors include age, socioeconomic level, pregnancy, tonsillectomy, and trauma. In addition, Nottay et al. (1981) reported that poliovirus undergoes continuous genomic mutation as it spreads from one individual to another. The influence of nongenetic factors coupled with this continuing in vivo mutation of the infecting poliovirus (with perhaps an associated change in neurovirulence of the virus) have made an evaluation of host genetics difficult.

Genetic factors are clearly important at many stages of the virus replicative cycle. The observation that poliovirus infection is restricted to primate cells in vitro has directed attention to genes coding for the virus receptor. Somatic cell hydridization localized susceptibility to poliovirus types 1 and 2 infection to human chromosome 19 (Miller et al., 1974). Susceptibility is presumably due to dominant phenotypic expression of the virus receptor since nonhuman cells are permissive to poliovirus replication if viral RNA is introduced within the cell (Neff and Enders, 1968) even if human chromosome 19 is absent. The poliovirus receptor has been demonstrated to differ from the echovirus and coxsackievirus receptor (Philipson and Bengtsson, 1962). In support of this observation Miller et al. (1974) found that the presence of chromosome 19 was not necessary for the expression of echovirus 7 and rhinovirus 1A receptors. A number of unanswered questions remain regarding the role of this receptor. For example, can one extrapolate from the results of in vitro fibroblast experiments to in vivo infection? By what mechanism has the Lansing strain of poliovirus type 2 become mouse adapted? Is the selective vulnerability of the motor neuron to poliovirus and certain features of the pathogenesis of poliomyelitis related to availability and expression of the virus receptor—or is there more likely a block at another stage in intracellular virus replication?

A number of twin or pedigree studies have suggested involvement of autosomal recessive genes in influencing susceptibility to paralytic poliomyelitis. Herndon and Jennings (1951) found a statistically significant difference in concordance rate for paralytic poliomyelitis in monozygous compared with dizygous twins. Penetrance figures reported have been 35 percent on the basis of a study of 47 twins (Herndon and Jennings, 1951), 70 percent on the basis of a study of one large pedigree with 29 cases of poliomyelitis (Addair and Snyder, 1942), and 80 percent in a study of 50 pedigrees including 153 cases (Reedy, 1957). These studies have been criticized because of small sample sizes, the lack of a matched control group, and the unavoidable influence of nongenetic familial determinants of susceptibility (for example, hygiene).

Wyatt (1975, 1978) has proposed that a dominantly inherited gene, *p*, confers susceptibility to poliomyelitis. He claims that studies of disease distribution agree well with the predicted presence of 2 percent homozygotes (p^+p^+) and 24 percent heterozygotes (p^+p^-) in the population. Phenotypic expression of the gene is age dependent, giving rise to the age variation in frequency of paralytic disease.

Several studies have analyzed the influence of HLA type on acquisition of poliomyelitis. Although Pietsch and Morris (1974) noted a significant association of HLA-3 and HLA-7 with paralytic poliomyelitis, three other studies were unable to confirm the findings (Dausset and Hors, 1975; Zander et al., 1976; Lasch et al., 1979). The differing results may reflect population differences or errors related to chance.

Many of these questions regarding poliomyelitis pathogenesis and the role of genetics will remain unanswered as a result of the introduction of poliovirus vaccination.

As a result of the introduction and use of poliovirus vaccination, questions regarding poliomyelitis pathogenesis and the role of host genetics may have to await studies involving mouse-adapted poliovirus.

Theiler's Murine Encephalomyelitis Viruses

Theiler's murine encephalomyelitis viruses (TMEV) are a group of mouse enteric picornaviruses which cause a variety of neurological diseases (reviewed by Roos, 1984). Members of the GDVII subgroup of TMEV cause an acute, self-limiting neuronal disease in weanling mice; members of the TO subgroup cause a biphasic disease characterized by an early, acute neuronal infection followed by a late, chronic demyelinating disease. Several studies suggest that the late demyelination may be immune mediated; for example, certain pathological features of this disease are said to be reminiscent of those seen in acute experimental allergic encephalomyelitis, an experimental demyelinating disease caused by T lymphocyte autoimmunity directed against myelin basic protein.

Since experimental allergic encephalitis is believed to be under control of at least two genes, one being H-2 linked (Knobler, Linthicum and Cohn, in press), Lipton and Melvold (1984) investigated the influence of H-2 genes on susceptibility to TMEV late demyelinating disease. One hundred percent of SJL strain mice developed clinical evidence of demyelination by 105 days after TMEV inoculation, while none of the Balb/c mice did so. By using data from inbred mouse strain crosses, they reported that susceptibility to demyelination was influenced in a gene dosage fashion by both a gene linked to H-2 and an independently segregating non-H-2 gene(s). Studies using recombinant inbred strains will clearly be of interest (Knobler, Linthicum and Cohn, in press).

Encephalomyocarditis Virus

Infection of mice with the M variant of encephalomyocarditis (EMC) virus causes a diabetic syndrome in mice resulting from infection of beta cells. Studies of crosses between strains of mice demonstrated that susceptibility to EMC virus is inherited as an autosomal recessive trait (Onodera et al., 1978). The data were most consistent with control from a single locus with two or more alleles. Similar genetic factors seem to determine susceptibility of mice for the pancreatropic endocrine effects of inoculation with human coxsackievirus group B, type 4 (Webb and Madge, 1980).

Yoon and Notkins (1976) attempted to delineate the level of EMC restriction. No difference between susceptible and resistant mice in the development of neutralizing antibody was noted. A higher titer of virus was found in the pancreas of susceptible mice as compared with resistant ones. Immunofluorescent and infectious center assays showed that a greater number of cells were infected in the susceptible animals. No difference in viral replication was noted in cultured kidney and embryo cell monolayers from susceptible and resistant mice, but there was 50 times less virus recovered from cultured purified beta cells of susceptible compared with resistant mice. The studies indicate that genetic factors that function at the level of the beta cells determine susceptibility to diabetes. These factors may be related to a specific block in intracellular replication of the virus or the response and repair of the beta cell to infection. Chairez, Yoon, and Notkins (1978) reported that twice as much virus binds to beta cells from susceptible mice than from resistant ones, suggesting that restriction is partly or completely expressed at the level of virus adsorption.

Orthomyxovirus

Orthomyxoviruses are enveloped viruses containing single-stranded, segmented RNA of negative polarity. This section will deal with one member of the orthomyxovirus subgroup, influenza virus. Although influenza virus is not usually associated with neurological disease, the extensive literature regarding *Mx*, a dominantly inherited resistance gene, has prompted its inclusion here (Lindenmann, Lane, and Hobson, 1963). This resistance gene is expressed in the neurotropic strain when inoculated by the IC route, the pneumotropic strain by the intranasal route, and the hepatropic strain by the IP route. The gene restricts virus replication 100-fold in resistant animals.

The mechanism of resistance has been under investigation for some time. Since the gene continues to act even when introduced into a nude mouse (without T cells) and in CBA/N mice with an X-linked recessive B cell defect, the mechanism of gene action is presumed to act by means other than the B or T cell immune system. The resistance phenotype is expressed in vitro in macrophages, aggregating brain cells, and hepatocyte cultures (Lindenmann et al., 1978; Arnheiter, Haller, and Lindenmann, 1980; Haller et al., 1980; Haller, 1981); virus yields can vary as much as 10^4 between resistant and sensitive cultures. Transplantation of macrophage precursors from resistant animals to susceptible irradiated ones and vice versa does not change the susceptibility of the recipient, implying that in vivo resistance is not merely a function of the macrophage.

Several observations suggest that *Mx* gene expression depends on, and possibly is equated with, the action of IFN (Arnheiter, Haller, and Lindenmann, 1980; Haller, 1981). Treatment of brain cells and hepatocytes in vitro from resistant animals or inoculation in vivo of resistant animals with anti-IFN globulin renders them fully susceptible and increases virus replication. The resistance phenotype of hepatocytes can then be restored by adding IFN. Although different strains of mice are known to produce different amounts of IFN to a specific virus (De Maeyer et al., 1973), the finding of similar levels of IFN in influenza virus–infected resistant and susceptible mice suggests that resistant strains have a greater sensitivity to IFN activity compared with susceptible strains against influenza virus. Sensitivity to the immunomodulating effect of IFN is known to vary in different mouse strains (De Maeyer and De Maeyer-Guignard, 1980). The importance of *Mx* in modulating the action of IFN against influenza virus infection was emphasized by Meyer and Horisberger (1984). They found that IFN treatment lowered the virus yield 800-fold in cultivated macrophages from mice homozygous for *Mx* and only sevenfold from macrophages lacking *Mx*.

The site of action of IFN on cultivated *Mx*-bearing macrophages is under study. IFN does not affect influenza virus attachment or penetration in mouse macrophage cultures (Horisberger, Haller, and Arnheiter, 1980), nor does it inhibit virus uncoating (Meyer and Horisberger, 1984). The transcription of the virus' negative-stranded RNA into messenger RNA is reduced twofold in macrophages from mice both which bear and lack *Mx*. Although there was no inhibition of influenza virus translation in macrophages lacking *Mx*, there was a marked (100-fold) reduction of virus polypeptide synthesis in *Mx*-bearing macrophages. This particular block in translation occurred despite the presence of reasonable levels of influenza virus messenger RNA which were functioning in in vitro translation systems.

Horisberger, Staeheli, and Haller (1983) have described a 72,000 molecular weight protein (P72) in *Mx*-bearing cells which is induced by IFN and is correlated with the antiviral state. The particular site of action of P72 in inhibiting influenza virus translation is under investigation. It will be of interest to determine the relationship and cosegregation of the *Mx* gene with the gene coding for P72, once it is identified.

Paramyxovirus

Paramyxoviruses are enveloped viruses containing single-stranded RNA of negative polarity. Measles virus, a member of the paramyxovirus group, causes the common exanthematous disease and two main central nervous system diseases, postinfectious encephalomyelitis (PIE) and subacute sclerosing panencephalitis (SSPE). PIE is believed to be an autoimmune demyelinating disease pathologically similar to experimental allergic encephalomyelitis. SSPE is a chronic, persistent measles virus encephalitis occurring in the pediatric age group after an interval that averages five years from the primary measles exanthematous disease. Virus is isolated from SSPE only occasionally and requires prolonged culturing and cocultivation. This difficulty is presumed to be due to the defective nature of the measles virus, which lacks M protein, a protein important in virus budding (Hall and Choppin, 1979; Lin and Thormar, 1980). The pathogenesis of SSPE is still poorly understood. Epidemiological studies have stressed the importance of primary measles virus infection before age two in those individuals destined to develop SSPE. The infecting virus, however, is believed to be normal wild type measles, which then undergoes mutation during the prolonged SSPE incubation period. Host genetic factors as well as immune parameters may play an important role in the genesis of the defective virus infection.

Animal studies have begun to examine genetic factors important in acute and chronic measles virus infections. Rager-Zisman et al. (1980) described a variation in mortality among mouse strains inoculated IC with measles virus. When inoculated three to six days after birth, 100 percent of C3H/HeJ succumbed, but no SJL mice died. Studies of F_1 hybrid mice and backcrosses showed that resistance was controlled by a dominant gene(s) that segregated independently from the H-2 complex. The mechanism of virus restriction is not known.

Massanari, Wilson, and Leininger (1980) inoculated hamster brain–adapted SSPE virus into four different highly inbred hamster strains. Different strains exhibited variations in onset of encephalitis, mortality, and ease of virus reisolation. Differences were also noted in the extent and timing of development of neutralizing antibody.

Herpesvirus

Herpesviruses are a group of DNA-containing viruses in which some members display the properties of latency and oncogenesis. This section will deal with three members of the herpes group that are associated with neurological disease: herpesvirus types 1 and 2 (HSV-1, HSV-2), cytomegalovirus (CMV), and Marek's disease virus (MDV). HSV-1, the cause of the common "cold sore," is one of the frequently identified etiologic agents of sporadic adult encephalitis. HSV-2 causes genital ulcers, aseptic meningitis, and neonatal encephalitis. CMV normally causes subclinical disease, although it can cause an atypical mononucleosis syndrome and an intrauterine infection with encephalitis. CMV can produce a chronic central nervous system infection primarily in the immunosuppressed (Dorfman, 1973). The CMV studies dealt with involve murine cytomegalovirus (MCMV). MDV causes a lymphoproliferative disease in chickens, with frequent neural involvement. The recurrent demyelination caused by MDV closely resembles the lesions seen in human Guillain-Barré syndrome (Prineas and Wright, 1972); susceptibility to Guillain-Barré syndrome may be influenced by histocompatibility antigen type (reviewed by Roos, 1981b), as in Marek's disease.

Herpes Simplex Virus

On the basis of genetic crosses, Lopez (1980, 1981) has described two major dominant genetic loci in mice that govern resistance to death from encephalitis produced after HSV-1 IP inoculation. For example, the LD_{50} for IP inoculation of a strain of HSV-1 varies from 10^6 plaque forming units PFU for an adult resistant mouse strain to 10^1–10^2 PFU for a susceptible one. The different mouse strains are all exquisitely sensitive to small amounts of this HSV-1 strain; however, there may be similar factors involved in IC inoculations. There are also "minor loci" that govern resistance.

Macrophages from susceptible mice are more efficient in replicating HSV-1 after in vitro infection than are macrophages from resistant mice; however, this in vitro macrophage permissiveness does not segregate with susceptibility in F_1 crosses, demonstrating that the mechanism for resistance is not controlled by macrophages (Lopez and Dudas, 1979).

Lopez (1981) has emphasized the similarities between HSV-1 resistance and resistance to bone marrow allografts. Treatment of resistant mice with strontium-89, a treatment known to ablate natural killer (NK) cells, renders resistant mice susceptible to HSV. NK cells are believed to mediate bone marrow allograft and tumor cell rejection and resistance to virus infections; they are able to cause cell lysis without previous exposure to an antigen and without dependence on the presence of antibody. In support of the findings of Lopez, Kirchner et al. (1980) detected greater NK cell activity in infected resistant mice. Still, Kirchner emphasizes the importance of IFN. Spleen cells from resistant mice produce more IFN when exposed to viral antigen than do cells from susceptible mice; Lopez (1981) reported similar in vitro studies. Different strains of mice have similar levels of IFN except under very specific conditions—that is, when inoculated with high doses of virus and when examined hours after infection (Kirchner et al., 1980). It may be that IFN activates NK cells or that NK cells produce IFN; however, Kirchner et al. (1980) claim that the spleen cell producing IFN appears to differ from a NK cell.

Mogensen (1977, 1980) described an X-linked dominant gene(s) controlled resistance to hepatitis following HSV-2 IP inoculation of mice. Four days after inoculation 100 percent of Balb/c mice developed microscopic liver lesions, but no GR mice did; in addition, virus titers were 100 times higher in Balb/c mice. In F_1 hybrids 100 percent of males and no females had microscopic lesions. The first backcross generation from resistant F_1 females mating with susceptible Balb/c mice showed a segregation ratio of about 1 : 1. This resistance was expressed at the level of the macrophage but not embryonic fibroblasts.

Cytomegalovirus

Resistance of adult mice to IP inoculation of murine cytomegalovirus (MCMV) is complicated but at least partly controlled by dominant susceptibility genes linked to the H-2 complex (Grundy, Mackenzie, and Stanley, 1981). These studies are difficult because of the multiplicity of genes involved and because of the slight differences in LD_{50} determining the resistant or susceptible phenotype. One gene maps in the *K/IA* subregion and the second in the *D* subregion. These genes are not phenotypically expressed till the first few weeks of life.

The mechanism of susceptibility exerted by the H-2-linked genes is unclear. These genes are not thought to be dominantly inherited immune response genes since one would normally expect the genes' actions to lead to resistance, not susceptibility. Lopez (1980), however, comments that mice with H-2–associated susceptibility have a heightened inflammatory response, suggesting immunopathological disease.

Resistance is also affected by interactions of these two genes with non-H-2-linked genes, for example, the C57B1 genetic background. The non-H-2-linked genes are probably dominantly expressed (Selgrade and Osborn, 1974; Chalmer, 1980). The non-H-2-associated resistance genes are correlated with increased in vitro cytotoxicity by spleen cells from infected mice against an NK-susceptible target cell (Bancroft, Shellam, and Chalmer, 1980). The cytotoxic cells appeared most similar to NK cells. Support for this observation was the finding that the "beige" mutant C57 B1/6J mice, which have a defect in NK activity, are more susceptible to MCMV infection than NK-competent heterozygote littermates. The role of IFN in this infection is yet to be delineated.

Embryo cell cultures and macrophages taken from different mouse stains replicate MCMV with similar efficiency, implying that susceptibility is not expressed at this level (Selgrade and Osborn, 1974; Grundy, Mackenzie, and Stanley, 1981). On the other hand, tracheal organ cultures taken from susceptible mice replicate 10- to 100-fold more MCMV than cultures from resistant mice (Nedrud, Collier, and Pagano, 1979). Nedrud and coworkers suggest that epithelial cells may be a critical target cell in vivo; for example, perhaps slowed replication in these cells of resistant animals allows the immune system to deal with the infection. The importance of the immune system is supported by the finding that cortisone treatment rendered resistant animals more susceptible in vivo (Grundy, Mackenzie, and Stanley, 1981). It is clearly necessary to study and compare the progress and the pathogenesis of disease in resistant and susceptible mice to delineate the initial target tissues.

Marek's Disease Virus

Strains of chickens vary considerably in susceptibility to Marek's disease (MD). There appear to be at least two genetic loci that confer resistance. One locus is associated with the major histocompatibility (MHC) locus of chickens (the B locus). For example, Longenecker et al. (1976) found that resistance to MD was associated with a particular allele of the B locus, perhaps through immunological mechanisms.

Briles et al. (1983) used a genetic recombinant between highly resistant B^{21} and B^{19} haplotypes to further delineate the subregion responsible for MD resistance. Their data suggest a gene(s) within or closely linked to the B-F regions seems to fully account for the MD resistance of the B^{21} haplotype. The B-F region determines the specificity of the histocompatibility antigens present on lymphocytes and erythrocytes, and may be important in the production of a T cell response against altered self-antigens during tumorigenesis.

The second locus that confers resistance to MD is not associated with the B locus, as evidenced by differing susceptibilities of line 6 and line 7 chickens, which do not differ at the MHC locus (Gallatin and Longenecker, 1979). These two lines differ at the Ly-4 locus, which codes for alloantigenic markers present on T cells. Lee et al. (1981) carefully compared Marek's disease virus (MDV) infections of resistant line 6 and susceptible line 7 chickens to clarify the mechanism of resistance. Line 6, but not line 7, chickens demonstrated late T cell–mediated viral immunity and significant cell-mediated cytotoxicity against MD tumor–associated surface antigen at several time points. In addition, titers of MDV were lower in line 6 than line 7 chickens as early as three days after infection. Lower titers were also present in spleens from infected line 6 embryos, suggesting the importance of cellular factors in resistance.

Other studies have emphasized the importance of cellular resistance specifically involving the target cell, that is, the T-lymphocyte. Studies have demonstrated a similar in vitro susceptibility of embryonic fibroblasts from line 6 and 7 chickens but a differential in vitro susceptibility of spleen and thymus cells from the two lines to turkey herpesvirus, a herpesvirus related to MDV (Gallatin and Longenecker, 1979). Spleen cells from susceptible chickens adsorbed more turkey herpesvirus and MDV than cells from resistant chickens (Powell et al., 1982). Suspension cultures of spleen cells from susceptible chickens, however, did not have more MDV antigen expression or production than cells from resistant chickens (Calnek et al., 1982). Calnek and coworkers explain this finding by claiming that they are observing an in vitro equivalent to early cytolytic MDV infection, which does not correlate with the later viral transformation and lymphoproliferative disease. Transplants of thymuses from susceptible to resistant chickens increased susceptibility of resistant chickens (Gallatin and Longenecker, 1979; Powell et al., 1982). Gallatin and Longenecker (1979) question whether a lower number of target cells, a lower number of receptors, or a lower affinity of receptors on target cells may be the critical determinant of resistance in line 6 chickens. Lee et al. (1981) verified that line 6 chickens did have fewer lymphocytes in their spleens and peripheral blood than line 7. In summary, resistant line 6 chickens had a greater immune response and fewer target cells available for transformation than line 7 chickens.

"HEREDITARY INFECTION" AND THE SUBACUTE SPONGIFORM ENCEPHALOPATHY AGENT

Probably one of the most provocative association of host genetic aspects with a "viral" disease is seen in connection with the subacute spongiform encephalopathies (SSEs) (reviewed by Roos, 1981a, 1981c). The SSEs comprise a group of two human diseases, kuru and Creutzfeldt-Jakob disease (CJD), and two animal diseases, scrapie and transmissible mink encephalopathy, which have similar clinical pathological characteristics and etiologies. All of these diseases cause a subacute, chronic, progressive, noninflammatory, fatal neurological syndrome. All have evidence of pathology confined to the central nervous system characterized by a noninflammatory neuronal loss, gliosis, spongiosis, scrapie associated fibrils (Merz et al., 1984), and, in some cases, amyloid plaques. The diseases are called *slow virus infections* because filtered material transmits the disease following a long incubation period, at times longer than two decades (Prusiner, Gajdusek, and Alpers, 1982).

The SSE transmissible agent has been of intense interest to molecular biologists because of its unusual resistance to physical and chemical agents and its apparent lack of antigenicity. Because the agent's behavior is unique among viruses, the possibility that the agent may be a *viroid* (a naked piece of RNA that transmits a variety of plant diseases) (Diener, 1981) or a non-nucleic acid-containing agent has been proposed. Although experiments have shown that the agent is not a viroid (Marsh et al., 1974; Ward, Porter, and Stevens, 1974), it is clear that it will prove to be unconventional in its structure.

Host genetic factors have been implicated in the three main SSE diseases studied—scrapie, kuru, and CJD. Some 10 percent of CJD cases are familial with a pattern suggesting dominant inheritance within the pedigrees. What is most remarkable and least understood is the finding that the familial cases are transmissible, a situation which has been referred to as "hereditary infection" (Harper, 1977).

Scrapie

Most studies on host genetics in the SSEs have been performed on scrapie, undoubtedly owing to its well-established transmission to rodents (Chandler, 1961). The natural disease in sheep and goats is characterized by ataxia and autonomic dysfunction. The disease is worldwide with over 218 outbreaks in the United States over the last 30 years (Hourrigan et al., 1979).

The means of spread of natural scrapie is unknown. Such factors as crosscontamination by scratching fence wires, bites, or the ingestion of contaminated grass have been implicated. There is a prominent "maternal effect" seen with scrapie; that is, offspring are more likely to be affected if a parental ewe rather than ram has scrapie (Dickinson, Stamp, and Renwick, 1974). This observation has raised the possibility that transmission to the lamb occurs by ingestion of infected placenta, which is known to be infectious. Support for this means of spread comes from recent studies by Hadlow et al. (1979) demonstrating that one of the earliest sites of infectivity in natural scrapie is the gastrointestinal tract.

Host genetic factors play a critical role in determining susceptibility and the length of the incubation period. The interpretation of these studies is unavoidably complicated by such factors as influence of virus strain, passage history, and route of inoculation. Dickinson and Fraser (1979) described a dominantly inherited gene—*Sip*—that determines susceptibility of sheep to subcutaneous inoculation with the SSBP/1 scrapie agent. Although homozygous *sip/sip* animals are resistant to subcutaneous inoculation, half of them succumb to IC challenge, suggesting that the gene may actually regulate the length of incubation period rather than absolute susceptibility; the incubation period of scrapie in these homozygous, resistant animals may exceed the life of the sheep. Different scrapie agent strains behave in different fashions with the same strains of sheep, suggesting that each strain of scrapie is unique because of its different passage history and that the agent-host interaction is important in disease development.

In mice Dickinson and Fraser (1979) described a *sinc* gene with two alleles, *S7* and *P7*, which determine the length of incubation period of scrapie in mice. Mice with the *S7/S7* genotype (for example, C57/B1) have a short incubation period for ME7 scrapie agent, whereas mice with *P7/P7* genotype have a prolonged incubation period. The heterozygote *S7/P7* has an incubation period of intermediate length. As with sheep scrapie, genetic influences change with different scrapie strains. The *S7/P7* heterozygote inoculated with 22A scrapie strain displays "overdominance" with a longer incubation period than with either homozygote. Dickinson and Outram (1979) suggest that there is a complex interaction of the gene products of S7/P7 with the agent itself.

Kingsbury et al. (1983) identified at least one and perhaps two other genetic loci which influence the length of the incubation period of scrapie (and CJD) in mice. Alleles coding for the longer incubation period are dominant over alleles coding for short incubation times. Alleles of the D subregion of the H-2 haplotype of the mice have a major control of the incubation period length.

Of great interest is the variation in pathology depending on the host genotype (and agent strain). The number of amyloid plaques and their distribution in inoculated mice (and presumably natural SSE disease) vary remarkably depending on the agent strain–host strain combination (Fraser and Dickinson, 1973; Outram, Fraser, and Wilson, 1973; Bruce, Dickinson, and Fraser, 1976).

Kuru

Kuru is a progressive cerebellar disease found in the highlands of Papua New Guinea among the Fore and tribes with which they intermarry (Gajdusek and Zigas, 1957). The clinical and pathological similarities between scrapie and kuru led to kuru transmission attempts, which succeeded in 1966 (Gajdusek, Gibbs, and Alpers, 1966). The incubation period of kuru is usually over one year when inoculated into subhuman primates.

The disease is primarily confined to adult females and children of both sexes. The epidemiological observation reflects the fact that these groups were given high-titered brain tissue during endocannibalistic rituals practiced by the Fore. The cessation of cannibalism has, as expected, resulted in the disappearance of kuru, with a steadily increasing age of remaining victims (Prusiner, Gajdusek, and Alpers, 1982).

The relationship between host genetic factors and kuru has been mentioned often in the scientific literature. Attention was originally focused on the frequent occurrence of kuru within pedigrees, prompting Bennett, Gray, and Auricht (1959) to propose that kuru was caused by an autosomal gene dominant in females and recessive in males. The hypothesis is no longer tenable with the successful transmission of kuru and with the disappearance of kuru coincident with the cessation of endocannibalism. The apparent vertical inheritance was a presumed artifact of frequent horizontal transmission. A number of genetic markers of kuru and nonkuru Fore were studied to determine any association with acquisition of kuru genetic marker (Kitchin et al., 1972). The only positive finding, of uncertain importance, was a significantly increased frequency of the group-specific aboriginal variant allele (Gc^{Ab}) among kuru Fore. The fact that kuru was a common cause of death among the Fore suggests that all Fore were in fact susceptible to the disease.

Creutzfeldt-Jakob Disease

Creutzfeldt-Jakob disease (CJD) is a rare disease found throughout the world with an average annual age-specific mortality rate of 0.26 per million in the United States (Roos, Gajdusek, and Gibbs, 1973; Masters, Harris,

Gajdusek, Gibbs, Bernoulli, and Asher, 1979). The average age of onset is in the early fifties, but cases from the teens to over 80 are described. The disease has varied clinical manifestations depending on the focus of the pathology. Most commonly one observes dementia, myoclonus, and a characteristic periodic, paroxysmal electroencephalogram leading to a mute akinetic state in 3 to 6 months, and death in 6 to 12 months. Ataxia, extrapyramidal signs, and amyotrophy are not infrequent.

CJD can be transmitted to subhuman primates and a variety of other species. The host susceptibility of certain species depends on the particular strain of the agent. For example, tissue from only some cases of CJD can be transmitted to the mouse (Kingsbury et al., 1982). In contrast, all typical CJD cases transmit to the chimpanzee. It may be that some cases do not transmit to the mouse because the incubation period of disease from some strains exceeds the life span of the mouse. The successful transmission to mice has offered the possibility of performing studies of host genetic effects such as Dickinson has employed in scrapie mice. Preliminary studies show that the incubation period does vary among different mouse studies (Kingsbury et al., 1982).

In a recent study about 15 percent of 1,435 cases of CJD were found to be familial (Masters, Harris, Gajdusek, Gibbs, Bernoulli, and Asher, 1979). One must be a bit cautious about this figure since, on one hand, familial cases may tend to be overreported and, on the other hand, CJD atypical clinical syndromes or early deaths in relatives destined to acquire CJD may make ascertainment incomplete. At least one initially unexplained focus of CJD, in Libya, was subsequently found to have unsuspected familial cases (Neugut et al., 1979). Masters, Gajdusek, Gibbs, Bernoulli, and Asher (1979) reviewed 37 pedigrees involving a total of 155 CJD patients. In most pedigrees there is evidence of affecteds in multiple generations, suggesting an autosomal dominant inheritance. The most noted example is the Backer family, which includes 15 individuals with possible, probable, or definite CJD over four generations. The familial cases are indistinguishable clinically from sporadic CJD except that they tend to have a younger age of onset. No maternal effect was observed.

Cases have rarely occurred among non–blood relatives. There are claims that three conjugal pairs succumbed to CJD, but documentation in the cases is not always convincing (Jellinger et al., 1972; Matthews, 1975; Masters, Gadjusek, Gibbs, Bernoulli, and Asher, 1979). In addition, one individual related by marriage to affected individuals (cousins and uncles) developed CJD (Masters, Gajdusek, Gibbs, Bernoulli, and Asher, 1979).

The familial cases remarkably transmit in a fashion indistinguishable from sporadic CJD. Tissues from over 17 of 26 patients and from more than 10 of 37 pedigrees with familial CJD have transmitted the disease. How can a disease that behaves like an autosomal dominantly inherited condition be transmissible? What is the means of spread of familial CJD? What is known concerning the spread of CJD in general?

The means of spread of CJD is generally not understood. The only mechanism of transmission presently known is an iatrogenic one; examples of iatrogenic spread have involved individuals who have received (1) contaminated, inadequately sterilized brain electrode implants and (2) a corneal transplant (from a CJD patient) (Duffy et al., 1974; Bernoulli et al., 1977). It is unclear how frequently iatrogenic spread of virus occurs, but given the extreme resistance of the agent to physical chemical agents, it is a cause of serious concern.

A variety of proposals have been entertained to explain the pathogenesis of familial (transmissible) CJD. This author feels that most likely the familial cases represent inadvertant horizontal transmission to a genetically susceptible individual because (1) we know the agent is extremely hardy and resistant to physical and chemical insults and (2) one lession of kuru has been that the seemingly vertical inheritance pattern actually represented inadvertant horizontal transmission. Given the extraordinary hardiness of the agent and the intimacy of relatives, one wonders why familial CJD is not more common. Most likely, the routes of transmission lead to extremely inefficient "takes," resulting in long incubation periods; perhaps the incubation periods extend longer than the life span of an individual. In addition, the genetic substrate of the host for these inoculation routes may influence the success and spread of transmission. Another essential component for successful transmission relates to the strain of the agent. There are many other examples in which transmission of SSE to the susceptible host depends on the agent strain. Perhaps families such as the Backer family, with four generations of CJD, possess a specific strain of CJD agent—that is, one that will successfully transmit parenterally, given the appropriate genetic host substrate. Our final understanding of these issues may be long in coming considering the rarity of the disease and the length of the incubation period.

The successful transmission of familial cases of CJD has directed attention to a variety of other heredofamilial neurological diseases. Support for such investigations has been gained from the observation that occasional pedigrees with familial CJD have had atypical clinical pictures. For example, one larger CJD pedigree included cases diagnosed as motor neuron disease (without dementia) and spinocerebellar degeneration (Rosenthal et al., 1976), whereas another had affected individuals originally diagnosed as Huntington's disease (Masters, Gajdusek, Gibbs, Bernoulli, and Asher, 1979). Somewhat confusing has been that apparent transmission of a spongiform encephalopathy from two cases of familial Alzheimer's disease. These transmissions may have been laboratory "errors" since the transmissions cannot be reproduced and since other familial and sporadic Alzheimer's disease cases have not transmitted (Goudsmit et al., 1980). At present no other clinical pathological group besides the SSEs has convincingly transmitted.

THE MECHANISMS OF GENES DETERMINING RESISTANCE

These many examples of host genetic factors affecting disease outcome probably only touch upon the plethora of genetic influences on virus infections that can be seen. The gene products can act at many levels to interfere with the virus replicative cycle; for example, they can affect immune response, interferon production, virus adsorption, penetration, transcription, translation, integration (in the case of retroviruses), assembly and budding, activation of macrophages or natural killer cells, and cellular repair mechanisms. In addition, gene products can interfere with the spread of virus to the central nervous system by mechanisms that may vary depending on the means by which the virus invades the nervous system (reviewed by Johnson, 1982). For example, genetic effects on the circulation of spinal fluid, axoplasmic flow, and permeability of blood vessels may change susceptibility to central nervous system disease.

The identification of genes within the MHC of the mouse, H-2, which controls the immune response to certain antigens, and the finding that cytotoxic T cells only kill target cells that share the same H-2 antigens (reviewed by Benacerraf, 1981) have suggested that resistance genes may be associated with the MHC. In fact, only a few resistance genes have been localized to the MHC, such as some of the host genes controlling retrovirus infections (Pincus, 1980), Marek's disease (Longenecker et al., 1976) and murine cytomegalovirus infection (Grundy, Mackenzie, and Stanley, 1981). Most of the resistance genes, however, do not seem to act by means of the immune system, as evidenced by the fact that their actions are frequently complete before the immune system is in evidence. In addition, the resistance phenotypes are frequently expressed at the cultured cell level.

A first line of resistance at the cellular level is at the level of the virus receptor. In some cases the possession of a virus receptor determines species specificity. For example, poliovirus types 1 and 2 receptor is coded by a gene in human chromosome 19 (Miller et al., 1974). The human coronavirus (229E) receptor is coded for by a gene in human chromosome 15 (Sakaguchi and Shows, 1981). The ecotropic murine leukemia virus receptor is coded for by a gene on chromosome 5 (Oie et al., 1978).

In many cases the resistance phenotype is only displayed by the target tissue. Quantitative differences in receptor availability, with subsequent changes in virus binding, may at times be responsible for the variation in resistance among strains, as with encephalomyocarditis virus (Chairez, Yoon, and Notkins, 1978). Usually, however, virus adsorption of cells from resistant strains is similar to adsorption to cells from susceptible animals, suggesting that virus restriction is not at this level.

It is clear that cells of resistant animals can produce specific quantitative and qualitative alterations in the viral nucleic acid produced. For example, Brinton (1981b) has demonstrated such changes in flavivirus RNA following infections in resistant cells; she postulated that an increased production of defective interfering particles may be a key mechanism in virus resistance (Brinton, 1981a, 1981b, 1983; Brinton and Fernandez, 1983). Chinsky and Soeiro (1982) have attempted to localize the Fv-1 block to the retrovirus integration stage. We eagerly await continuing molecular studies in this area and suspect they will help elucidate the level of virus restriction.

The rapidity of phenotypic expression of the resistance gene has alerted investigators to the possible involvement of natural host defenses, such as NK cells, IFN, and macrophages. There is increasing evidence that NK cells, a subpopulation of lymphoid cells, mediate resistance to virus infections (reviewed by Herberman and Ortaldo, 1981). Lopez (1981) has implicated NK in host genetic resistance to herpesvirus infections. IFN may also play a role in these infections since IFN is known to activate NK cells. Macrophages, lymphocyte-macrophage interactions, and lymphocyte-macrophage-IFN interactions may all be important, as has been suggested in mouse hepatitis type 2 infection (Bang and Cody, 1980). Multiple interactions may make delineation of a single critical resistance event very difficult. It may be that some of the genes are pleiotropic or that a complex of events is inextricably involved in resistance. For example, the effect of a gene coding for a virus receptor as well as for proteins that alter lymphocyte membranes and thereby change immune defenses against the infection will be difficult to predict. In addition, one must be cautious of incomplete data. Lopex and Dudas (1979), for example, mention that herpes simplex virus susceptibility phenotype is present on macrophages from susceptible mice; however, this in vitro susceptibility does not segregate with in vivo susceptibility in F_1 crosses.

Most likely the complexity of the interactions is a reflection of the complexity of the infection itself. One must bear in mind that the race between host defenses and virus intracellular infection is always ongoing. An additional complication is the fact that many of these infections involve cells important to the host defense, such as the macrophage and the NK cell. For example, the outcome of infection will be difficult to predict if a cell important in virus clearance has an enhanced permissiveness to virus replication. Conversely, exogenous or endogenous factors such as antibody or IFN may change intracellular virus replicative events. These interactions stress the dynamic changing character of a viral infection and emphasize the dangers in extrapolating in vitro data to disease in the whole animal.

It seems likely that an understanding of genetic factors altering resistance may lead to new therapeutic approaches to disease prevention and treatment. For example, if the action of the *Mx* resistance gene is mediated by interferon, then administration of IFN in orthomyxovirus infection may be indicated. An understanding of the mechanism of resistance of certain sheep to scrapie would obviously be of great value in therapy of CJD. Our knowledge of important inherited immune defenses may lead to targeted immunotherapy. The delineation of the molecular structure of cell receptors for viruses also suggests novel antiviral tactics, such as the production of antibodies or specific viral polypeptides to cover the receptors.

What new research directions will help us to unravel the complex host genetic-viral interactions? The use of recombinant inbred strains will probably have an increasing role in the analysis of susceptibility to virus infection. Recombinant inbred strains are derived from inbreeding of two unrelated inbred strains (Bailey, 1971; Knobler, Linthicum, and Cohn, in press). Mice from two inbred strains are crossed and their progeny are independently inbred by successive generations of brother-sister matings. The inbreeding leads to the establishment of strains with a recombination of linked alleles from the progenitor strains. Particular traits found in one of the progenitor strains will segregate into some of the recombinant strains. Examination of traits in the recombinant inbred strains may allow determination of genetic linkage of the traits and the number of loci determining a particular trait. The recombinant inbred strains will have increasing importance as a linkage map is accumulated. For example, the strain distribution pattern of nine recombinant inbred strains used to determine susceptibility to MHV-4 encephalomyelitis, acute experimental allergic encephalomyelitis and vasoactive amine sensitivity showed that the susceptibility to MHV-4 was not linked to susceptibility to experimental allergic encephalomyelitis (Knobler, Linthicum, and Cohn, in press).

Other directions include the use of mutants (as Brinton [1981b] exploited) and virus variants (for example, through recombinant DNA technology) that are no longer under the control of the host resistance gene. Transplants, as in the case of thymus transplants in MD, could be an important research tool, especially with the success of "brain transplants" (reviewed by Marx, 1982). For example, the brain tissue from fetal or adult resistant strains of mice could be transplanted to brains of susceptible strains; infection of the transplant would suggest that the cells themselves are not sufficient for maintenance of the resistant gene phenotype. Finally, eventual manipulation of the host resistance genes themselves (with recombinant DNA technology) will provide a powerful molecular tool to elucidate many of the issues.

REFERENCES

Addair, J., and L. H. Snyder. 1942. Evidence for an autosomal recessive gene for susceptibility to paralytic poliomyelitis. J. Hered. 33:307–309.

Andrews, J. M., and M. B. Gardner. 1974. Lower motor neuron degeneration associated with type C RNA virus infection in mice: neuropathological features. J. Neuropathol. Exp. Neurol. 33:285–307.

Arnheiter, H., T. Baechi, and O. Haller. 1982. Adult mouse hepatocytes in primary monolayer cultures express genetic resistance to mouse hepatitis virus type 3. J. Immunol. 129: 1275–1281.

Arnheiter, H., O. Haller, and J. Lindenmann. 1980. Host gene influence on interferon action in mouse hepatocytes: specificity for influenza virus. Virology 103:11–20.

Bailey, D. W. 1971. Recombinant-inbred strains. Transplantation 11:325–327.

Bancroft, G. J., G. R. Shellam, and J. E. Chalmer. 1980. Association of host genotype with the augmentation of natural killer cells and resistance to murine cytomegalovirus. In Genetic Control of Natural Resistance to Infection and Malignancy. E. Skamene, P. A. L. Kongshavn, and M. Landy (Eds.), pp. 227–282, New York, Academic.

Bang, F. B., and T. S. Cody. 1980. Genetic resistance to mouse hepatitis virus. In Genetic Control of Natural Resistance to Infection and Malignancy, E. Skamene, P. A. L. Kongshavn, and M. Landy (Eds.), pp. 215–226, New York, Academic.

Benacerraf, B. 1981. Role of MHC gene products in immune regulation. Science 212:1229–1238.

Bennett, J. H., A. J. Gray, and C. O. Auricht. 1959. The genetical study of kuru. Med. J. Austr. 2:505–508.

Bernoulli, C., J. Siegfried, G. Baumgarten, F. Regli, T. Rabinowicz, D. C. Gajdusek, and C. J. Gibbs, Jr. 1977. Danger of accidental person-to-person transmission of Creutzfeldt-Jakob disease by surgery. Lancet i:478–479.

Bhatt, P. N., and R. O. Jacoby. 1976. Genetic resistance to lethal flaviruses encephalitis. II. Effect of immunosuppression. J. Inf. Dis. 134:166–173.

Briles, W. E., R. W. Briles, R. E. Taffs, and H. A. Stone. 1983. Resistance to a malignant lymphoma in chickens mapped to subregion of Major Histocompatibility (*B*) Complex. Science 219:977–979.

Brinton, M. A. 1980. Genetically controlled resistance to togaviruses. In Genetic Control of Natural Resistance to Infection and Malignancy, E. Skamene, P. A. L. Kongshavn, and M. Landy (Eds.), pp. 297–303, New York, Academic.

Brinton, M. A. 1981a. Genetically controlled resistance to flavivirus and lactate-dehydrogenase-elevating virus-induced disease. In Current Topics in Microbiology and Immunology: Natural Resistance to Tumors and Viruses, O. Haller (Ed.), pp. 1–14, Berlin, Springer-Verlag.

Brinton, M. A. 1981b. Isolation of a replication-efficient mutant of West Nile virus from a persistently infected genetically resistant mouse cell culture. J. Virol. 39:413–421.

Brinton, M. A. 1983. Analysis of extracellular West Nile virus particles produced by cell cultures from genetically resistant and susceptible mice indicates enhanced amplification of defective interfering particles by resistant cultures. J. Virol. 46:860–870.

Brinton, M. A., and Fernandez, A. V. 1983. A replication-efficient mutant of West Nile viruses insensitive to DI particle interference. Virology 129:107–115.

Brooks, B. R., J. R. Swarz, and R. T. Johnson, 1980. Spongiform polioencephalomyelopathy caused by a murine retrovirus. I. Pathogenesis of infection in newborn mice. Lab. Invest. 43:480–486.

Bruce, M. E., A. G. Dickinson, and H. Fraser. 1976. Cerebral amyloidosis in scrapie in the mouse: effect of agent strain and mouse genotype. Neuropathol. Appl. Neurobiol. 2:471–478.

Burks, J. S., B. L. DeVald, L. D. Jankovsky, and J. C. Gerdes. 1980. Two coronaviruses isolated from the central nervous system tissue of two multiple sclerosis patients. N. Engl. J. Med. 209:933–934.

Calnek, B. W., K. A. Schat, W. R. Shek, and C.-L. H. Chen. 1982. In vitro infection of lymphocytes with Marek's disease virus. J. Natl. Canc. Inst. 69:709–713.

Chairez, R., J.-W. Yoon, and A. L. Notkins. 1978. Virus-induced diabetes mellitis. X. Attachment of encephalomyocarditis virus and permissiveness of cultured pancreatic B cells to infection. Virology 85:606–611.

Chalmer, J. E. 1980. Genetic resistance to murine cytomegalovirus infection. In Genetic Control of Natural Resistance to Infection and Malignancy, E. Skamene, P. A. L. Kongshavn, and M. Landy (Eds.), pp. 283–290, New York, Academic.

Chandler, R. L. 1961. Encephalopathy in mice produced with scrapie brain material. Lancet i:1378–1379.

Chattopadhyay, S. K., W. P. Rowe, N. M. Teich, and D. R. Lowy. 1975. Definitive evidence that the murine C-type virus inducing locus *Akv-1* is viral genetic material. Proc. Natl. Acad. Sci. 72:906–910.

Chinsky, J., and R. Soeiro. 1982. Studies with aphidicolin on the *Fv-1* host restriction of Friend murine leukemia virus. J. Virol. 43:182–190.

Darnell, M. B., and H. Koprowski. 1974. Genetically determined resistance to infection with Group B arboviruses. II. Increased production of interfering particles in cell cultures from resistant mice. J. Inf. Dis. 129:248–256.

Dausset, J., and J. Hors. 1975. Some contributions of the HL-A complex to the genetics of human diseases. Transplant Rev. 22:44–74.

De Maeyer, E., and J. De Maeyer-Guignard. 1980. Host genotype influences immunomodulation by interferon. Nature 284:173–175.

De Maeyer, E., J. De Maeyer-Guignard, W. T. Hall, and D. W. Barley. 1973. A locus affecting circulating interferon levels induced by mouse mammary tumour virus. J. Gen. Virol. 23:209–211.

Desgroseillers, L., and P. Jolicoeur. 1983. Physical mapping of the Fv-1 tropism host range determinant of Balb/c murine leukemia virus. J. Virol. 48:685–696.

Dickinson, A. G., and H. Fraser. 1979. An assessment of the genetics of scrapie in sheep and

mice. In Slow Transmissible Diseases of the Nervous System, Vol. 1, S. J. Prusiner and W. J. Hadlow (Eds.), pp. 367–385, New York, Academic.
Dickinson, A. G., and G. W. Outram. 1979. The scrapie replication-site hypothesis and its implications for pathogenesis. In Slow Transmissible Diseases of the Nervous System, Vol. 2, S. J. Prusiner and W. J. Hadlow (Eds.), pp. 13–31, New York, Academic.
Dickinson, A. G., J. T. Stamp, and C. C. Renwick. 1974. Maternal and lateral transmission of scrapie in sheep. J. Comp. Path. 84:19–25.
Diener, T. O. 1981. Viroids Adv. Virus Res. 28:244–283.
Dorfman, L. J. 1973. Cytomegalovirus encephalitis in adults. Neurology 23:136–144.
Duffy, P., J. Wolf, G. Collins, A. G. Devoe, B. Streeten, and D. Cowen. 1974. Person-to-person transmission of Creutzfeldt-Jakob disease. N. Engl. J. Med. 299:692–693.
Dupuy, C., D. Lafforet-Cresteil, and J. M. Dupuy. 1980. Genetic study of MHV3 infection in mice: in vitro replication of virus in macrophages. In Genetic Control of Natural Resistance to Infection and Malignancy, E. Skamene, P. A. L. Kongshavn, and M. Landy (Eds.), pp. 241–246, New York, Academic.
Fraser, H., and A. G. Dickinson. 1973. Scrapie in mice. Agent-strain differences in the distribution and intensity of gray matter vacuolation. J. Comp. Path. 83:29–40.
Gajdusek, D. C., C. J. Gibbs, Jr., and M. Alpers. 1966. Experimental transmission of a kuru-like syndrome to chimpanzees. Nature 209:794–796.
Gajdusek, D. C., and V. Zigas. 1957. Degenerative disease of the central nervous system in New Guinea: the endemic occurrence of "kuru" in the native population. N. Engl. J. Med. 257:974–978.
Gallatin, W. M., and B. M. Longenecker. 1979. Expression of genetic resistance to an oncogenic herpesvirus at the target cell level. Nature 280:587–589.
Gardner, M. G., B. E. Henderson, J. E. Officer, R. W. Rongey, J. C. Parker, C. Oliver, J. D. Estes, and R. J. Huebner. 1973. A spontaneous lower motor neuron disease apparently caused by indigenous type C RNA virus in wild mice. J. Natl. Canc. Inst. 51:1243–1254.
Gardner, M. B., V. Klement, B. E. Henderson, H. Meier, J. D. Estes, and R. J. Huebner. 1976. Genetic control of type C virus of wild mice. Nature 259:143–145.
Gazdar, A. F., H. Oie, P. Lalley, W. W. Moss, J. D. Minna, and U. Francke. 1977. Identification of mouse chromosomes required for murine leukemia virus replication. Cell 11:949–956.
Goudsmit, J., C. H. Morrow, D. M. Asher, R. T. Yanagihara, C. L. Masters, C. J. Gibbs, Jr., and D. C. Gajdusek. 1980. Evidence for and against the transmissibility of Alzheimer disease. Neurology 30:945–950.
Grundy (Chalmer), J. E., J. S. Mackenzie, and N. F. Stanley. 1981. Influence of H-2 and non-H-2 genes on resistance to murine cytomegalovirus infection. Infect. Immun. 32:277–286.
Hadlow, W. J., R. E. Race, R. C. Kennedy, and C. M. Eklund. 1979. Natural infection of sheep with scrapie virus. In Slow Transmissible Diseases of the Nervous System, Vol. 2, S. J. Prusiner and W. J. Hadlow (Eds.), pp. 3–12, New York, Academic.
Hall, W. W., and P. W. Choppin. 1979. Evidence for lack of synthesis of the polypeptide of measles virus in brain cells in subacute sclerosing panencephalitis. Virology 99:443–447.
Haller, O. 1981. Inborn resistance of mice to orthomyxoviruses. In Current Topics in Microbiology and Immunology: Natural Resistance to Tumors and Viruses, O. Haller (Ed.), pp. 25–52, Berlin, Springer-Verlag.
Haller, O., H. Arnheiter, J. Lindenmann, and J Gresser. 1980. Host gene influences sensitivity to interferon action selectively for influenza virus. Nature 283:660–662.
Harper, P. S. 1977. Mendelian inheritance or transmissible agent?—The lesson of kuru and Australia antigen. J. Med. Genet. 14:389–398.
Herberman, R. B., and J. R. Ortaldo. 1981. Natural killer cells: their role in defenses against disease. Science 214:24–30.
Herndon, R. M., D. E. Griffin, U. McCormick, and L. P. Weiner. 1975. Mouse hepatitis virus-induced recurrent demyelination. Arch. Neurol. 32:32–35.
Herndon, C. N., and R. G. Jennings. 1951. A twin-family study of susceptibility to poliomyelitis. Amer. J. Hum. Genet. 3:17–46.
Horisberger, M. A., O. Haller, and H. Arnheiter. 1980. Interferon-dependent genetic resistance to influenza virus in mice: virus replication in macrophages is inhibited at an early step. J. Gen. Virol. 50:205–210.
Horisberger, M. A., Staeheli, P., and Haller, O. Interferon induces a unique protein in mouse cells bearing the gene for influenza virus resistance. In the Biology of the Interferon System, E. De Mayer, and H. Schellekens (Eds.), pp. 251–256, Amsterdam, Elsevier.

Hourrigan, J., A. Klingsporn, W. W. Clark, and M. de Camp. 1979. Epidemiology of scrapie in the United States. In Slow Transmissible Diseases of the Nervous System, Vol. 1, S. J. Prusiner and W. J. Hadlow (Eds.), pp. 331–356, New York, Academic.

Jacoby, R. O., and P. N. Bhatt. 1976. Genetic resistance to lethal flavivirus encephalitis. I. Infection of congenic mice with Banzi virus. J. Inf. Dis. 134:158–173.

Jellinger, K., F. Seitelberger, W. Hess, and W. Holczabek. 1972. Subakuten spongiöse Encephalopathie. Wiener. Klin. Wochenshr. 84:245–249.

Jenkins, N. A., N. G. Copeland, B. A. Taylor, and B. K. Lee. 1981. Dilute *(d)* coat colour mutation of DBA/2J mice is associated with the site of integration of an ecotropic MuLV genome. Nature 293:370–374.

Johnson, R. T. 1982. Viral infections of the nervous system. New York, Raven Press.

Jolicoeur, P. 1979. The *Fv-1* gene of the mouse and its control of murine leukemia virus replication. Curr. Top. Microbiol. Immunol. 86:67–122.

Kai, K., H. Ikeda, Y. Yussa, S. Suzuki, and T. Okada. 1976. Mouse strain resistant to N-, B-, and NB-tropic murine leukemia viruses. J. Virol. 20:436–440.

Kingsbury, D. T., K. C. Kasper, D. P. Stites, J. D. Watson, R. N. Hogan, and S. B. Prusiner. 1983. Genetic control of scrapie and Creutzfeldt-Jakob disease in mice. J. Immunol. 131: 491–496.

Kingsbury, D. T., D. A. Smeltzer, H. L. Amyx, C. J. Gibbs, Jr., and D. C. Gajdusek. 1982. Evidence for an unconventional virus in mouse-adapted Creutzfeldt-Jakob disease. Infect. Immun. 37:1050–1053.

Kirchner, H., H. Engler, R. Zawatsky, and C. H. Schroder. 1980. Studies of resistance of mice against herpes simplex virus. In Genetic Control of Natural Resistance to Infection and Malignancy, E. Skamene, P. A. L. Kongshavn, and M. Landy (Eds.), pp. 267–276, New York, Academic.

Kitchin, F. D., A. G. Bearn, M. Alpers, and D. C. Gajdusek. 1972. Genetic studies in relation to kuru. III. Distribution of the inherited serum group-specific protein (Gc) phenotypes in New Guineans: an association of kuru with the Gc Ab phenotype. Amer. J. Hum. Genet. 24(suppl.):S72–S85.

Knobler, R. L., M. V. Haspel, and M. B. A. Oldstone. 1981. Mouse hepatitis virus type 4 (JHM strain) induced fatal central nervous system disease. I. Genetic control and the murine neuron as the susceptible site of disease. J. Exp. Med. 153:832–843.

Knobler, R. L., D. S. Linthicum, and M. Cohn. Host genetic regulation of acute MHV-4 viral encephalomyelitis and acute experimental autoimmune encephalomyelitis in (Balb/c X SJL/J) recombinant-inbred mice. J. Neuroimmunology (in press).

Kumar, V., and M. Bennett. 1981. Genetic resistance to Friend-virus induced erythroleukemia and immunosuppression. In Current Topics in Microbiology and Immunology: Natural Resistance to Tumors and Viruses, O. Haller (Ed.), pp. 65–82, Berlin, Springer-Verlag.

Kumar, V., L. Goldschmidt, J. W. Eastcott, and M. Bennett. 1978. Mechanisms of genetic resistance of Friend virus leukemia in mice. IV. Identification of a gene *(Fv-3)* regulating immunosuppresion in vitro, and its distinction from *Fv-2* and genes regulating marrow allograft reactivity. J. Exp. Med. 147:422–433.

Kumar, I., P. Resnick, J. W. Eastcott, and M. Bennett. 1978. Mechanism of genetic resistance to Friend virus leukemia in mice. V. Relevance of *Fv-3* gene in the regulation of in vivo immunosuppression. J. Natl. Canc. Inst. 60:1117–1123.

Lasch, E. E., H. Joshua, E. Gazit, M. El-Massri, O. Marcus, and R. Zamir. 1979. Study of the HLA antigen in Arab children with paralytic poliomyelitis. Isr. J. Med. Sci. 15:12–13.

Lee, L. F., P. C. Powell, M. Rennie, L. J. N. Ross, and L. N. Payne. 1981. Nature of genetic resistance to Marek's disease in chickens. J. Natl. Canc. Inst. 66:789–796.

Lévy-Leblond, E., D. Oth, and J. M. Dupuy. 1979. Genetic study of mouse sensitivity to MHV3 infection: influence of the H-2 complex. J. Immunol. 122:1359–1362.

Lilly, F. 1966. The inheritance of susceptibility to the Gross leukemia virus in mice. Genetics 53:529–539.

Lilly, F. 1968. The effect of histocompatability-2 type on response to the Friend leukemia virus in mice. J. Exp. Med. 127:465–473.

Lin, F. H., and H. Thormar. 1980. Absence of M protein in a cell-associated subacute sclerosing panencephalitis virus. Nature 285:490–491.

Lindenmann, J., E. Deuel, S. Fanconi, and O. Haller. 1978. Inborn resistance of mice to myxoviruses: macrophages express phenotype in vitro. J. Exp. Med. 147:531–540.

Lindenmann, J., C. A. Lane, and D. Hobson. 1963. The resistance of A2G mice to myxoviruses. J. Immunol. 90:942-951.

Lipton, H. L., and R. Melvold. 1984. Genetic analysis of susceptibility to Theiler's virus-induced demyelinating disease in mice. J. Immunol. 132:1821-1825.

Longenecker, B. M., F. Pazderka, J. S. Gavora, J. L. Spencer, and R. F. Ruth. 1976. Lymphoma induced by herpesvirus: resistance associated with a major histocompatibility gene. Immunogenetics 3:401-407.

Lopez, C. 1980. Resistance to HSV-1 in the mouse is governed by two major, independently segregating, non-H-2-loci. Immunogentics 11:87-92.

Lopez, C. 1981. Resistence to herpes simplex virus type 1 (HSV-1). In Current Topics in Microbiology and Immunology: Natural Resistance to Tumors and Viruses, O. Haller (Ed.), pp. 15-24, Berlin, Springer-Verlag.

Lopez, C., and G. Dudas. 1979. Replication of herpes simplex type 1 in macrophages from resistant and susceptible mice. Infect. Immun. 23:432-437.

MacNaughton, M. R., and S. Patterson. 1980. Mouse hepatitis virus strain 3 infection of C57, A/Sn and A/J strain mice and their macrophages. Arch. Virol. 66:71-75.

Marsh, R. F., J. S. Semancik, K. C. Medappa, R. P. Hanson, and R. R. Rueckert. 1974. Scrapie and transmissible mink encephalopathy: search for infectious nucleic acid. J. Virol. 13: 993-996.

Martinez, D., M. A. Brinton, T. G. Tachovsky, and A. H. Phelps. 1980. Identification of lactate dehydrogenase-elevating virus as the etiologic agent of genetically restricted, age-dependent polioencephalomyelitis of mice. Infect. Immun. 27:979-987.

Marx, J. L. 1982. Transplants as guides to brain development. Science 217:340-342.

Massanari, R. M., M. L. Wilson, and J. R. Leininger. 1980. Subacute sclerosing panencephalitis in inbred hamsters. In Genetic Control of Natural Resistance to Infection and Malignancy, E. Skamene, P. A. L. Kongshavn, and M. Landy (Eds.), pp. 353-359, New York, Academic.

Masters, C. L., D. C. Gajdusek, C. J. Gibbs, Jr., C. Bernoulli, and D. M. Asher (1979). Familial Creutzfeldt-Jakob disease and other familial dementias: An inquiry into possible modes of transmission of virus-induced familial diseases. In Slow Transmissible Diseases of the Nervous System, Vol. 1, S. J. Prusiner and W. J. Hadlow (Eds.), pp. 143-194, New York, Academic.

Masters, C. L., J. O. Harris, D. C. Gajdusek, C. J. Gibbs, Jr., C. Bernoulli, and D. M. Asher. 1979. Creutzfeldt-Jakob disease: patterns of worldwide occurrence and the significance of familial and sporadic clustering. Ann. Neurol. 5:177-188.

Matthews, W. B. 1975. Epidemiology of Creutzfeldt-Jakob disease in England and Wales. J. Neurol. Neurosurg. Psychiatr. 38:210-213.

Merz, P. A., R. G. Rohwer, R. Kascsak, H. M. Wisniewski, R. A. Somerville, C. J. Gibbs, Jr., and D. C. Gajdusek. 1984. Infection specific particle from the unconventional slow virus diseases. Science 225:437-440.

Meyer, T., and M. A. Horisberger. 1984. Centralized action of mouse α and β interferon in influenza virus-infected macrophages carrying the resistance gene *Mx*. J. Virol. 49:709-716.

Miller, D. A., O. J. Miller, V. G. Dev, S. Hashmi, R. Tantravahi, L. Medrano, and H. Green. 1974. Human chromosome 19 carries a poliovirus receptor gene. Cell 1:167-173.

Mogensen, S. C. 1977. Genetics of macrophage-controlled resistance to hepatitis induced by herpes simplex virus type 2 in mice. Infect. Immun. 17:268-273.

Mogensen, S. C. 1980. Genetics of macrophage-controlled natural resistance to hepatitis induced by herpes simplex type 2 in mice. In Genetic Control of Natural Resistance to Infection and Malignancy, E. Skamene, P. A. L. Kongshavn, and M. Landy (Eds.), pp. 291-296, New York, Academic.

Mohan, S., and B. K. Pal. 1982. Binding characteristics of wild mouse type C virus to mouse spinal cord and spleen cells. Infect. Immun. 37:532-538.

Murphy, W. H., M. R. Tam, R. L. Lanzi, M. R. Abell, and C. Kauffman. 1970. Age dependence of immunologically induced central nervous system disease in C58 mice. Canc. Res. 30: 1612-1622.

Nedrud, J. G., A. M. Collier, and J. S. Pagano. 1979. Cellular basis for susceptibility to mouse cytomegalovirus: evidence from tracheal organ culture. J. Gen. Virol. 45:737-744.

Neff, J. M., and J. F. Enders. 1968. Poliovirus replication and cytopathology in monolayer hamster cell cultures fused with beta propiolactone-inactivated Sendai virus. Proc. Soc. Exp. Biol. Med. 127:260-267.

Neugut, R. H., A. I. Neugut, E. Kahana, Z. Stein, and M. Alter. 1979. Creutzfeldt-Jakob disease: familial clustering among Libyan-born Israelis. Neurology 29:225-231.

Nottay, B. K., O. M. Kew, M. H. Hatch, J. T. Heyward, and J. F. Obijeski. 1981. Molecular variation of type 1 vaccine-related and wild polioviruses during replication in humans. Virology 108:405–423.

Oie, H. K., A. F. Gazdar, P. A. Lalley, E. K. Russell, J. D. Minna, J. DeLarco, G. J. Todaro, and U. Francke. 1978. Mouse chromosome 5 codes for ecotropic murine leukemia virus cell-surface receptor. Nature 274:60–62.

Onodera, T., J.-W. Yoon, K. S. Brown, and A. L. Notkins. 1978. Evidence for a single locus controlling susceptibility to virus-induced diabetes mellitus. Nature 274:693–696.

Outram, G. W., H. Fraser, and D. T. Wilson. 1973. Scrapie in mice. Some effects on the brain lesion profile of ME 7 agent due to genotype of donor, route of infection and genotype of recipient. J. Comp. Path. 83:19–28.

Pease, L. R., G. D. Abrams, and W. H. Murphy. 1982. Fv-1 restriction of age-dependent paralytic lactic dehydrogenase virus infection. Virology 117:29–37.

Philipson, L., and S. Bengtsson. 1962. Interaction of enteroviruses with receptors from erythrocytes and host cells. Virology 18:457–469.

Pietsch, M. C., and P. J. Morris. 1974. An association of HL-A3 and HL-A7 with paralytic poliomyelitis. Tissue Antigens 4:50–55.

Pincus, T. 1980. The endogenous murine type C viruses. In Molecular Biology of RNA Tumor Viruses, J. R. Stephenson (Ed.), pp. 77–130, New York, Academic.

Pincus, T., J. W. Hartley, and W. P. Rowe. 1971. A major genetic locus affecting resistance to infection with murine leukemia virus. I. Tissue culture studies of naturally occurring viruses. J. Exp. Med. 133:1219–1233.

Powell, P. C., L. F. Lee, B. M. Mustill, and M. Rennie. 1982. The mechanism of genetic resistance of Marek's disease in chickens. Int. J. Canc. 29:169–174.

Prineas, J. W., and R. G. Wright. 1972. The fine structure of peripheral nerve lesions in a virus-induced demyelinating disease in fowl (Marek's disease). Lab. Invest. 26:548–557.

Prusiner, S. B., D. C. Gajdusek, and M. P. Alpers. 1982. Kuru with incubation periods exceeding two decades. Ann. Neurol. 12:1–9.

Quint, W., H. van der Putten, F. Janssen, and A. Berns. 1982. Mobility of endogenous ecotrophic murine leukemia viral genomes within mouse chromosomal DNA and integration of a mink cell focus-forming virus-type recombinant provirus in the germ line. J. Virol. 41: 901–908.

Rager-Zisman, B., P. A. Neighbour, G. Ju, and B. R. Bloom. 1980. The role of H-2 in resistance and susceptibility to measles virus infection. In Genetic Control of Natural Resistance to Infection and Malignancy, E. Skamene, P. A. L. Kongshavn, and M. Landy (Eds.), pp. 313–320, New York, Academic.

Rasheed, S., E. Toth, and M. B. Gardner. 1977. Characterization of purely ecotropic and amphotropic naturally occurring wild mouse leukemia virus. Intervirol. 8:323–335.

Reedy, J. J. 1957. Recessive inheritance to susceptibility to poliomyelitis in fifty pedigrees. J. Hered. 48:37–44.

Riley, V. 1974. Persistence and other characteristics of the lactate dehydrogenase-elevating virus (LDH-virus). In Progress in Medical Virology, Vol. 18, J. L. Melnick (Ed.), pp. 198–213, Basel, S. Karger.

Roos, R. P. 1981a. Alzheimer's disease and the lessons of transmissible virus dementia. In Epidemiology of Dementia, J. A. Mortimer, L. M. Schuman, and A. Lilienfeld (Eds.), pp. 73–86, New York, Oxford.

Roos, R. P. 1981b. Genetics of Guillain-Barré syndrome. In Handbook of Clinical Neurology, G. Bruyn, P. Vinken, and N. C. Myrianthopoulos (Eds.), pp. 645–647, Amsterdam, North-Holland.

Roos, R. P. 1981c. Genetics of subacute spongiform encephalopathies. In Handbook of Clinical Neurology, G. Bruyn, P. Vinken, and N. C. Myrianthopoulos (Eds.), pp. 655–659, Amsterdam, North-Holland.

Roos, R. P. 1983. Viruses and demyelinating disease of the central nervous system. In Neurologic Clinics—Symposium on Multiple Sclerosis, J. P. Antel (Ed.) Vol. I, pp. 681–700, Philadelphia, Saunders, Co.

Roos, R. P., D. C. Gajdusek, and C. J. Gibbs. 1973. The clinical characteristics of transmissible Creutzfeldt-Jakob disease. Brain 96:1–20.

Rosenthal, N. P., J. Keesey, B. Crandall, and W. J. Brown. 1976. Familial neurological disease associated with spongiform encephalopathy. Arch. Neurol. 33:252–259.

Ruddle, N. H., B. S. Conta, L. Leinwand, C. Kozak, F. Ruddle, P. Besmer, and D. Baltimore.

1978. Assignment of the receptor for ecotrophic murine leukemia virus to mouse chromosome 5. J. Exp. Med. 148:451–465.
Sakaguchi, A. Y., and T. B. Shows. 1981. Coronavirus 229E susceptibility in man-mouse hybrids is located on human chromosome 15. Somat. Cell Genet. 8:83–94.
Selgrade, M. K., and J. E. Osborn. 1974. Role of macrophages in resistance to murine cytomegalovirus. Infect. Immun. 10:1383–1390.
Shif, I., and F. B. Bang. 1970. In vitro interaction of mouse hepatitis virus and macrophages from genetically resistant mice. J. Exp. Med. 131:843–850.
Smith, A. L., R. O. Jacoby, and P. N. Bhatt. 1980. Genetic resistance to lethal flavivirus infection: detection of interfering virus produced in vivo. In Genetic Control of Natural Resistance to Infection and Malignancy, E. Skamene, P. A. L. Kongshavn, and M. Landy (Eds.), pp. 305–312, New York, Academic.
Sorensen, O., R. Dugre, D. Percy, and S. Dales. 1982. In vivo and in vitro models of demyelinating disease: endogenous factors influencing demyelinating disease caused by mouse hepatitis virus in rats and mice. Infect. Immun. 37:1248–1260.
Steeves, R., and F. Lilly. 1977. Interactions between host and viral genomes in mouse leukemia. Ann. Rev. Genet. 11:277–296.
Stroop, W. G., and M. A. Brinton. 1983. Mouse-strain-specific central nervous system lesions associated with lactate dehydrogenase-elevating virus infection. Lab. invest. 49:334–345.
Stohlman, S. A., and J. A. Frelinger. 1978. Resistance to fatal central nervous system disease by mouse hepatitis virus, strain JHM. I. Genetic analysis. Immunogenetics 6:277–281.
Taguchi, F., R. Yamaguchi, S. Makino, and K. Fugiwara. 1981. Correlation between growth potential of mouse hepatitis viruses in macrophages and their virulence for mice. Infect. Immun. 34:1059–1061.
Virelizier, J.-L. 1981. Role of macrophages and interferon in natural resistance to mouse hepatitis virus infection. In Current Topics in Microbiology and Immunology: Natural Resistance to Tumors and Viruses, O. Haller (Ed.), pp. 53–64, Berlin, Springer-Verlag.
Virelizier, J.-L., A. D. Dayan, A. C. Allison. 1975 Neuropathological effects of persistent infection of the mouse by mouse hepatitis virus (MHV-3). Infect. Immun. 12:1127–1140.
Virelizier, J.-L., and I. Gresser. 1978. Role of interferon in the pathogenesis of viral diseases of mice as demonstrated by the use of anti-interferon serum. V. Protective role in mouse hepatitis was type 3 infection of susceptible and resistant strains of mice. J. Immunol. 120:1616–1619.
Ward, R. L., D. D. Porter, and J. G. Stevens. 1974. Nature of the scrapie agent: evidence against a viroid. J. Virol. 14:1099–1103.
Webb, S. R., and G. E. Madge. 1980. The role of host genetics in the pathogenesis of coxsackievirus infection in the pancreas of mice. J. Inf. Dis. 141:47–54.
Weiner, L. P. 1973. Pathogenesis of demyelination induced by a mouse hepatitis virus (JHM virus). Arch. Neurol. 28:298–303.
Weiser, W., and F. B. Bang. 1976. Macrophages genetically resistant to mouse hepatitis virus converted in vitro to susceptible macrophages. J. Exp. Med. 143:690–695.
Wyatt, H. V. 1975. Is poliomyelitis a genetically-determined disease? Med. Hypotheses 1:35.
Wyatt, H. V. 1978. Abortive poliomyelitis or minor illness as a clue to genetic susceptibility. Med. Microbiol. Immunol. 166:29–36.
Yamada, A., F. Taguchi, and K. Fujiwara. 1979. T lymphocyte–dependent difference in susceptibility between DDD and C3H mice to mouse hepatitis virus MHV-3. J. Exp. Med. 49: 413–421.
Yoon, J.-W., and A. L. Notkins. 1976. Virus-induced diabetes mellitus. J. Exp. Med. 143: 1170–1185.
Zander, H., H. Grosse Wolde, S. Scholz, B. Kuntz, B. Netzel, and E. D. Alberts. 1976. Poliomyelitis: An analysis of HLA-A, -B and -D alleles (Abst.). Presented in a Symposium on HLA and Disease: Predisposition to Disease and Clinical Implications, p. 90, Paris, INSERM.

7 Genetic Jeopardy and the New Clairvoyance

Nancy S. Wexler

Hereditary Disease Foundation
Santa Monica, California.

FOREWORD

December, 1984. This chapter was finished in summer, 1983. Barely a month following its completion, the world of Huntington's disease research was revolutionized. A DNA marker was discovered, closely linked to the gene for Huntington's disease.[1] Through *in situ* hybridization of the marker, the gene was localized to chromosome four.[1] The marker, known as G8, is a unique 17.6 kilobase piece of anonymous DNA, a restriction fragment length polymorphism (RFLP) approximately 5 centimorgans from the disease gene.[1,2] Linkage between marker and gene was discovered in a huge kindred from Venezuela and has been confirmed in three families with Huntington's disease in the United States.[3] In these families, if the affected parent is heterozygous for the marker (there are four possible haplotypes), presymptomatic and prenatal diagnosis can be accomplished with 95% accuracy (there is a 5% chance of recombination).[3,4,5]

Work is now in progress to determine if the Huntington's disease gene is linked to RFLP marker G8 in families with Huntington's disease throughout the world and from all different racial, ethnic, and geographic backgrounds. As no new mutation has ever been confirmed in this disorder, it is hoped that one or possibly a few mutations in antiquity were preserved and spread throughout the world by trade and migration. If the question of genetic heterogeneity can be eliminated, families insufficiently large to prove linkage can still be tested on the assumption that the gene is on chromosome four. These families must have enough genetically informative relatives to determine which haplotype of the marker is diagnostic of the Huntington's disease gene in that family, and the affected relative must be heterozygous for the marker.[6]

I first wrote this chapter as a review of current attitudes toward prediction and speculations about the future some ten or more years hence when a marker would be found. Had a marker for the gene been in existence, I would have written this chapter from an entirely different perspective. I cannot pretend now to have known about the marker while writing, making judicious edits here and there to update the material. The structure of the paper would have been different, the level of urgency higher, and the conclusions more directive. Some sections would have been eliminated and others added. This paper is a leisurely walk through "premarker" attitudes toward predictive testing and the beginning of preparation for when the marker is in hand. I cannot revise it to mask the fact that the marker did not exist while I was writing without starting substantially from the beginning. And the reader deserves to know that the chapter was written before the marker was discovered. Some portions are out of date. The discussion about research to generate a predictive test is now moot. But I choose to leave it in because it gives the reader an historical perspective on the type of discussion held in the past regarding predictive testing and the deep need some feel for a test, even if it were a provocative, intrusive test.

My hypotheses regarding the psychological needs and reactions of those undergoing testing should be just as relevant today because they were predicated on a test being available. It is even more critical now that we meet these needs for trained, excellent psychotherapeutic support. No one has yet received presymptomatic or prenatal testing, but soon (perhaps even before this chapter is published), testing will begin. The accuracy of the test will climb to close to 100 percent with the discovery of a flanking marker — a discovery which may well be made by publication date. Its applicability will also improve as new restriction enzymes are used to increase heterozygousity. I have never been so overjoyed to be out of date and I beg the reader's indulgence with parts of this chapter that are passe. I look forward to the day when the rest of this chapter regarding the emotional and other turmoil of families undergoing testing will be equally obsolete because a cure for Huntington's disease will be at hand.

The ultimate goal of current research is to move from the marker to the isolation and characterization of the Huntington's disease gene and the eventual interdiction of the illness through prevention, treatment or gene repair. Presymptomatic diagnosis is only a way-station on the road to treatment and cure. The searing psychological, ethical and legal stresses generated by this ability to diagnose but not treat will be relieved by the further advances of science.

NOTES

1. Gusella J. F., Wexler N. S., Conneally P. M., Naylor S. L., Anderson M. A., Tanzi R. E., Watkins, P. C., Ottina K., Wallace M. R., Sakaguchi A. Y., Young A. B., Shoulson I., Bonilla E., Martin J. B., 1983. A polymorphic DNA marker genetically linked to Huntington's disease. *Nature* 306:234.
2. Gusella J. F., Tanzi R. E., Anderson, M. A., Hobbs W., Gibbons K., Raschtchian R., Gilliam T. C., Wallace M. R., Wexler N. S., Conneally P. M., 1984. DNA markers for nervous system diseases. *Science* 225:1320.
3. Gusella J. F., Gibbons K., Hobbs W., Hift R., Anderson M., Raschtchian R., Folstein S., Wallace P., Conneally P. M., Tanzi T., 1984. The G8 locus linked to Huntington's Disease. *Am J Hum Genet* 36:139S.
4. Conneally P. M., Gusella J. F., Wexler N. S., 1985. Huntington's Disease: Linkage with G8 on Chromosome 4 and Its Consequences, in *Genetic Disorders. Medical Genetics: Past, Present, Future*, K. Berg and Alan R. Liss, (eds.), New York, 1975.
5. Wexler N. S., Conneally P. M., Gusella J. F., 1984. Huntington's Disease "Discovery" Fact Sheet, Hereditary Disease Foundation, Santa Monica, CA.
6. Wexler N. S., Conneally P. M., Housman D., Gusella J. F., in press. A DNA Polymorphism for Huntington's Disease Marks the Future, *Archives of Neurology*.

If you ask, I will say
I was not born
but assembled by blind men
from scraps.

Kate Braverman, *Lullaby For Sinners*

INTRODUCTION

Huntington's disease (HD) is an autosomal dominant disorder characterized by abnormal involuntary movements involving all parts of the body, emotional disturbance, and intellectual decline. Onset is typically mid-life, but 20 percent of cases manifest below the age of 20 or above 60. There is no diagnostic test to identify the presymptomatic gene carrier, so that those "at risk" for the disease, carrying a 50 percent probability, must live in uncertainly until symptoms appear. The gene is fully penetrant. Treatment is palliative and does nothing to halt the 10- to 20-year progressive course to death, usually from aspiration pneumonia or choking (See Chapter 4 for a thorough review).

Along the shores of Lake Maracaibo, Venezuela, a very large kindred of Huntington's disease families can be found in small fishing villages close to the water's edge. Traceable to a woman living in the early 1800s in a *pueblo de agua*, or "water village," this kindred has grown to over 3,500 individuals; 3,000 of whom are living. There are almost 100 living Huntington's disease patients, over 1500 persons at 50 percent and 25 percent risk.

An interdisciplinary group of scientists, including myself, has been involved in research on this community for the last four years. One day I was sitting in a remote village with a 25-year-old at-risk girl on the rough wood porch of her one-room house, perched on stilts over the water. "Yes, I have inherited the disease," she cheerfully told me. I looked at her carefully to spot any tell-tale symptoms I had missed before. "What problems do you have that make you think you have the illness?" I asked. "Oh, I don't *have* it, God willing, I've only *inherited* it."

We quickly learned that for this Venezuelan Huntington's disease family *to inherit* Huntington's disease and *to have* the disorder are two entirely different phenomena. Our style of collecting pedigree information changed as we asked about each individual, "Did X inherit the illness? And was he or she sick with the disease?" To the Venezuelans, every descendant of a Huntington's disease patient inherits the illness, unless the connection is very remote, but only certain people become sick, for more or less mysterious reasons.

Upon reflection, the Venezuelans may distort a medical fact, but they express a psychological truth. The experience of being at risk is a separate and distinct state, with its own unique psychology. Those who "inherit" Huntington's disease are not entirely normal because they may have an especially vulnerable physiology or fatal flaw in their constitution, as yet unrevealed. Neither are they diseased. Their lot is one of chronic, unremitting ambiguity—neither healthy nor sick, neither confident that health will continue nor certain that illness will intervene.

RISK

The Psychology of Risk

In studies of the psychology of being at risk, the ambiguity of the situation seems one of the most difficult aspects to bear.[1] This would apply to any late onset autosomal dominant disease, such as dominant polyposis, some dystonias, or spinocerebellar atrophies. Particularly when there is no early diagnostic test, as in Huntington's disease, and when the initial signs of the illness are insidious—small memory lapses, irritability, tiny movements, and decreasing coordination—the uncertainty can be overwhelming.

Exacerbating the chronic psychological tension involved in maintaining an ambiguous health status, at-risk individuals have no control over the means of resolving the situation. Only time will reveal the answers; all decisions in the meantime must be made in uncertainty. The experiments of Weiss and others[2,3] using "executive" animals who make decisions and suffer the consequences of only their own actions and other animals passively "yoked" to them, unable to control their own fate but forced to suffer the consequences of the "executive" animals' actions would suggest that lack of control and unpredictability, with no means of escape, produce maximal stress. All three conditions—lack of control, unpredictability, and no means of escape—characterize the state of being at risk.[4]

Most at-risk individuals adjust and accommodate in a productive, healthy fashion to what would otherwise be unrelenting stress. But for many, a low-grade *basso continuo* of tension underlies much of life, rising and falling in pitch as decisions are made and life crises confronted. By acknowledging the separate status of the experience of being at risk, the Venezuelans validate a syndrome of which chronic stress is a primary component. The main precipitant of this stress is lack of control. Treatment, therefore, must center around providing a means of controlling the anxiety generated by ambiguity—not only concerning the odds of inheriting Huntington's disease but also all of the choices such as childbearing, marriage, or financial planning, to name only a few, that are influenced by these odds.

The Age of Risk: Onset Curves

Fortunately, the peak age of onset of Huntington's disease in the third and fourth decades allows at-risk individuals to feel increasing relief as they grow older, symptom free. Most, however, do not know that their statistical probability of inheriting the illness actually declines, and would probably be comforted by this information. Providing reduced risks based on age is complicated by variable ages of onset within and among families. Determining accurate age of onset within a family is problematic because of the long, gradual initiation of symptoms. Using age at diagnosis as an indicator of symptom onset has proved even more error-ridden. Patients frequently will not visit a physician until symptoms are flagrant, and retrospective histories by relatives are influenced by their familiarity with the illness or are determined only by a particular event, such as losing a job or divorcing. A number of prospective studies are now in progress that should provide more accurate data (Shoulson, Folstein, Young, Venezuela Collaborative Project, to mention a few).

Data from the National Research Roster for Huntington's Disease Patients and Families at Indiana University on 999 affected individuals provide a mean age of onset of 36.11 years, an age also found in other studies.[5] A number of investigators have been interested in age of onset correlations among relatives to determine the genetic contribution to onset age so that these findings might be utilized in genetic counseling.[5] Parent-child correlations vary between 0.50 to 0.78 in different studies.[5]Pericak-Vance at al. analyzed separately age of onset data from families with a juvenile onset proband (65 families) and those with an adult onset proband (89 families).[6,7] They found that the variation in age of onset among families within each group far exceeded variation in age of onset between the two—juvenile and adult—onset family groups. In light of the magnitude of differences among families in age of onset, they recommend that sufficiently large families be analyzed individually to determine age of onset and within-member correlations for the kindred.

A peculiarity of Huntington's disease, as yet unexplained, is the finding that the majority of juvenile cases, both male and female, inherit their illness from their fathers. In general, the mean age of onset in all offspring of affected males is 3.5 years earlier than in offspring of affected females, a statistically significant difference.[5] Probabilities for being a carrier must therefore be analyzed separately for offspring of affected males and females.[5]

Given the complexity of factors influencing age of onset, extreme caution must be exercised in using within-family correlations for genetic counseling, particularly with regard to family planning.

PRESYMPTOMATIC PREDICTION

Predictive Testing

Every aspect of Huntington's disease symptomatology has been scrutinized in detail for ways to measure its minutest expression in at-risk persons. Electroencephalography,[8] electromyography,[9] accelerometer studies,[10] neuropsychological testing,[11,12] eye movement monitoring[13,14]—all have proved inconclusive. Biophysical studies measuring electron spin resonance[15] and nuclear magnetic resonance[16] of tissue membranes remain unconfirmed. Biochemical measurements of cerebrospinal fluid gamma-aminobutyric acid and other compounds failed to show a bimodal difference in those at risk,[17] and other studies of cell properties such as growth,[18] nutritional status,[17] and vulnerability to toxins[20] remain controversial. The so-called levodopa loading test, provocative in many respects, proved invalid on follow up.[21,22] Recent studies of altered red blood cell osmotic fragility in HD patients and some at-risk individuals must be verified.[23]

Computerized axial tomography has been demonstrated to be insensitive to the earliest signs of illness,[24,25] but positron emission tomography offers some hope of being informative presymptomatically.[26,27] In the only study to date, 15 at-risk persons were tested and 6 showed an abnormal decrement in metabolism similar to those definitively diagnosed.

Recombinant DNA and Gene Linkage

Linkage analysis using classic markers such as red cell antigens and protein polymorphisms has been pursued as a means of developing a diagnostic test and localizing the mutant gene on a chromosome. Using these techniques, the Huntington's disease gene has been excluded from approximately 20 percent of the genome.[5] But classical linkage reached an impasse with the exhaustion of existing markers and slow discovery of new ones.

Recent advances in recombinant DNA technology promise to alter radically the field of genetic linkage. Restriction fragment length polymorphisms (RFLPs) represent unique sequences of varying lengths in the DNA and are usually inherited as Mendelian codominant traits.[28] They provide the possibility of an almost unlimited number of new linkage markers.

Scientists have estimated that approximately 200 RFLPs, evenly spaced every 10 centimorgans throughout the genome, will permit the detection of linkage with any gene of interest. More than twice this number of markers may need to be generated to select the requisite evenly spaced ones.[29] Generating these RFLPs is estimated to require between 3 to 10 years at a cost of approximately $10 million.* Hard work and serendipity may deliver a marker for Huntington's disease early in the decade, or it may await the marker saturation of the map; but its impact, whenever it is found, will be profound for research and clinical care. Pre- and postnatal diagnosis will be possible. In addition, investigators can focus their energies on studying the chromosome actually containing the gene. Concentrated effort can be directed toward understanding the nature of the gene defect and developing therapeutic interventions.

The parameters of any marker for the Huntington's disease gene will have to be carefully described. It may be a private marker, unique to only one family, or a more generalized marker, of use to most or all Huntington's

*Ray L. White and David Housman, 1983 : personal communication.

disease and the fact that a new mutation has never been documented, it is quite possible that the vast majority of families do share the same Huntington's disease allele and common markers.[5]*

Predictive Planning

Although a biochemical indicator may be discovered before an RFLP marker is developed, the likelihood is high that some diagnostic test of the gene's presence will be available within the next decade. We would be reckless not to plan in advance for such an occurrence. In trying to conceptualize the problem, the following questions can be posed: What are current attitudes of at-risk individuals and families toward presymptomatic detection tests? To whom should the test be available and under what circumstances? What will be the psychological impact on those discovered to have the Huntington's disease gene and on their families? What will be the psychological impact on those found to be free of the gene and on their families? How can we plan services accordingly to assist?

A Debate among Few Debaters

In reviewing the state of predictive art, the British Medical Journal called for a debate regarding presymptomatic testing:

> There remains the difficult problem of the ethics of such testing. When a definitive test is eventually introduced—as will surely happen—its application will cause controversy. Perhaps the time has come when we should debate possible guidelines for handling the results before the test is introduced.[30]

Subsequent authors took up the debate. Segal raised the pertinent question of who will tell the individual and poignantly described the dilemma a practitioner faces:

> Considering the possible mental anguish and exceptionally high suicide in Huntington's families, the dilemma is whether it is humane and ethical to direct a young person's thoughts to the, at best (or worse), fifty-fifty chance that he or she might deteriorate into the condition witnessed in a relative, and should govern life accordingly. Thus, an individual who might never have consciously or otherwise contemplated the possibility of personal involvement (in this disease) might have life and hope blighted, and survival without development of the disease might be little compensation for a lifetime of anxiety, dread, and single status.[31]

*Now that a marker of the Huntington's disease gene has been found, presymptomatic and prenatal diagnosis is possible in those families in which the disease gene and the RFLP marker, G8, have been proven to be linked. Testing is now in progress to exclude genetic heterogeneity; families of all racial groups and ethnic backgrounds are being tested throughout the world to determine if Huntington's disease gene and G8 are linked.

It may be only the most enlightened health and allied health professionals who are struggling with these issues. The genetics of Huntington's disease are unknown to many practitioners around the world, so that their ethical sensibilities in this area remain virgin. Caro et al. complained vociferously in 1976 that neither neurologists nor family doctors always ensure that a new diagnosis of Huntington's disease is communicated to all of a patient's children who may be grown and scattered around the world. Families, once alerted, are appreciative: "We knew that Huntington's chorea was in our family but not what it meant for us and our children."[32]

Caro was not being unduly harsh to the British medical profession, as a 1983 study by Martindale and Yale demonstrated. They surveyed all health professionals within a provincial health district of approximately 160,000 that contained a teaching hospital. All neurologists, psychiatrists, general practitioners, social workers, district nurses, and health visitors in the area with knowledge of a Huntington's disease family were contacted and interviewed. The results were appalling:

> In no case did the neurologist making the diagnosis point out the implications for the patient's family to the appropriate general practitioners who, with one exception, remained in ignorance of the genetic risk. Serious opportunities for the prevention of Huntington's chorea are being neglected.[33]
>
> In one instance, a married daughter of a Huntington's disease patient was seen once by a geneticist on the GP's advice, "but all the health professionals concerned thought it inappropriate to approach the 28-year-old son because he was single!"[33]

If the inadequacy of good genetic information continues, it is unlikely that the development of a presymptomatic test will make Huntington's disease disappear in a generation, as some have suggested.[34]

In the United States the provision of genetic counseling can be equally inconsistent and sporadic. Johnston and Seitz, in 1983, surveyed staff on 94 wards in 78 Veterans Administration Medical Centers across the country. Although 82 percent indicated that they provide genetic information to HD families, less than 1 percent of the hospitals had a formal genetic counseling policy. In 38 percent of the hospitals an individual, usually a neurologist or social worker, was specially designated to counsel patients and families. Despite these arrangements, genetic counseling was not always provided in a consistent and formal fashion. In many hospitals it depended on the interests of current ward personnel, often medical students who frequently change rotations. In some instances information provided by untrained personnel was inaccurate. If staff assumed that families knew the genetics, they often did not give information or check the family's understanding. Referrals for additional help received minimal follow-up.[35]

Despite the enormity of its implications, Huntington's disease families are surprisingly unanimous in wanting accurate diagnosis and genetic counseling as quickly as possible.[36] The most frequent reactions when information was inaccurate or withheld, particularly until after marriages were made and children born, were rage, bitterness, frustration, grief, and betrayal.[1,36]

FAMILY ATTITUDES

> *My name is Lynnette. I'm in the fifth grade. My father has had Huntington's disease all his life that I can remember. We had to stay by the house to watch him. We had to give him medicine every night. Every year we go around for Huntington's disease. It can defect the brain. It can make you bad bad sick, think wrong, and it makes you mean. You pull off covers and pour water on people's head. It is a very bad disease. It make nervous; you have to leave the house sometimes. My mother would get griped out for being late. He lied on the couch day by day.*
>
> *One day they might get a cure for it.*
>
> *Ten years old.*[36]

Attitudes toward Reproduction

The hereditary aspect of Huntington's disease is particularly pernicious. One of the most common immediate reactions among newly diagnosed parents is concern about inflicting the illness on the next generation. An installation engineer, upon diagnosis, said, "Oh my God, what have I passed on to my children?" He never mentioned the hereditary part again.[37]

The trauma of guilt and worry can be so great that patients choose to break with reality rather than face it. One man, when asked how he felt about his children developing his illness, denied vociferously that he had Huntington's disease, so he could not pass it on.[37] A female patient, a geneticist in full possession of all the medical facts, suddenly announced that the illness was not hereditary.

A recent survey of 420 members of the Association to Combat Huntington's Chorea (Combat), a lay organization in Great Britain, netted 153 completed questionnaires from spouses, children, and siblings of Huntington's disease patients.[38] Struggles between sympathetic but paternalistic practitioners and families wanting accurate information were evident in the responses. Some families were informed only about the nature of the illness and not its genetics, particularly when young offspring already existed. The hereditary aspect was only discussed with childless couples. In other instances, only couples with children were informed: " . . . some family doctors who have seen a generation or more with the disease in a given family may choose not to tell young women, especially until after they were married and had about 2 children, thus, giving them the benefit of the doubt. . . . Those who produced children in ignorance of the genetic consequences bitterly resented not being informed at the time and usually regretted having children once they discovered the true facts."[38]

It is difficult to know if those currently with children would have really foregone childbearing, had they known. Once children are present, the most crucial concern is fear of hurting them. It may be somewhat easier for some to make a theoretical decision in hindsight rather than the actual sacrifice. Childless couples, in full possession of the genetic facts, often find childlessness too painful a deprivation to bear, particularly in addition to coping with the disease and its threat. Accurate information regarding attitudes toward procreation must be collected prospectively from currently well-informed couples of childbearing age, prior to reproduction.

Frustrated by the perpetuation of the illness, some have advocated directive genetic counseling.[39] Others find that directiveness only leaves clients feeling isolated and alone, frightened to return for help if their decisions contradict their physician's.[40]

Using a novel approach of nondirective counseling and active, sustained availability and contact, including annual or biannual home visits, Harper and coworkers have demonstrated that a program of nondirective support, education, crisis intervention, and empathic rapport between health care providers and families can decrease the number of births to at-risk individuals.[41,42] (Some cynics have speculated that only non-gene-carrying at-risk individuals are refraining from reproduction. Procreation in the face of genetic facts is such an error of judgment, they argue, that it must be pathognomonic of the illness.)[43]

Attitudes toward Presymptomatic Testing

These findings suggest that if a reliable predictive test were available, the birthrate might fall even more precipitously. According to the Combat questionnaire, 80 percent favored at-risk persons taking such a test, 68 percent thought it should be administered before the age of 18, and 22 percent preferred informing themselves and their children between the ages of 19 and 40. Asked what they would do if the test proved positive, the overwhelming majority would limit children either by contraception or, more frequently, by sterilization, and as many as 70 percent said they would avail themselves of an abortion if pregnancy did occur. These questions were obviously painful and difficult to answer because the response rate declined to a little over 60 percent. The majority also wished to take part in initial research trials on a predictive test; and 75 percent of these wanted to know the results, even if they were uncertain.[38]

The statistics from the British questionnaire regarding prediction are almost identical to figures gathered elsewhere. In 1972 Stern and Eldridge conducted the largest attitudinal survey ever undertaken in the United States.

From over 2,600 members of the U.S. Lay Organization Committee to Combat Huntington's Disease. 1,065 questionnaires were completed. Some 94 percent of the 31 affected individuals would have wanted a presymptomatic test, although only 80 percent of all respondents would—77 percent of those at high risk and 86 percent at low risk. Only 15 percent overall were uncertain, but 17 percent of those at high risk were unsure. And 82 percent of respondents indicated they would limit family size if discovered to have the gene, but those at highest risk were somewhat less likely to make this response.[44]

This seemingly illogical finding could be explained by viewing it as another indication of the high risk individual's need for denial and insistence on normalcy, i.e., raising a family, to bolster this denial that having the gene is anything very bad or should influence one's behavior in any way. On the other hand, it may be that those at low or no risk, spouses for example, might more glibly give the conventionally "rational" response—limiting family size—since it is less likely that they will have to carry out this choice. Those at high risk have had to think realistically about the possibility that they do bear the gene. Their hypothetical choice to have children despite presymptomatic identification as gene carriers may speak to the little understood, but intense power of the urge to procreate. The high risk respondents are replying honestly with what they would do, while the low risk respondents are conveying what they feel they ought to do.

Wexler interviewed in-depth 36 individuals at 50 percent risk for Huntington's disease, subjects garnered through voluntary health organizations but many not particularly active members. All agreed that a predictive test should be developed for purposes of science and for those who wish to utilize it. Although all acknowledged that they would be terrified to take a test, two thirds said they would actually want to be tested. Ranging from realism to bravado, reactions varied from those who thoughtfully decided to know to those who felt they ought to want to be tested. All asked for counseling to be coupled with testing.[1]

In the Australian state of Victoria, where subjects were somewhat more representative than those selected from the membership of voluntary health organizations, 84 percent of at-risk respondents were positive toward a safe, simple, and 100 percent certain test, while 64 percent had a generally positive response to predictive testing.[44]

Bates wrote one of the most eloquent pleas in the literature for a safe and reliable test,[46] arguing that information is necessary to give people time to plan marriage, reproduction, career, and finances and to prepare emotionally for the inevitable. A representative of a lay health organization, she is strongly in favor of the development of a presymptomatic test.

The medical profession, perhaps uneasy with its responsibility as bearer of bad news, continues to temporize. In an editorial solicited by *Neurology*, Emery writes, "My feeling is that (to pursue research on predictive tests) would only be justified once an effective treatment has been found."[47] Emery appears to be one of the few who report that the majority of at-risk individuals with whom he has discussed the matter would prefer to live in uncertainty rather than risk taking a presymptomatic test.

It is unlikely that we will have the luxury to pursue Emery's suggestion. Any finding that purports to reveal a difference between symptomatic and normal individuals is immediately tested to determine its discriminatory powers in those at risk. Research on the etiology of the illness will inevitably lead to the development of a presymptomatic test, probably prior to the discovery of adequate treatment.

TEST DEVELOPMENT*

Research on Tests versus Delivery of Validated Tests

Almost all surveys soliciting reactions to presymptomatic testing ask if subjects are willing to take a 100 percent-certain test. Unfortunately, a misunderstanding has arisen in the literature that threatens to obscure important information. Statements pertaining to *research* on the *development* of a test are confused with remarks on the *use* of a *perfected* test. Fahn and others have advocated that an ideal test would be one in which the subject would be unaware of the results.[20] The major disadvantage of the levodopa and other provocative tests is that during the research to develop the test symptoms are obvious to subjects, leading them to leap to unwarranted conclusions. Fahn is not suggesting that the results of a valid and reliable presymptomatic test should be withheld from subjects, or even that a provocative test that is 100 percent certain should not be administered. Fahn and colleagues are arguing that a diagnostic test would be ideal if it could be tested extensively without the individual seeing possibly spurious results.

McCormick and Lazzarini sought to dispel the notion that provocative testing has no role to play in presymptomatic diagnosis. In a survey of 40 at-risk individuals with a mean age of 35, 40 percent expressed a willingness to undergo provocative testing with a chance that transient chorea might appear. An additional 5 percent would take provocative testing if no movements were evoked.[48] The survey stipulates, however, that this putative test is 100 percent certain. The high enthusiasm for taking a provocative test speaks mainly to the urgency with which a test is desired and the price people are willing to pay for definitive information. For research purposes—designing and validating the test—a nonobvious test has less risk of provoking misinterpretations.

The levodopa loading test is a case in point. Although the authors are not explicit about the psychological reactions of those 11 of 28 testing positive-

*This section is now out of date since a specific test is being developed. It sets an historical context for the discussion of presymptomatic testing which is relevant to understanding current concerns.

ly, at least one individual attempted suicide.* On long-term follow-up three false negatives have been diagnosed.[19,20] Another drawback of the testing is that the positives must be followed until they are over 80 and examined pathologically to confirm any false positives.

The levodopa test is not reliable, and the extent of its unreliability is unknown. People in general have variable sensitivity to dopamine, as evidenced by some individuals' susceptibility to develop tardive dyskinesia when exposed to neuroleptic medication. Yet those witnessing transient chorea develop in themselves during the L-dopa loading test must have been hard-pressed to explain their symptoms other than as premonitory for the disease, particularly after reading published reports of the project stating that none of the normal controls and approximately two thirds of those at risk developed chorea.

Research Risks: The Dangers of Partial Information

Scientists internationally are concerned about the dangers of imparting partial and potentially misleading information to research subjects. In 1982 the World Federation of Neurology Research Group on Huntington's Chorea published the following Letter to the Editor of the New England Journal of Medicine:

> The World Federation of Neurology Research Group on Huntington's Chorea reviewed and discussed the current status of possible predictive tests of this disease at its 9th biannual meeting in Japan in September, 1981. The members have concluded that so far no proposed "predictive test" has been validated. They recommend that until the predictive findings of investigations have been confirmed neither the results of the tests nor predictive opinion based on this test should be disclosed to the subjects at risk who were tested. A committee of the Research Group is going to consider this issue further. Suggestions are welcome.[48]

Under the Freedom of Information Act, individuals who participate in federally funded research can request their test results. They cannot sign away these rights in an informed consent form even if the form specifies that test results will not be given nor should they be requested. Even if test results must be provided under the Freedom of Information Act, the hypothesized interpretation of such results need not be included. The majority of the time, results cannot be interpreted because their accuracy is the essence of the research. Investigators are not required to speculate. For example, if an at-risk subject in a positron emission transaxial tomography study requests his or her scan, computations about the scan can be given with no interpretation

*1979, Klawans, H. : personal communication.

regarding the scan or its possible implications. Of course, most subjects inquisitive enough to demand test results are likely to find some means of interpreting them, even if not provided by the investigator.

The gravest danger regarding research on individuals at risk is that they be given partial and incomplete information on research results. A new presymptomatic diagnostic test requires extensive statistical and longitudinal testing for specificity, reliability, and validity. Partial information stating that a subject performed abnormally on a test is misleading and dangerous to a subject until the value of the test for diagnostic purposes is clearly known. Test performance can be so influenced by anxiety and other factors as to render the test useless. Partial information can also be destructive because it is given out by individual investigators with no control over how information is communicated and no organized follow-up plan.

Investigators can protect themselves to some extent by coding all test results and giving the code to a safekeeper. If an individual presses to see results, the investigator would have to break the code, but most subjects are reluctant to push a scientist to do something injurious to the quality of the science. Investigators should not feel compelled to give out any information since their experimental findings are neither valid nor reliable. Rather, they should discuss the possible meanings of various outcomes with at-risk subjects and help them live their lives of uncertainty with greater equanimity.

Investigators should not be seduced by their own wishful thinking that they have discovered an early diagnostic test into divulging meaningless and possibly damaging information to research subjects as if it were meaningful data. No matter how many caveats a scientist might add to the information, subjects hear "normal" or "abnormal" and draw their own deductions accordingly. At-risk individuals are so eager for, and fearful of, information that will end their ambiguity that even the most sophisticated may jump to conclusions, try as they may to maintain intellectual caution. Investigators should not be so psychologically naive as to believe that if they tell a subject, "You were abnormal, in the Huntington's disease performing range, but don't take it to heart because we don't know yet what it means—it may not be specific to Huntington's disease," that subject would not interpret, "I have Huntington's disease."

TEST UTILIZATION

Questions Regarding the Use of a Safe, Reliable, and Valid Predictive Test

It is probable that a marker indicating the gene's presence will be discovered before an effective treatment or prophylactic intervention. If an RFLP is the first diagnostic marker developed, it will require arduous work to extract the mutant gene from a strand of adjacent DNA, sequence it, and try to develop some corrective. The abnormal gene product may be totally novel. The marker will tell us nothing about the nature of the defective gene except its chromosomal location and some estimate as to the proximity between marker and gene. It is likely that there will be a long hiatus between the discovery of a diagnostic test and the development of a safe and effective treatment. This means that we will shortly have the capacity to tell people that they or their children or spouses will develop Huntington's disease sometime in the future but that there is no treatment for it.

Quality Controls on Testing

Given all that depends on this information, we must have utmost confidence in the accuracy of the test. This may require that testing should be done in only one or two laboratories in the country to ensure quality control. Perhaps individuals should be tested several times to confirm the findings. Periodic checking of laboratories with known positive samples may be advisable.

Impact on the Information Provider

Modern genetic counselors are taught to be nondirective in their counseling. This orientation has all the earmarks of enlightenment and progressive thinking. Individuals and couples are responsible for themselves, after all, and the counselor's job is to make sure they know the correct information and to help them make decisions most suitable for them.

In certain respects this solicitous but laissez-faire attitude on the part of counselors permits them to avoid grappling with questions such as whether couples should expose children to the risk of genetic disease. If a couple gambles and loses, the counselor is not held responsible for the life, suffering, costs, or joys of that child.

But what if a counselor is responsible for telling an individual that he or she is carrying the Huntington's disease gene, and that person commits suicide? Can the counselor maintain the same equanimity? How should the information provider be protected, as well as the information receiver?

A Test for All or Some?

Should a person of any age be able to be tested? How can a 10-year-old, for example, cope with knowing that in 30 or 40 years the illness will probably start? If a minor does not receive the test, can parents ask for test results on their minor children? Surely this information would be communicated nonverbally or otherwise to the child. Since there is nothing prophylactically that parents can do, except maintain the good health and nutrition of their children, which they should do under all circumstances, of what use is it to provide parents with this information? Would parents be any more thoughtful, sensitive, or emphathetic to their children knowing they carry the gene than if they were merely at risk? Is early diagnosis necessary for financial planning? Is this planning sufficient reason to test minor children? The parents may have to save less, knowing their children are free, and more could be spent on the childrens' upbringing and education. But is it worth the emotional cost of knowing if a child will be affected?

In families with several or more children, it would be exceedingly disruptive for parents to know that one or more children are carriers and the others are not. Would parents be overly protective of a gene-carrying child or subtly disengage from an emotional bond that will inevitably prove painful when the child falls ill? Will the same financial and time investments be made in fostering the career of a child destined to have shorter productive years? Will parents steer their presymptomatic children, overtly or covertly, away from careers requiring coordination and judgment, such as medicine, finances, aviation, or sports? Will they encourage or discourage interpersonal relationships, marriage, or children? Should parents be permitted to know this information without their children's consent, whatever the cost or benefits may be?

There appears to be no medical precedent for withholding a diagnostic test until a person reaches a certain age. It may even be illegal to withhold. Whatever the age of the information recipient, there are important questions to answer regarding delivery of this information. Who should give the news? Should neurologists or family doctors communicate the information? Genetic counselors? Anybody with whom an individual has an ongoing medical relationship? If an at-risk person walks into a doctor's office for the first time and requests such a test, should it be done, information provided, and the individual left to adjust as best as possible?

THE IMPACT OF KNOWLEDGE

Impact of a Predictive Test on Those with the Gene

The ability to effect a sudden transition from hopeful uncertainty to catastrophic clairvoyance makes medical science quail at its power. Bates is one of the few who is sanguine and optimistic that those discovered to have the gene will use the information constructively.[46] Others paint a scenario of suicide and despair.[49]

Probably both visions are correct. Our challenge is to maximize the productive outcome. A key concept in counseling for Huntington's disease is time. Much depends on the age at which an individual discovers he or she is a gene carrier. Consider first the reactions of those discovered to carry the gene. There may be some relief that the uncertainty is finally resolved. But the most expectable reaction would be disappointment, grief, rage, frustration, despair, depression, and possibly suicidal thoughts and actions.

A New Waiting Game

The long, insidious onset of the illness also poses particular difficulties. In telling people they have a malignancy, the illness is already in progress when the diagnosis is made. The tumor is palpable and its response to treatment can be measured. If the cancer goes into remission, the waiting game begins, as in Huntington's disease, but there is always some hope that the cancer will never reappear.

In Huntington's disease, onset is extraordinarily difficult to determine. In prospective studies minimal neurological abnormalities or emotional disturbances can be present for years before reaching the level of diagnostic certainty.* It is even possible that symptoms do begin early in life but are subthreshold.

"Freed" from the uncertainty of hope that the gene has been escaped, the presymptomatic person moves into a new state of limbo. When will the disease appear? How many good years are left? At-risk individuals chronically "symptom seek" when their odds are 50/50. Knowing that they are harboring the gene but uncertain as to when its manifestations will begin, presymptomatic individuals will not know how to interpret clumsiness, irritability, emotional instability, lapses in memory or judgment, or other everyday events. Healthy, productive years may be lost by a premature decision that the disease has begun. Presymptomatic persons may turn themselves into patients merely to resolve the uncertainty and ambiguity of this new waiting game. Closure is better, even if it means adopting a patient status.

On the other hand, denial may set in, as it does in many early patients, and some might cling to their presymptomatic status in the face of obvious illness.

One solution is to turn over responsibility for determining disease onset to a skilled physician familiar with the disorder who is available on a long-term, consistent basis. The presymptomatic person could delegate all symptom searching to the physician, secure in the knowledge that the physician will be honest when unequivocal signs begin. Presymptomatic persons can share the ambiguity of their situation with another responsible individual. In this way 20 or 30 years of productive life might be protected.

Interpersonal Relationships

If gene carriers are single, how can they make a commitment to others, knowing the future? Prospective spouses of at-risk individuals might be willing to take a risk with their partner, but would they still marry knowing the certainty of illness? Can marriages already made withstand the strain of knowledge? With prenatal prediction possible, gene carriers will have the option of bearing gene-free children, but would they reproduce knowing that their parenting role would be abnormal? Could commitments to careers be

*Venezuela Collaborative HD Study, 1983 : personal communication.

made with the same vigor and enthusiasm, knowing that they will most likely be sharply curtailed?

One of the most pernicious aspects of a late onset disorder such as Huntington's disease is that parents are often in the terminal phases of the disease when their own offspring are diagnosed. It may be very difficult for those found to be gene carriers to be around their ill relatives, seeing their future enacted before them. Responsibilities for established caretaking relationships within a family might alter after a presymptomatic diagnosis is made.

Stigmatization and Discrimination

Stigmatization of asymptomatic gene carriers could become a severe problem. If gene carriers are discriminated against in hiring practices by employers receiving even minimal federal funds, they are protected under Section 503 of the Rehabilitation Act of 1973. This act specifies that persons cannot be discriminated against if they are handicapped or are perceived to be handicapped.[50] It is unclear whether or not an employer could insist on a prospective at-risk employee taking a test as a condition of hiring. Insurance companies would be influenced by the availability of the test and might adjust their benefits and costs accordingly. Following a positive test, serious occupational and financial repercussions could ensue. The Commission for the Control of Huntington's Disease and Its Consequences received many letters of testimony on current problems in this area:

> My youngest nephew has had his goal set on being a doctor all of his life and has worked toward that aim. He is an excellent student and has worked for the finances involved in such an education. He is now faced with problems in being accepted in medical school because he is a member of an HD family. He is attempting to lead as normal a life as possible but society will not allow it. He is being made to feel "guilty" about having a father who died of HD when the committee who interviewed him for entrance into medical school admitted that the reason for his not being accepted was HD. Such barriers cannot be broken down in a short time, but some measures need to be instituted to educate people, even people associated with the medical field.[50]

> Our son had an ROTC scholarship at Colby and was to be a pilot in the USAF; after he finished basic and went to flight school some doctor discovered on his medical record his father had HD. He was immediately grounded and treated as if he had nothing more to live for. Finally after months of telling him they were going to discharge him, they changed their minds and sent him to radar school instead.[50]

Counseling Suggestions

One approach could be to engage the person immediately in planning for the future. Many HD patients have had satisfying careers and good interpersonal relationships before the illness begins that continue as the disease progresses. Presymptomatic individuals can be helped to have the same. Life options should not be closed just because they carry the gene. Financial and medical insurance planning should have top priority. A counselor, be it a psychotherapist, physician, or other professional, is needed to help the person make realistic preparations for life, maintaining a delicate balance between optimism and practicality. Hope for medical advances should be constantly maintained. Time is on their side in permitting science years to work.

Many at-risk individuals feel insecure, inferior, and damaged just by virtue of being at risk. They believe their at-risk status makes them less desirable as marriage partners, particularly if prospective spouses must forgo their own biological children as a condition of the union. The self-esteem of the presymptomatic person suffers an even more serious blow. The counselor's positive and supportive attention can do much to bolster crumbling self-regard. Counselees often take their cues from how the counselor perceives them. Family members may react to the early diagnosis with withdrawal from the individual due to their own grief and loss. In some instances parental guilt is unleashed as anger at the new patient for being sick causing them anguish, even if it is the parents who "caused" this illness." The counselor can protect the presymptomatic person from feeling like a pariah at home and in society. Families will also need extensive help in coping with the impact of one or possibly multiple early diagnoses.

Supportive counseling is time-consuming and expensive. In a survey of neurological disorders seen in the medical genetics clinic at the University of Washington, Huntington's disease represented over 11 percent of all clinic visits. The authors conclude: "This high figure undoubtedly reflects the serious emotional, social and financial difficulties weighing on families with Huntington's disease, rather than its uncommon prevalence."[51] Barette and Marsden were also struck by the "widespread anxiety and suffering among Huntington's chorea families" and by their repeatedly expressed need for expert medical and social advice.[38]

This need will be magnified dramatically when a predictive marker is discovered. Most insurance plans have limited outpatient and psychiatric coverage. Extensive genetic counseling services are often not reimbursable. Covering the cost of vital long-term counseling might prove to be a major financial burden on those who need most to garner their funds for the future.

In discussions of presymptomatic testing, suicide is usually depicted as the ultimate negative consequence of a positive test. Suicide has been found to be at least six times more common in Huntington's disease patients than in normals.[52] It occurs most frequently in young, often not-yet-diagnosed patients—exactly the status of the presymptomatic individual. For many at risk, however, the option of suicide permits them to live a freer life. One at-risk man described his gun as his "insurance policy" in case he should develop the disease. Knowing that one can take an active stance to relieve the burden of illness resolves the ambiguity somewhat. Surely many at risk question why they should bother to invest themselves in life when it may be interrupted in a particularly disagreeable fashion. Those with the gene may be even more tempted to end their lives. Some counselors have been effective in telling clients that they always have the option of committing suicide in the future, so why not just try living day by day, thereby coaxing them away from a precipice of despair. Maintaining the choice of suicide gives people a psychological sense of control and mastery, even of the most negative kind, which can be helpful to some in an out-of-control situation. Suicidal threats need not be met with moralistic horror but rather should be responded to with other options for control.

Impact on Those Not Carrying the Gene

The easy assumption is that those found to be free of the gene will be elated, relieved, and proceed to live a normal life. Certainly exquisite joy and relief are to be expected. But this group has special needs and concerns as well which should not be ignored in developing counseling programs.

Those no longer at-risk must also adjust to a major identity upheaval. For many, being at-risk for Huntington's disease has been part of their identity for most or all of their lives. They have accommodated to uncertainty and learned to live with it, some even to use it creatively to spur them into new activities or achievements. The sense of perspective which being at risk can bring, the acknowledgement of the possible compression of life, can be an energizing source to propel people toward activities or relationships which would otherwise intimidate or frighten them. Life takes on a certain immediacy, a preciousness when tinged with its possible loss.

People at risk can also feel themselves to be somewhat out of the ordinary, called upon (for whatever reason), to confront an unusual life situation. They can gain a special status with family and friends. Within voluntary health organizations for Huntington's disease families, those at risk feel a certain bond, a special comradery. Siblings at risk can draw more closely together as they share an experience unique from their friends.

Some families may also make special dispensations for those at risk. Children are not teased for being clumsy or uncoordinated. Temper outbursts may be more easily tolerated with the explanation that the person at risk may be under a great deal of stress, or perhaps the temper is the first manifestation of illness. Some at risk may play on these "secondary gains" of their risk status, but most, if they recognize them, appreciate the consideration of others in allowing them a little special license.

If presymptomatic testing turns out negative, terror is gone but so is specialness, so is license. The "formerly at risk" must adapt to a new identity in which they are just the same as everyone else and accountable for all behavior. If they had decided not to pursue some goals—marriage, children, education—because they felt the future to be too uncertain, they may feel bitter and resentful. Particularly those who chose not to procreate due to the genetic risk and who are no longer able to do so may feel that their sacrifice was in vain and they are left childless for no reason. They may regret choices viewed with hindsight.

Relationships among family members can also experience a tremendous wrenching. Siblings united by the bond of risk may suddenly find themselves separated by the outcome of testing. It is not unusual for even the closest siblings occasionally to wish the disease on the others instead of themselves. If another sibling is diagnosed, the guilt is intense. Many may suffer from "survivor guilt," particularly characteristic of wartime soldiers who live while their buddies are killed. Survivor guilt may be generated not only from their own nuclear or extended family, but also from the Huntington's disease community at large. Within the family, siblings may draw apart, resulting in the loss of an important relationship for all of them. A well sibling may feel compelled, from love, guilt, or whatever the reason, to dedicate him or herself to caring for a presymptomatic sib, while that sib may find it difficult to be near the one who is lucky. The jealousy and enmity of less fortunate siblings may be difficult to bear.

People found to be free of the gene may find it embarrassing or "greedy" to ask for attention for themselves in the face of the obvious needs of those diagnosed positively. Programs must be developed specifically for them. The counselor's acknowledgement of the needs of these individuals will help in validating them, and providing counseling as an integral part of the overall program will legitimize their need for help.

Those Who Prefer Not to Know

Approximately a quarter of those surveyed in the United State, the United Kingdom, and Australia indicated that they would not want to avail themselves of a presymptomatic test.[38,44,45] These individuals will need to be protected from undue pressure to take the test. Some of this pressure will no doubt come from themselves.

> God, get it over with. I'm so tired of wondering. If they would tell me that I wasn't going to get it, they could take my arm off! What if they do tell you you've got HD, how do you live with that? Like, if they were to say to me today, "You're going to get HD when you are 30," do you know what every day would be like? Every day would not be a real life. . . . I just couldn't live with that. Now, at least I have a 50-50 chance; knowing and not knowing. I can live with that. Now, I have optimism. Then it would be real.[1]

> Can you imagine knowing that there's some place you can just walk into? Every day, that would prod your mind, "I'm going. I'm going. Just for the sake of science, I'm going. For the sake of knowing, I'm going. I can't stand it anymore." Then all of a sudden, having somebody tell you you're going to have HD. I think this would prey on their minds and everybody would probably go. But they really wouldn't want to.[1]

Family members may also insist that an at-risk person take the test, particularly for family planning purposes or on behalf of 25 percent at-risk individuals. If an at-risk parent is found to be free, then his or her children are also free. The fate of spouse and children are dramatically affected by test results; do they have a right to force the test? Direct or covert pressure may also come from employers, insurance companies, or other institutions. Protecting an individual's right not to know is as important as providing support for those who choose to know.

CONCLUSION

Recommendations

These recommendations are meant primarily to stimulate thought and discussion. Assuming that an RFLP will be the first gene marker and that we will have our current state of treatment, the following suggestions are offered:

1. All diagnostic testing must be voluntary.
2. No individual should be denied the opportunity to be tested.
3. All diagnostic tests should be done in one or only a few laboratories in which quality control can be strictly maintained and a protocol followed in handling test reults.
4. The test should be available to those age 18 or older who can choose independently. For those below the age of 18, the test should only be used in special circumstances, such as childbearing or marriage, which should be evaluated on a case by case basis. Minor children should not be tested. Children should be encouraged to have some years of growth and development without their future compromised.
5. The tests should be given by a physician with a good relationship with the client, if possible, and by someone very familiar with Huntington's disease and Huntington's disease family care.
6. An individual should have at least four or more interviews, if possible, prior to taking the test to discuss it. Sessions both singly and jointly with other family members should be encouraged.
7. A protocol should be developed including some questions to ask routinely of all prospective test takers. Good clinicians will vary the questions according to their own understanding of the client and the situation, but guidelines might prove useful. Questions should "walk" the client through the experience of taking the test, including projecting reactions to different outcomes. Time should be spent discussing the person's motivation for taking the test and why at that particular moment. The physician should inquire about family members' reactions to possible test results. Through role playing, and emphasizing specific details, the clinician should try to make all possible outcomes vivid to the client. At-risk individuals should

not feel pressured to take the test because other family members have taken it or because their families want them to. Clinicians should legitimize *not* taking it, as well.

8. At-risk individuals should, if at all possible, take time to reflect between talking to the physician or counselor and taking the test. They should know when the initial appointment is made that the test will not be given on the first visit to avoid undue anticipatory stress.
9. If a client is pregnant and waiting is not an option, time should be spent in defining and developing the client's support system.
10. Test results should NEVER be given over the telephone. No matter what the cost or distance, unless there are extraordinary circumstances, all results, positive and negative, should be discussed in person between the client (and family) and the clinician making the initial assessment. Giving a diagnosis to gene carriers over the telephone is tantamount to malpractice. And those discovered not to have the gene may have guilts and concerns as compelling and worthy of attention as those in whom the gene is revealed.
11. A regular program of telephone and in-person follow-up should be initiated with all persons posttesting. Follow-up groups for those receiving positive and negative results could provide peer support. Occasional joint sessions would allow each to hear the reactions of the opposite group, which may contain or represent other family members, rather than to imagine what the others might feel. In this way sibs who have different outcomes on the test are immediately put into a "sib" group of people sharing their own outcome. Follow-up of some sort should be daily at first, then weekly, then no less than monthly for the first year posttest. Support groups should also be organized for non-at-risk family members—parents, spouses, and so on, others—to help them cope.
12. Particularly intensive care will need to be provided for those having prenatal testing. If the test is negative, the at-risk parent can be assured of having a child free of Huntington's disease, regardless of whether he or she develops the illness. The parent is free to choose to live in personal uncertainty, secure in the knowledge that the child is safe from genetic jeopardy. But if the test is positive, two massive losses are suffered simultaneously. The child will most likely be aborted, or else the test would not have been taken initially. And the parent is immediately diagnosed as an obligate carrier. At-risk parents hear their own death knell with that of their child. And the spouse suffers the same devastating double loss of spouse and offspring. Knowledge that the couple can try again for a Huntington's disease gene-free baby might provide some solace. On the other hand, a couple might choose not to try again, knowing that one parent will be ill.

Some imaginative ways around this problem have been proposed. If the gene is localized to a specific chromosome through gene mapping, one could determine whether the at-risk person's fetus carries either the normal or affected grandparent's chromosome in question. If inherited from the normal grandparent, the baby is free of risk. But if the affected grandparent's chromosome is passed on, the fetus has a 50/50 risk, since it is unknown whether the chromosome is carrying the gene. A baby safe from Huntington's disease can be virtually assured, while protecting the at-risk parent from devastating news.*

Common Problems—Cooperative Solutions

There is a more extensive literature on predictive testing for Huntington's disease than for other similar disorders, perhaps because Huntington's disease is more common. Cautionary tales about other illnesses do exist, however. Widespread confusion greeted the early screening programs for sickle-cell disease. Failing to differentiate between sickle-cell trait and disease, sickle-cell carriers were sometimes denied employment and insurance. Differentiating between a clinical patient and a presymptomatic but predestined patient will require even more sophistication.

Results from a prospective study of genetic screening and counseling for sickle-cell disease in the Greek rural community of Orchemenos are no more heartening. Marriages were equally frequent between carriers, but a new stigmatized "carrier class" had developed. Follow-up studies on Tay-Sachs screeening programs have been conducted in urban U.S. and Canadian cities, but even among this highly educated, sophisticated population, about one half of the carriers expressed anxiety and unease after learning of their carrier status.[53]

The paucity of literature on predictive testing may attest to our lack of preparedness. Fortunately, the experiences of one disease group will be highly informative for the others. In lieu of Huntington's disease, this chapter could as easily have referred to problems associated with predictive screening for a variety of spinocerebellar degenerations (Joseph's disease or olivopontocerebellar ataxia); or dystonia, myotonic dystrophy, hereditary polyposis, familial hypercholesterolemia; or any late onset, autosomal dominant disorder. Familial Alzheimer's disease patients may share appallingly similar difficulties.

Although the problems are formidable, the creativity of the professional and lay community in designing and financing medical, emotional, economic, and legal support systems for families jeopardized by any genetic illness will have broad applications. The day may be close when individuals can request a personalized genetic "map" with all of their health "land mines" demarcated. Geneticists may replace astrologers in predicting strengths, weaknesses, predilections, and predispositions. They can know our future by reviewing our

---

*David Housman, 1979 : personal communication.

past inheritance. But we must determine if this new knowledge is to be life enriching or destructive.

> Men at some time are masters of their
> fates:
> The fault, dear Brutus, is not in our
> stars,
> But in ourselves, that we are underlings.
>
> *Julius Caeser*, act 1, sc. 2, line 134

NOTES

1. Wexler, N.: "Genetic Russian Roulette: The Experience of Being at Risk for Huntington's Disease," *in* Genetic Counseling: Psychological Dimensions, Academic Press, New York, 1979, pp. 199–220.
2. Weiss, J. M.: "A Model for Neurochemical Study of Depression," presentation before the American Psychological Association, Los Angeles, CA.
3. Weiss, J. M.: "Coping Behavior: Explaining Behavioral Depression Following Uncontrollable Stressful Events," *in* Behavioral Research Therapeutics, 1980, 18:485–504.
4. Wexler, N.: "Huntington's Disease and Other Late Onset Genetic Disorders," *in* Psychological Aspects of Genetic Counseling, Academic Press, London, 1984, pp. 125–146.
5. Conneally, P. M.: "Huntington's Disease: Genetics and Epidemiology," *in* American Journal of Human Genetics, 1984, 36:506–526.
6. Pericak-Vance, M. A.; Conneally, P. M.; Merritt, A. D.:"Genetic Linkage in Huntington's Disease," *in* Advances in Neurlogy. Vol. 23, Huntington's Disease, eds. T. N. Chase, N. S. Wexler, and A. Barbeau, Raven Press, New York, 1979, pp. 59–72.
7. Pericak-Vance, M. A.; Elston, R. C.; Conneally, P. M.; Dawson, D. V.: "Age of Onset Heterogeneity in Huntington's Disease Families," *in* American Journal of Medical Genetics, 1983, 14:49–59.
8. Patterson, R. M; Bagchi, B. K.; Test, A.: "The Prediction of Huntington's Chorea: An Electroencephalographic and Genetic Study," *in* American Journal of Psychiatry, 1948, 104: 786–797.
9. Baro, F.: "A Neuropsychological Approach to Early Detection of Huntington's Chorea," *in* Advances in Neurology, Vol. 1, Huntington's Chorea, eds. A. Barbeau, T. N. Chase, G. W. Paulson, Raven Press, New York, 1973, p. 329.
10. Falek, A.: "Preclinical Detection of Huntington's Chorea: A Preliminary Report," *in* Progress in Neurogenetics, 2nd International Congress, eds. A. Barbeau, J. R. Brunette, Excerpta Medica Foundation, Amsterdam, 1969, p. 529.
11. Lyle, O. E.; Gottesman, I. I.: "Subtle Cognitive Deficits as 15- to 20-Year Precursors of Huntington's Disease," *in* Advances in Neurology, Vol. 23, Huntington's Disease, eds. T. N. Chase, N. S. Wexler, A. Barbeau, Raven Press, New York, 1979, pp. 227–238.
12. Wexler, N. S.: "Perceptual-Motor, Cognitive, and Emotional Characteristics of Persons at Risk for Huntington's Disease," *in* Advances in Neurology, Vol. 23, Huntington's Disease, eds. T. N. Chase, N. S. Wexler, A. Barbeau, Raven Press, New York, 1979, pp. 257–271.
13. Starr, A.: "A Disorder of Rapid Eye Movement in Huntington's Chorea," *in* Brain, 1967, 90:545.
14. Petit, H.; Milbled, G.: "Anomalies of Conjugate Ocular Movements in Huntington's Chorea: Application to Early Detection," *in* Advances in Neurology, Vol. 1, Huntington's Chorea, eds. A. Barbeau, T. N. Chase, G. W. Paulson, Raven Press, New York, 1973.
15. Butterfield, D. A.; Markesbery, W. R.: "Erythrocyte Membrane Alterations in Huntington's Disease," *in* Advances in Neurology, Vol. 23, Huntington's Disease, eds. T. N. Chase, N. S. Wexler, A. Barbeau, Raven Press, New York, 1979, pp. 397–408.
16. Appel, S. H: "Membrane Defects in Huntington's Disease," *in* Advances in Neurology, Vol.

23, Huntington's Disease, eds. T. N. Chase, N. S. Wexler, A. Barbeau, Raven Press, New York, 1979, pp. 387–396.
17. Manyam, N. V. B.; Hare, T. A.; Katz, L.: "Cerebrospinal Fluid GABA Levels in Huntington's Disease, 'At-Risk' for Huntington's Disease, and Normal Controls," *in* Advances in Neurology, Vol. 23, Huntington's Disease, eds. T. N. Chase, N. S. Wexler, A. Barbeau, Raven Press, New York, 1979, pp. 547–556.
18. Goetz, I.; Roberts, E.; Warren, J.; Comings, D. E.: "Growth of Huntington's Disease FIbroblasts during Their in Vitro Lifespan," *in* Advances in Neurology, Vol. 23, Huntington's Disease, eds. T. N. Chase, N. S. Wexler, A. Barbeau, Raven Press, New York, 1979, pp. 351–359.
19. Tourian, A.; Hung, W.: "Huntington's Disease Fibroblasts: Nutritional and Protein Glycosylation Studies," *in* Advances in Neurology, Vol. 23, Huntington's Disease, eds. T. N. Chase, N. S. Wexler, A. Barbeau, Raven Press, New York, pp. 371–386.
20. Gray, P. N.; May, P. C.; Mundy, L.; Elkins, J.: "L-glutamate Toxicity in Huntington's Disease Fibroblasts," *in* Biochemistry Biophysics Research Communication, 1980, 95:707–714.
21. Klawans, H. L.; Goetz, C. G.; Paulson, G. W.; Barbeau, A.: "Levodopa and Presymptomatic Detection of Huntington's Disease: Eight-Year Follow up Letter," *in* New England Journal of Medicine, 1980, 302:1090.
22. Fahn, S.: "Levodopa Provocative Test for Huntington's Disease," *in* New England Journal of Medicine, 1980, 303:884.
23. McCormack, M. K.; Ullman, J.; Lazzarini, A.: "Altered Red Cell Osmotic Fragility in Huntington's Disease," *in* American Journal of Medical Genetics," 1982, 11:53–59.
24. Terrence, C. F.; Delaney, J. F.; Alberts, M. C.: "Computerized Tomography for Huntington's Disease," *in* Neuroradiology, 13:173–175.
25. Neophytides, A. N.; Di Chiro, G.; Barron, S. A.; Chase, T. N.: Computed Axial Tomography in Huntington's Disease and Persons At-Risk for Huntington's Disease," *in* Advances in Neurology, Vol. 23, Huntington's Disease, eds. T. N. Chase, N. S. Wexler, A. Barbeau, Raven Press, New York, 1979, pp. 185–191.
26. Kuhl, D. E.; Phelps, M. E.; Markham, C. H.; Metter, J.; Riege, W. H.; Winter, J.: "Cerebral Metabolism and Atrophy in Huntington's Disease Determined by 18 FDG and Computed Tomography Scan," *in* Annals of Neurology, 1983, 12:425–434.
27. Kuhl, D. E.; Metter, E. J.; Riege, W. H.; Markham, C. H.: "Patterns of Cerebral Glucose Utilization in Parkinson's Disease and Huntington's Disease," *in* Annals of Neurology, 1984, 15:S119–S125.
28. Botstein, D.; White, R. L.; Skolnick, M.; Davis, R. W.: "Construction of a Human Genetic Linkage Map in Man Using Restriction Fragment Length Polymorphisms," *in* American Journal of Human Genetics, 1980, 32:314–331.
29. Housman, D.; Gusella, J.: "Molecular Genetic Approaches to Neural Degenerative Disorders," *in* Molecular Genetic Neuroscience: A New Hybrid. eds. F. O. Schmitt, S. J. Bird, F. E. Bloom, Raven Press, New York, 1982, pp. 415–424.
30. Editorial: "Predictive Tests in Huntington's Chorea," *in* British Medical Journal, March 4, 1978, pp. 528–529.
31. Segal, M.: "Predictive Tests in Huntington's Chorea," *in* British Medical Journal, April 1, 1978, p. 859.
32. Caro, A.; Jones, M.; Stephens, J.; Evans, K.; Walmsley, W.; Randall, D.; Johnston, A.; Heald, J.: "Genetic Counseling in Huntington's Chorea," *in* British Medical Journal, August 14, 1976, p. 420.
33. Martindale, B.; Yale, R.: "Huntington's Chorea: Neglected Opportunities for Preventive Medicine," *in* Lancet, 1983, 634–636.
34. Hemphill, M.: "Pre-testing for Huntington's Disease: An Overview," *in* Hastings Center Reports, 1973, 3:12–13.
35. Kodanaz, A.; Ziegler, D.: "Genetic Counseling in Huntington's Disease," unpublished manuscript.
36. Report: Commission for the Control of Huntington's Disease and Its Consequences, Vol. IV, Parts 1-6, DHEW Publication No. NIH 78-1505.
37. Rubin, R.: "Shouldering Telephone Poles: The Experiences of Families Coping with Huntington's Disease," unpublished manuscript, New Jersey, 1977.
38. Barette, J.; Marsden, C. D.: "Attitudes of Families to Some Aspects of Huntington's Chorea," *in* Psychological Medicine, 1979, 9:327–336.

39. Perry, T. L.: "Some Ethical Problems in Huntington's Chorea," *in* Canadian Medical Association Journal, 1981, 125:1098–1100.
40. Shoulson, I.: "Care of Patients and Families with Huntington's Disease," *in* Movement Disorders, eds. C. D. Marsden, S. Fahn, Butterworth Scientific (International Medical Review Series), London, 1982, pp. 277–290.
41. Harper, P. S.; Walker, D. A.; Tyler, A.; Newcombe, R. G.; Davie, S. K.: "Huntington's Chorea: The Basis for Long-term Prevention," *in* Lancet, 1979, 2:346–349.
42. Harper, P. S.; Tyler, A.; Smith, S.; Jones, P.; Newcombe, R. G.; McBroom, V.: Decline in the Predicted Incidence of Huntington's Chorea Associated with Systematic Genetic Counseling and Family Support," *in* Lancet, 1981, 2:411–413.
43. Thomas, S.: "Ethics of a Predictive Test for Huntington's Chorea," *in* British Medical Journal, 1982, 284:1383–1385.
44. Stern, R.; Eldridge, R.: "Attitudes of Patients and Their Relatives to Huntington's Disease," *in* Journal of Medical Genetics, 1975, 12:217–223.
45. Teltcher, B.; Polgar, S.: "Objective Knowledge about Huntington's Disease and Attitudes toward Predictive Testing of Persons at Risk," *in* Journal of Medical Genetics, 1981, 18:31–39.
46. Bates, M.: "Ethics of Provocative Test for Huntington's Disease," Letter to the Editor, New England Journal of Medicine, 1981, Vol. 307, 304:175–176.
47. Emery, A. E. H.: "Whether or Not Predictive Test?" *in* Neurology, 1980, 30:345–346.
48. McCormick, M. K.; Lazzarini, A.: "Attitudes of Those at Risk for Huntington's Disease toward Presymptomatic Provocative Testing," *in* New England Journal of Medicine, 1982: 1406.
49. Stevens, D. L.: "Tests for Huntington's Chorea," *in* New England Journal of Medicine, 1971, 285:413–414.
50. Report: Commission for the Control of Huntington's Disease and Its Consequences, Vol. 1: Overview, DHEW Publication N. NIH 78–1501.
51. Bird, T. D.; Hall, J. G.: "Clinical Neurogenetics," *in* Neurology, 1977, 27:1057–1060.
52. Myers, R. H.: Schoenfeld, M.; Bird, E. D.: Progress in Medical Genetics, Vol. 6, Chap. 4.
53. Kenen, R. H.; Schmidt, R. M.: "Stigmatization of Carrier Status: Social Implications of Heterozygote Screening Programs," *in* American Journal of Public Health, 1978, 68:1116–1120.

Index